**INTRODUCTION
TO
ABSTRACT
ALGEBRA**

INTERNATIONAL SERIES IN PURE AND APPLIED MATHEMATICS

William Ted Martin, E. H. Spanier, G. Springer, and P. J. Davis. Consulting Editors

**McGRAW-HILL
BOOK COMPANY**

New York
St. Louis
San Francisco
Auckland
Düsseldorf
Johannesburg
Kuala Lumpur
London
Mexico
Montreal
New Delhi
Panama
Paris
São Paulo
Singapore
Sydney
Tokyo
Toronto

LOUIS SHAPIRO
Howard University

Introduction to Abstract Algebra

This book was set in Times New Roman.
The editors were A. Anthony Arthur and David Damstra;
the production supervisor was Dennis J. Conroy.
The drawings were done by Santype Ltd.
R. R. Donnelley & Sons Company was printer and binder.

QA
162
S47

Library of Congress Cataloging in Publication Data

Shapiro, Louis, date
    Introduction to abstract algebra.

    (International series in pure and applied mathematics)
    Bibliography: p.
    1. Algebra, Abstract. I. Title.
QA162.S47   512'.02   74-11448
ISBN 0-07-056415-9

**INTRODUCTION
TO
ABSTRACT
ALGEBRA**

1 2 3 4 5 6 7 8 9 0 DO DO 7 9 8 7 6 5

CONTENTS

Note: Explanations of the asterisk and dagger symbols are given in the preface.

PREFACE

The emphasis in this book is on examples and exercises, and they provide much of the motivation for the material. I have also tried to provide some historical comment and to examine the connections between modern algebra and other fields. (These comments and connections often appear in the exercises.) In particular I have tried to point out the connections with elementary mathematics as they occur. A special cross-reference index has also been included for this purpose.

The basic plan of the book is as follows: The first two chapters cover groups, homomorphisms, and rings. Set theory, functions, equivalence relations, the integers, and the complex numbers are developed when needed, rather than in a chapter. A trimester or semester course can cover the first two chapters, especially if the sections marked † are omitted. The rest of the book is devoted to various topics which might be covered in a second trimester or semester. The main options are:

1 Groups acting on sets
2 Integral domains, number theory
3 Vector spaces (Secs. 3.1 and 3.2), further group theory (Sec. 5.1), field theory
4 Vector spaces, Wedderburn theorems, group representations
5 Wedderburn theorems

Chapter 4, "Groups Acting on Sets," is somewhat novel. It provides a chance to do some significant applications, such as counting benzene-based molecules. It also provides a more unified approach to the Sylow theorems, the class equation, Cayley's theorem with respect to a subgroup, and permutation groups. The first two sections are very easy, and I hope the student will read them.

Another novelty is the three concluding survey chapters, designed to introduce the reader to some more advanced topics. The approach to this material is again through examples and exercises, and here many of the proofs are omitted. Thorough references to the literature are given, and I hope many readers will find these chapters useful as a first introduction to the material.

Sections marked * in the first two chapters may be omitted or treated lightly by anyone already acquainted with their contents. Sections marked † are not mandatory for a reading of Chapters 1 and 2 or for a first course.

Many exercises, are included. I have tried to place the exercises strategically, hoping the pedagogical gain outweighs the loss of smoothness of exposition.

Answers are provided for some of the exercises, hints for some others; overall, the answer section is extensive. The answers and hints should only be used as a last resort, or to check an answer. Mathematics is learned through idea formation and, though this may take different forms, looking up answers in the back of the book is not one of these.

The pace of Chapters 1 and 2 is leisurely, and the pace of the remaining chapters moderate. I have attempted to write a long, easy book rather than a short, hard one.

I would also like to thank the following people for their assistance in developing and preparing this book: William Blair, Robert Shapiro, Jacob Goldhaber, Ralph Greenberg, Brooke Stephens, Alan MacConnell, David Schneider, Wen-Tin Woan, James Donaldson, Horace Komm, Evelyn Roane, Patricia Berg, Julie Smith as well as Jack Farnsworth, Anthony Arthur, Edward Millman, and David Damstra from McGraw-Hill. I'm also grateful to many friends and colleagues for their continuing interest and support. The many students who worked through the preliminary versions have my deepest thanks for both their patience and many improvements. In particular I want to thank Lucious Anderson, Margarita Calhoun, Saul Claton, Valerie Oldwine, Theresa Manning, Peter Philip, Evelyn Roane, Vernise Steadman, and Cynthia Storrs.

I would be happy to receive comments from interested readers. I am, in addition, willing to supply those proofs that, because of the limitations of space and the example-exercise approach taken in the later part of the book, had to be omitted.

LOUIS SHAPIRO

*1.1 SETS AND BINARY OPERATIONS

In this book, definitions will be given in terms of elements, sets, and elements belonging to sets. In this section we set up the basic definitions, with the hope that most readers will be able to proceed directly to the next section. We assume some acquaintance with the ideas of function and ordered pair. We shall introduce additional material on functions (Sec. 1.5), permutations (Sec. 1.7), the division algorithm (Sec. 1.3), and complex numbers (Sec. 2.2) as it is needed.

If an element t is in a set S, we denote this by $t \in S$.

Definition If A and B are sets, then the **intersection** of A and B, denoted $A \cap B$, is the set of all elements that are in both A and B.

As an example, assume that A consists of the elements 1, α, 10, r, and 37π. Suppose B consists of the elements -1, 1, -3, r, s, and $\frac{4}{3}$. Then $A \cap B$ is the set consisting of 1 and r.

We describe a small set by listing all the elements. For instance, we could rewrite the example above as: If $A = \{1, \alpha, 10, r, 37\pi\}$ and $B = \{-1, 1, -3, 4, r, \frac{4}{3}\}$, then $A \cap B = \{1, r\}$.

We adopt the standard convention of using braces { }, only when listing sets. For larger sets we often write

$$S = \{x \mid \text{defining properties for } x \text{ to be an element of } S\}$$

For instance, if S is the set of all positive real numbers we can abbreviate this description to

$$S = \{x \mid x \text{ is a real number and } x > 0\}$$

The vertical bar is often read as "such that."

One often-encountered set is the natural numbers, $\mathbb{Z}^+ = \{1,2,3,...\}$, where the three dots indicate that we continue on in the same manner. If $T = \{x \mid x \text{ is a real number and } x \leq \pi\}$, then we have $\mathbb{Z}^+ \cap T = \{1,2,3\}$.

As another example consider

$$A = \{(x,y) \mid x^2 + y^2 = 1\} = \text{unit circle in the plane}$$

and

$$B = \{(x,y) \mid y = x\} = \text{the } 45° \text{ line through the origin}$$

Then $A \cap B = \{(1,1),(-1,-1)\}$. The term **intersection** originally comes from this kind of example, where the sets involved were geometric figures. The standard example for third graders is the intersection of two streets.

Definition If A and B are sets, then $A \cup B$, the **union** of A and B, is the set of all elements in A or in B or in both.

Let $A = \{1,\alpha,10,r,37\pi\}$ and $B = \{-1,1,-3,r,s,\frac{4}{3}\}$; then

$$A \cup B = \{-1,1,-3,10,\tfrac{4}{3},37\pi,r,s,\alpha\}$$

Definition A set T is said to be a **subset** of a set S if every element of T is an element of S. This is written $T \subseteq S$ or $S \supseteq T$. The **empty set,** denoted $\varnothing$ and containing no elements, is considered to be a subset of every set. If $x \in S$, then the subset of S consisting of just the element x is denoted $\{x\}$. In fact, the statements $\{x\} \subseteq S$ and $x \in S$ are equivalent Any set is also considered to be a subset of itself. A subset of a set S that is neither $\varnothing$ nor S is said to be a **proper subset** of S. If S contains T but $T \neq S$, this is denoted as $T \subset S$ or, for emphasis $T \subsetneq S$.

Several sets that we will use frequently are denoted as follows:

$\mathbb{Z}^+ = \{1,2,3,...\} = \text{all positive integers}$
$\mathbb{Z} \ = \{0,1,-1,2,-2,...\} = \text{all integers}$
$\mathbb{Q} \ = \{a/b \mid a,b \in \mathbb{Z}, b \neq 0\} = \text{all the rational numbers}$
$\mathbb{R} \ = \text{the set of all real numbers}$

$\mathbb{Q}^+$ and $\mathbb{R}^+$ are the set of all positive rational numbers and all positive real numbers, respectively. We thus have

$$\mathbb{Z}^+ \subset \mathbb{Q}^+ \subset \mathbb{Q} \subset \mathbb{R} \qquad \mathbb{Z} \subset \mathbb{R} \qquad \text{and so on}$$

Exercises

1 Show, for two sets A and B, that $A \subseteq B$ if and only if $A \cap B = A$. [*Hint*: Two sets S and T can be shown to be equal by showing every element of S is in T (that is, $S \subseteq T$) and every element of T is in S (that is, $T \subseteq S$).]

2 Show, for two sets C and D, that $C \subseteq D$ if and only if $C \cup D = D$.

3 If $S = \{a,b,c\}$, how many subsets does S have? If $S = \{a_1,a_2,\ldots,a_n\}$, how many subsets does S have?

4 Suppose $A = \{1,2,a,b,s,t\}$, $B = \{(0,0),q,a,b\}$, and $C = \{2,q,b\}$.
  (*a*) What are $A \cap B$, $A \cap C$, $A \cup B$, and $B \cup C$?
  (*b*) Is $C \subseteq A \cap B$?
  (*c*) Is $C \subseteq A \cup B$?

5 (*a*) † Let $S = \{1,2,3,4,5\}$. How many subsets of S consist of two elements?
    (*b*) † Repeat part *a* for $S = \{1,2,3,4,5,6\}$ and $S = \{1,2,3,\ldots,n\}$.
    (*c*) † How many lines are needed to connect every pair of points if n points are placed about the circumference of a circle?
    (*d*) † How many subsets of $S = \{1,2,3,\ldots,n\}$ consist of k elements?

Definition A **binary operation** from S to T is a function from ordered pairs (s_1,s_2) of elements in S to T. If $T = S$, the binary operation is said to be **closed**.

    We now illustrate with some examples. We shall discuss functions more thoroughly in Sec. 1.5. Often, if our binary operation is f, we write $s_1 * s_2$ instead of $f((s_1,s_2))$.

EXAMPLE A Addition is a closed binary operation on $\mathbb{Z}^+$ because the sum of two positive integers is a positive integer.

EXAMPLE B Subtraction is not a closed binary operation on $\mathbb{Z}^+$. Although both 5 and 7 are in $\mathbb{Z}^+$, $5 - 7 = -2$, and -2 is not positive, so not in $\mathbb{Z}^+$.

EXAMPLE C Subtraction is a closed binary operation on $\mathbb{Z}$ because the difference between any two integers is again an integer. Historically, it was the necessity for subtraction to be a closed binary operation that led to the introduction of negative numbers and zero.

EXAMPLE D Multiplication is a closed binary operation on any of the sets $\mathbb{Z}^+$, $\mathbb{Z}$, $\mathbb{Q}^+$, $\mathbb{Q}$, $\mathbb{R}^+$, and $\mathbb{R}$.

EXAMPLE E $X * Y = 3X/(7Y^2 + 1)$ is a closed operation on $\mathbb{Q}$ or $\mathbb{R}$, but not on $\mathbb{Z}$, since $1 * 1 = 3 \cdot 1/(7 \cdot 1^2 + 1) = \frac{3}{8}$ which is not an integer.

EXAMPLE F $X * Y = $ the larger of X and Y is a closed binary operation on $\mathbb{Z}$ (or on $\mathbb{Z}^+$, $\mathbb{Q}$, or $\mathbb{R}$).

Exercise

6   Which of the following binary operations are closed?

(a) $f(a,b) = a/b$ (division) on $\mathbb{Z}^+$
(b) $f(a,b) = a/b$ on $\mathbb{Q}$
(c) $f(a,b) = a/b$ on $\mathbb{Q}^+$
(d) $f(a,b) = a + b + 5$ on $\mathbb{Z}$
(e) $f(a,b) = a + \sin b$ on $\mathbb{R}$

Definition   A binary operation $*$ on a set S is **commutative** if $a * b = b * a$ for all elements a and b in S. The operation is **associative** if $(a * b) * c = a * (b * c)$ for all elements a, b, and c in S.

EXAMPLE G   Consider the operation of subtraction on $\mathbb{Z}$. We can see that it is not commutative since, say, $3 - 1$ is not equal to $1 - 3$. Also, subtraction is not associative since, for instance, $(7 - 3) - 1$ is not equal to $7 - (3 - 1)$.

EXAMPLE H   Addition on $\mathbb{Z}$ (or $\mathbb{Q}$ or $\mathbb{R}$) works much better. When we were first taught addition, we learned that $a + b = b + a$ and that $(a + b) + c = a + (b + c)$. The associative law in this context just says that, when adding three numbers, we can add either the first two or the second two together first; either way, we get the same answer. (Adding up a column of three numbers should yield the same answer as adding down the column.)

EXAMPLE I   If $\mathbb{R}$ is our set, and we define an operation $*$ by $a * b = 5ab$, then $*$ is associative, since $(a * b) * c = (5ab) * c = 25abc$ and $a * (b * c) = a * 5bc = 25abc$. Also, $a * b = 5ab = 5ba = b * a$, so $*$ is commutative.

EXAMPLE J   Let $\mathbb{R}$ be our set, and let $a * b = a^3 + b^3$. Then $a * b = a^3 + b^3 = b^3 + a^3 = b * a$, and $*$ is commutative. However, $a * (b * c) = a * (b^3 + c^3) = a^3 + (b^3 + c^3)^3$, and $(a * b) * c = (a^3 + b^3) * c = (a^3 + b^3)^3 + c^3$. So $(a * b) * c \neq a * (b * c)$, and $*$ is not associative.

† **EXAMPLE K** (*Optional*)   Let S be the set of all 2×2 matrices with entries in $\mathbb{R}$, and let $*$ be matrix multiplication. Then $*$ is associative but not commutative.

Exercises

7   Which of the following are associative?

(a) $S = \mathbb{R}^+$ and $a * b = a/b$
(b) $S = \mathbb{Z}$ and $a * b = a + b + b^2$

(c) $S = \mathbb{Z}$ and $a * b = a + b + ab$
(d) $S = \mathbb{R}$ and $a * b = |a|b$
(e) $a * b = b$

8 Which of the operations in Exercise 4 are commutative?
9 Let S be a set, and let $\mathscr{S}$ be the set of all subsets of $\mathscr{S}$. Is $\cap$ a binary operation on $\mathscr{S}$? Is it closed? Commutative? How about $\cup$?

1.2 GROUPS, DEFINITIONS AND EXAMPLES

We now come to the central character of this book and one of the main unifying themes in mathematics.

Definition A **group** $(G,*)$ is a set G together with any binary operation $*$ such that

G1 $*$ is closed; i.e., for every ordered pair (g_1, g_2) of elements of G there is a unique element $g_1 * g_2$ also in G.
G2 $*$ is associative.
G3 There is an element e in G such that $x * e = e * x = x$ for all x in G.
G4 For every element x in G, there is an element x' in G such that $x * x' = x' * x = e$.

REMARKS The element e is called the *identity* element of G. (The use of e originates from *einheit*, the German term.) Since e is defined by Axiom G3, Axiom G4 must follow Axiom G3. $*$ need not be commutative, and x' is called the *inverse* of x. Although the same e works for all of G, each x has its own private inverse. The effect of Propositions 2 and 3 in the following section is that we can talk about "the identity" and "the inverse of x" instead of "an identity" and "an inverse."

EXAMPLE A $(\mathbb{Z}, +)$ is a group. We have already shown that $+$ is associative and closed on $\mathbb{Z}$. Since $0 + x = x + 0 = x$, 0 is an identity for $(\mathbb{Z}, +)$. Also, $x + (-x) = (-x) + x = 0$, so $x' = -x$, and the inverse Axiom G4 holds. Therefore $(\mathbb{Z}, +)$ is a group. This group, since it is so familiar, is the first one to think about if you wish to check some statement against a known example.

EXAMPLE B Is $(\mathbb{Z}^+, \cdot)$, where $\cdot$ denotes ordinary multiplication, a group? We have already established that $\cdot$ is closed and associative on $\mathbb{Z}^+$. Since $1 \cdot x = x \cdot 1 = x$ for any x in $\mathbb{Z}^+$, and 1 is in $\mathbb{Z}^+$, we see that 1 is an identity. We only have left to verify Axiom G4. The possible candidate for the inverse of 5 is $\frac{1}{5}$. since $5 \cdot \frac{1}{5} = 1$, but $\frac{1}{5}$ is not in $\mathbb{Z}^+$. Thus G4 does not hold, and $(\mathbb{Z}^+, \cdot)$ is not a group.

EXAMPLE C $(2\mathbb{Z}, +)$ is a group, where $2\mathbb{Z}$ denotes the even integers 0, 2, -2, 4, -4, ...

EXAMPLE D $(2\mathbb{Z}, \cdot)$ is not a group.

EXAMPLE E $(2\mathbb{Z} + 1, +)$ is not a group, where $2\mathbb{Z} + 1$ denotes the odd integers.

† EXAMPLE F (*Optional*) The 2×2 nonsingular matrices with real entries under the operation of matrix multiplication form a group. This operation is not commutative.

Definition A group $(G, *)$ where $*$ is commutative, is called an **abelian group**. (After Niels Henrik Abel, a brilliant Norwegian mathematician who died of poverty and tuberculosis at age 27.)

Exercises

1 Verify the statements made in Examples **B, C, D, E** (and **F**).

2 Let G be the real numbers, except -1, where $a * b = a + b + ab$. Is $(G, *)$ a group? Solve $2 * x * 3 = 35$ for x.

3 With $G = \mathbb{R} - (0)$, and $*$ defined by $a * b = 3ab$, find an identity for $(G, *)$. Find an inverse for 8, and then for any x in G. Is $(G, *)$ a group?

4 Which of the following are groups, where $+$ and $\cdot$ are ordinary addition and multiplication?

   (*a*) $(\mathbb{R}, +)$
   (*b*) $(\mathbb{R}, \cdot)$
   (*c*) $(S, \cdot)$, where $\{1, -1\} = S$
   (*d*) $(S, +)$, where $\{1, -1\} = S$
   (*e*) (*Optional*) † Let $\{1, -1, i, -i\} = S$, where $i = \sqrt{-1}$. Is $(S, \cdot)$ a group?

5 (*Optional*) † Let $C[0,1]$ be the set of all real-valued functions continuous on $[0,1] = \{x \mid 0 \le x \le 1\}$. (Included in $C[0,1]$ are most of the functions that come up in calculus, such as $\sin x$ and $x^2 - 2$.) If f and g are in $C[0,1]$, then we define $f + g$ by $(f + g)(x) = f(x) + g(x)$, which is the usual definition of addition of functions. Using any theorems from calculus you wish, show that $(C[0,1], +)$ is a group.

6 Let S be a set, and let G be the set of subsets of S. If A and B are subsets of S, then define $A * B = (A \cup B) - (A \cap B)$, that is, the subset of all elements in A or in B, but not in both.

   (*a*) Show, using Venn diagrams if you wish, that $(A * B) * C = A * (B * C)$.
   (*b*) Show that $(G, *)$ is a group.
   (*c*) Show that $(G, *)$ is an abelian group.

(d) If S has two or three elements, how many elements will be in G?

(e) If S has n elements, how many elements will be in G?

7 Show, step by step, that $(a * (b * c)) * d = (a * b) * (c * d)$ in any group $(G,*)$.

8 This exercise is highly recommended if you want or need to practice induction proofs. We define a^2 as $a * a$, $a^3 = a * a * a$, and for a positive integer n, $a^n = a * a * a * \cdots * a$ (n copies of a).

(a) If $(G,*)$ is an abelian group, then show that $(a * b)^n = a^n * b^n$ for all positive integers n.

(b) If n is a positive integer, then show that $(a^n)^{-1} = (a^{-1})^n$.

† Some historical comment: Until around 1890, many different definitions of group were in use. Some authors did not require anything but a closed binary operation. The three volumes of Sophus Lie on transformation groups are known for both their importance and their unreadability, and one trouble is that the definition of group is nonconstant. Some of the early definitions just left out the associative law, but this didn't do much damage since this law was assumed whenever needed.

1.3 ELEMENTARY PROPERTIES OF GROUPS

Proposition 1 If $(G,*)$ is a group, then $a * b = c * b$ implies $a = c$. Also, $a * b = a * c$ implies $b = c$. These are called the *right* and *left cancellation laws*.

PROOF Assume $a * b = c * b$. Then

$$(a * b) * b' = (c * b) * b'$$

$$\| \qquad \qquad \|$$ by G2, the associative law

$$a * (b * b') \quad c * (b * b')$$

$$\| \qquad \qquad \|$$ by the definition of inverse in G4

$$a * e \qquad c * e$$

$$\| \qquad \qquad \|$$ by the definition of identity in G3

$$a \qquad \qquad c$$

The second assertion is proved similarly. ////

Proposition 2 A group $(G,*)$ has a unique identity.

PROOF If e and f are both identities of G, then

$$x * e = x \qquad \text{for all } x \text{ in } G \qquad \text{(A)}$$

and

$$f * y = y \qquad \text{for all } y \text{ in } G \qquad \text{(B)}$$

so

$$e = f * e \qquad \text{by (B)}$$

$$= f \qquad \text{by (A)} \qquad ////$$

Proposition 3 An element x in a group $(G,*)$ has a unique inverse.

PROOF Let x' and x'' both be inverses of x. Then

$$
\begin{aligned}
x' &= x' * e & \text{definition of } e \\
&= x' * (x * x'') & \text{definition of inverse} \\
&= (x' * x) * x'' & \text{associative law} \\
&= e * x'' & \text{definition of inverse} \\
&= x'' & \text{definition of } e \qquad \text{////}
\end{aligned}
$$

Exercise

1 Show that, if x is an element of $(G,*)$ and $x * x = x$, then $x = e$.

REMARK From now on we shall use a^{-1} to denote the inverse of a where a is in a group $(G,*)$. The one (very important) exception is when dealing with abelian groups in which $+$ is the operation. There, we shall use $-a$ as the inverse of a.

Proposition 4 $(a^{-1})^{-1} = a$ in any group $(G,*)$.

PROOF Since a is in G, so is a^{-1}. Every element of the group, including a^{-1}, has an inverse, so $a^{-1} * (a^{-1})^{-1} = e$. However, it is also true that $a^{-1} * a = e$, and thus, since the inverse of a^{-1} is unique (Proposition 3), we must have $a = (a^{-1})^{-1}$. ////

Let us denote the set of all real numbers except 0 by $\mathbb{R}^*$. It is easy to show that this is a group under multiplication, where 1 is the identity and $\dfrac{1}{x} = x^{-1}$ is the inverse of x. Proposition 4 applied to this example shows that $\dfrac{1}{1/x} = x$. Similarly, $(\mathbb{R}, +)$ is a group and, using additive notation, we have $-(-x) = x$. Thus one of the by-products of proving Proposition 4 is that we have established two of the cardinal rules of high-school algebra.

NOTATION If S is a finite set, we denote the number of elements in S by $|S|$. If S has an infinite number of elements, we can write $|S| = \infty$.

Definition A group $(G,*)$ is finite if it has a finite number of elements. This number is the **order** of G.

REMARK If $(G,*)$ is finite, we can construct a multiplication table as follows:

*			y	
x			$x * y$	

The left cancellation law says that $x * y = x * z$ implies $y = z$. What does this mean in terms of a multiplication table?

*		y		z	
x		$x * y$		$x * z$	

Since $y \neq z$ we must have $x * y \neq x * z$; that is, no two entries in a row can be the same. Similarly, from the right cancellation law we find that no two entries in a column can be the same.

Let us attempt to figure out a multiplication table for a group G with three elements, which we may as well label e, a, and b (e being the identity). We start with

*	e	a	b
e			
a			
b			

Since $e * x = x * e = x$, we can fill in the top row and first column:

*	e	a	b
e	e	a	b
a	a		
b	b		

We now look at $a * b$, which has to equal e, a, or b. Since we can't have repetition in any row or any column, $a * b$ can't be a or b, and by elimination must be e. Similarly, $b * a = e$, so we have

*	e	a	b
e	e	a	b
a	a		e
b	b	e	

Now $a * a$ must equal e, a, or b; since e and a are impossible, $a * a = b$ and, similarly, $b * b = a$.

Finally we have the complete table:

*	e	a	b
e	e	a	b
a	a	b	e
b	b	e	a

When writing out a group multiplication table we almost always list the elements in the same order going across the top as along the side. Thus, if a and b are two distinct elements of $(G,*)$, then we have

*	a	b
a		$a * b$
b	$b * a$	

If $(G,*)$ is abelian, we have $a * b = b * a$ for all a and b in $(G,*)$. This means that flipping the multiplication table through the diagonal axis from upper left to lower right will not change the table. Thus, a quick look at the last table tells us that the three-element group is abelian. Since we had no free choices in constructing this table, there is essentially only one group of order 3.

Proposition 5 If a and b are any two elements in a group $(G,*)$, then $(a * b)^{-1} = b^{-1} * a^{-1}$.

PROOF We shall produce two inverses for the element $a * b$. Since inverses are unique by Proposition 3, they must be equal. $(a * b)^{-1}$ is the inverse of $a * b$ by definition. Also,

$$
\begin{aligned}
(a * b) * (b^{-1} * a^{-1}) &= a * [b * (b^{-1} * a^{-1})] && \text{associative law} \\
&= a * [(b * b^{-1}) * a^{-1}] && \text{associative law} \\
&= a * (e * a^{-1}) && \text{definition of inverse} \\
&= a * a^{-1} && \text{definition of identity} \\
&= e && \text{definition of inverse}
\end{aligned}
$$

So $b^{-1} * a^{-1}$ is also an inverse of $a * b$, and thus $b^{-1} * a^{-1} = (a * b)^{-1}$.

////

Although the only groups we have mentioned so far are abelian (except for the group of 2×2 matrices with real entries), in some sense most groups are nonabelian. For instance, there are essentially 14 different groups of order 16, 9 of which are nonabelian. Therefore it is very important to remember that, in general,

$$a * b \neq b * a \qquad \text{and} \qquad a^2 * b^2 \neq (a * b)^2$$

where $a^2 = a * a$. Also

$$a^n * b^n \neq (a * b)^n$$

where $a^3 = a * a * a = a^2 * a$, $a^n = a^{n-1} * a$, and $a^{-n} = (a^{-1})^n$.

Proposition 6 A group $(G,*)$ is abelian if and only if $(a * b)^{-1} = a^{-1} * b^{-1}$, for all elements a, b in $(G,*)$.

PROOF We know from Proposition 5 that $(a * b)^{-1} = b^{-1} * a^{-1}$. Thus, if $(G,*)$ is abelian,

$$(a * b)^{-1} = b^{-1} * a^{-1} = a^{-1} * b^{-1}$$

Conversely, if

$$(a * b)^{-1} = a^{-1} * b^{-1}$$

then
$$b^{-1} * a^{-1} = a^{-1} * b^{-1}$$

and
$$a * b^{-1} * a^{-1} = a * a^{-1} * b^{-1} = b^{-1}$$
$$b * a * b^{-1} * a^{-1} = b * b^{-1} = e$$
$$b * a * b^{-1} = b * a * b^{-1} * a^{-1} * a = e * a = a$$
$$b * a = b * a * b^{-1} * b = a * b$$

Since this is true for all a, b in $(G,*)$, $(G,*)$ is abelian.   ////

We now make an abbreviation in notation. If there is no ambiguity, we use G instead of $(G,*)$ and ab instead of $a * b$.

Exercises

2 Show that G is abelian if and only if $(ab)^2 = a^2 b^2$ for all a, b in G.
3 (*Optional*) † Find 2×2 nonsingular matrices A and B where $AB \neq BA$. Show also that $A^2 B^2 \neq (AB)^2$.

We shall discuss our first examples of nonabelian groups after we have discussed subgroups. Let us now discuss possible groups of order 4. Let the group G have elements e, a, b, and c. As before, we must have

*	e	a	b	c
e	e	a	b	c
a	a			
b	b			
c	c			

We look at $a * b$. Since no element repeats in any row or column, $a * b$ must equal e or c, and we consider these two cases separately.

CASE I:

*	e	a	b	c
e	e	a	b	c
a	a		e	
b	b			
c	c			

Now $a * c$ must be b, which forces $a * a$ to be c. Thus we have

CASE I:

*	e	a	b	c
e	e	a	b	c
a	a	c	e	b
b	b			
c	c			

So now $c * b = a$, $b * b = c$, and then $b * a = e$, $b * c = a$, $c^2 = e$, and $c * a = b$. Finally, we have

CASE I:

*	e	a	b	c
e	e	a	b	c
a	a	c	e	b
b	b	e	c	a
c	c	b	a	e

We note by inspecting the multiplication table that this group is abelian. We also see that $a^2 = b^2 = c$, $b^4 = a^4 = e$, and $a^3 = a^2 \cdot a = c \cdot a = b$. So, re-

arranging the third and fourth rows and columns and substituting, we have
CASE I:

	e	a	$a^2 = c$	$a^3 = b$
e	e	a	a^2	a^3
a	a	a^2	a^3	e
a^2	a^2	a^3	e	a
a^3	a^3	e	a	a^2

Let us go back and check the second possibility.
CASE II:

	e	a	b	c
e	e	a	b	c
a	a		c	b
b	b			
c	c			

We also assume $ac \neq e$, or else interchanging c and b would take us back to case I. Allowing no duplication in any row or column, we find $a^2 = e$. Similarly, $ca = b$, $ba = c$, $bc = a$, and so on. This gives, finally,
CASE II:

	e	a	b	c
e	e	a	b	c
a	a	e	c	b
b	b	c	e	a
c	c	b	a	e

Were $bc = e$, we again would revert to case I.

It must be pointed out that all we have here is two *possible* group tables. It is easy to check that the identity axiom is all right by looking at the first row and column. Since e shows up once in each row and column, inverses are all right. The associative law, however, is unchecked. To check it completely would require 4^2 multiplications to find all ab, and thus 4^3 multiplications to find all $(ab)c$. Another 4^3 multiplications would give us all $a(bc)$, so a small four-element group would require $2 \cdot 4^3$ multiplications. An n-element group would require $2 \cdot n^3$ multiplications to verify the associative law. Being clever, and using the commutative law and the identity, might cut the number of computations by two-thirds, but this still is awesome. What are the effects of this computational hazard? One is that we now merely accept that we do have essentially one group of order 3 and two of order 4. Second is that we become very interested in any method that guarantees us associativity. This is why composition of functions becomes so vital to us.

Exercises

4 If G is a group of even order, show that there is an element x in G other than e such that $x^2 = e$. In fact there are an odd number of such elements (which are called *involutions*).

5 Fill in the rest of the following table so as to make it a group table. The associative law will be useful. Is this group abelian?

	e	a	b	c	d	f
e	e	a	b	c	d	f
a	a		e	f	c	d
b	b	e	a			
c	c			a		
d	d	f				
f	f	c		a		

6 If G is a group such that $x^2 = e$ for all elements x in G, then show that G is abelian. (Compare with Exercise 2 of Sec. 5.1.)

   The following is a definition of finite group used by Frobenius in 1879. (You may have heard of Frobenius if you have studied power-series solutions of differential equations.)

7 If $(G,*)$ is a finite set G together with a closed associative binary operation $*$ such that $x * a = x * b$ implies $a = b$, and $a * x = a * y$ implies $x = y$, then $(G,*)$ is a group. Show this.

8 Using one of our earlier examples, show that the $(G,*)$ of Exercise 7 need not be a group if G is infinite.

9 Show that $x^n = e$ if and only if $(y^{-1}xy)^n = e$. [*Hint:* Show first that $(y^{-1}xy)^n = y^{-1}x^ny$. Don't use commutativity.]

   To conclude this section we introduce one new class of groups. Let Z_n be the set $(\bar{0},\bar{1},\bar{2},\ldots,\overline{n-1})$, and let $\oplus$ be defined by $\bar{a} \oplus \bar{b} = \bar{c}$ if $a + b = c + kn$, where a, b, c, k, and n are integers, and $0 \leq c < n$. For example, in Z_{12},

$$\bar{7} + \bar{8} = \bar{3} \quad \text{since} \quad 7 + 8 = 15 = 3 + 1 \cdot 12$$
$$\bar{3} + \bar{2} = \bar{5} \quad \text{since} \quad 3 + 2 = 5 + 0 \cdot 12$$

Proposition 7 The $(\mathbb{Z}_n, \oplus)$ are groups for all positive integers.

Division algorithm If a and b are integers, b positive, then there are two integers c and r such that $a = bc + r$ and $0 \leq r < b$. This pair, c and r, is unique.

PROOF The multiples of b are $\ldots, -2b, -b, 0, b, 2b, \ldots$, and between some pair of consecutive multiples we have $cb \leq a < (c + 1)b$ for some integer c. Subtracting, we have $0 \leq a - cb < (c + 1)b - cb = b$. Set $a - cb = r$, and then obviously $0 \leq r < b$ and $a = cb + r$. To show that the pair c and r are unique, we assume $cb + r = c'b + r'$ with, say, $0 \leq r \leq r' < b$. Then $(c - c')b = r' - r$ so that $r' = r$ since b divides $r' - r$ but $0 \leq r' - r < b$. Then we must have $c - c' = 0$ or $c = c'$.   ////

For example, if $a = 304$ and $b = 24$, then $a = 304 = 12 \cdot 24 + 16$ and $0 \leq 16 < 24$. We shall come back to this lemma later, but now we use it to prove Proposition 7.

PROOF OF PROPOSITION 7 $\oplus$ is closed on $\mathbb{Z}_n$ because, for any $\bar{x}, \bar{y}$ in $\mathbb{Z}_n$, we can take $x + y = cn + r$ (by the lemma) so that $\bar{x} \oplus \bar{y} = \bar{r}$ and $0 \leq r < n$, and thus we have closure.

We would also like to show that $\oplus$ is well-defined. This means that if $\bar{a} = \overline{a'}$ and $\bar{b} = \overline{b'}$, we still want $\bar{a} \oplus \bar{b} = \overline{a'} \oplus \overline{b'}$. First note that $\bar{a} = \overline{a'}$ if and only if $a = a' + dn$, and that $\bar{b} = \overline{b'}$ if and only if $b = b' + fn$. Thus $a + b = a' + b' + (d + f)n$. Now if $a + b = cn + r$, we shall have $a' + b' = (c - d - f)n + r$, and thus $\overline{a'} \oplus \overline{b'} = \bar{r} = \bar{a} \oplus \bar{b}$. Since r is unique, we couldn't have $a' + b'$ also equal to $c'n + r'$, and the uniqueness of r really translates into $\oplus$ being a function. (See Sec. 1.5 for more material on functions).

Since $(x + y) + z = x + (y + z)$ for any integers x, y, and z, we must have $(\bar{x} \oplus \bar{y}) \oplus \bar{z} = \bar{x} \oplus (\bar{y} \oplus \bar{z})$.

$\bar{0}$ is the identity for $\mathbb{Z}_n$, and $\overline{n - a}$ is the inverse for a if $1 \leq a \leq n - 1$. $\bar{0}$ is its own inverse.

Thus $(\mathbb{Z}_n, \oplus)$ is a group, and we are done.   ////

Exercises

10 Write down group tables for $(\mathbb{Z}_5, \oplus)$ and $(\mathbb{Z}_6, \oplus)$.

11 In $(\mathbb{Z}_{12}, \oplus)$, what is $\bar{9} \oplus \bar{7}$?

12 If it is now 9 o'clock, what will the time be in 7 hours?

13 We can also define a multiplication for $\mathbb{Z}_n$ by $\bar{a} \odot \bar{b} = \bar{r}$, where $ab = cn + r$ and $0 \leq r < n$. The following are tables for $(\mathbb{Z}_4, \odot)$, $(\mathbb{Z}_5, \odot)$ and $(\mathbb{Z}_6, \odot)$. Fill in the rest of the table for $(\mathbb{Z}_6, \odot)$. Are any of these groups? Which are commutative? What modification in $(\mathbb{Z}_5, \odot)$ would make it into a group? What about $(\mathbb{Z}_4, \odot)$? $(\mathbb{Z}_6, \odot)$?

$(\mathbb{Z}_4, \odot)$:

$\odot$	$\bar{0}$	$\bar{1}$	$\bar{2}$	$\bar{3}$
$\bar{0}$	$\bar{0}$	$\bar{0}$	$\bar{0}$	$\bar{0}$
$\bar{1}$	$\bar{0}$	$\bar{1}$	$\bar{2}$	$\bar{3}$
$\bar{2}$	$\bar{0}$	$\bar{2}$	$\bar{0}$	$\bar{2}$
$\bar{3}$	$\bar{0}$	$\bar{3}$	$\bar{2}$	$\bar{1}$

$(\mathbb{Z}_5, \odot)$:

$\odot$	$\bar{0}$	$\bar{1}$	$\bar{2}$	$\bar{3}$	$\bar{4}$
$\bar{0}$	$\bar{0}$	$\bar{0}$	$\bar{0}$	$\bar{0}$	$\bar{0}$
$\bar{1}$	$\bar{0}$	$\bar{1}$	$\bar{2}$	$\bar{3}$	$\bar{4}$
$\bar{2}$	$\bar{0}$	$\bar{2}$	$\bar{4}$	$\bar{1}$	$\bar{3}$
$\bar{3}$	$\bar{0}$	$\bar{3}$	$\bar{1}$	$\bar{4}$	$\bar{2}$
$\bar{4}$	$\bar{0}$	$\bar{4}$	$\bar{3}$	$\bar{2}$	$\bar{1}$

$(\mathbb{Z}_6, \odot)$:

$\odot$	$\bar{0}$	$\bar{1}$	$\bar{2}$	$\bar{3}$	$\bar{4}$	$\bar{5}$
$\bar{0}$	$\bar{0}$	$\bar{0}$	$\bar{0}$	$\bar{0}$	$\bar{0}$	$\bar{0}$
$\bar{1}$	$\bar{0}$	$\bar{1}$	$\bar{2}$	$\bar{3}$	$\bar{4}$	$\bar{5}$
$\bar{2}$	$\bar{0}$	$\bar{2}$	$\bar{4}$			
$\bar{3}$	$\bar{0}$	$\bar{3}$				
$\bar{4}$	$\bar{0}$	$\bar{4}$				
$\bar{5}$	$\bar{0}$	$\bar{5}$				

14 Compute:

(a) $\bar{0} + \bar{1}$ in $\mathbb{Z}_2$

(b) $\bar{0} + \bar{1} + \bar{2}$ in $\mathbb{Z}_3$

(c) $\bar{0} + \bar{1} + \bar{2} + \bar{3}$ in $\mathbb{Z}_4$

(d) $\bar{0} + \bar{1} + \bar{2} + \bar{3} + \bar{4}$ in $\mathbb{Z}_5$

(e) $\bar{0} + \bar{1} + \bar{2} + \cdots + \overline{(n-1)}$ in $\mathbb{Z}_n$

1.4 SUBGROUPS AND CYCLIC GROUPS

Definition A subset H of a group $(G, *)$ is a **subgroup** if the elements of H form a group under $*$.

The single-element subset $\{e\}$ is always a subgroup of G. So is G. These two subgroups are called *improper* and the other subgroups of G are called *proper subgroups*. To check that a subset is a subgroup, it is sufficient to check three things.

Proposition 1 To check that a subset H is a subgroup of $(G, *)$, it suffices to check that

S1 H is closed under $*$.

S2 e is in H.

S3 If x is in H, then x^{-1} is in H.

In fact, if H is nonempty, it suffices to check S1 and S3.

PROOF $(H,*)$ satisfies the associative law automatically, because $a * (b * c) = (a * b) * c$ for all a, b, and c in G, and H is a subset of G. So $(H,*)$ is a group if it satisfies S1, S2, and S3. Also, if H is nonempty then x is in H. By S3, x^{-1} is in H. By S1, $x * x^{-1} = e$ is in H. Thus, if H is nonempty and S1 and S3 hold, then H is a subgroup.   ////

EXAMPLE A $(2\mathbb{Z},+)$, the even integers under addition, is a subgroup of $(\mathbb{Z},+)$. Both in turn are subgroups of $(\mathbb{Q},+)$.

EXAMPLE B $(\mathbb{R}^+,\cdot)$, the reals under multiplication, is not a subgroup of $(\mathbb{R},+)$, the reals under addition. Both are groups and $\mathbb{R}^+$ is a subset of $\mathbb{R}$, but $(\mathbb{R}^+,\cdot)$ is not a subgroup of $(\mathbb{R},+)$ since the operations are different.

EXAMPLE C   The subset $S = \{\bar{0},\bar{2}\}$ of $(\mathbb{Z}_4,\oplus)$ is a subgroup. In fact, it is the only proper subgroup. $(S,\oplus)$ is a group because it is closed (that is, $\bar{0} \oplus \bar{0} = \bar{0}$, $\bar{0} \oplus \bar{2} = \bar{2} \oplus \bar{0} = \bar{2}, \bar{2} \oplus \bar{2} = \bar{0}$) and contains inverses (that is, $\bar{0} \oplus \bar{0} = \bar{2} \oplus \bar{2} = \bar{0}$). If $(H,\oplus)$ is any other subgroup of $(\mathbb{Z}_4,\oplus)$ and $\bar{1}$ is in $(H,\oplus)$, then so are $\bar{1} \oplus \bar{1}$, $\bar{1} \oplus \bar{1} \oplus \bar{1}$, and $\bar{0}$, so $H = \mathbb{Z}_4$. Similarly, if $\bar{3}$ is in H, so are $\bar{3} \oplus \bar{3}$, $\bar{3} \oplus \bar{3} \oplus \bar{3}$, and $\bar{0}$, so $H = \mathbb{Z}_4$. Again, H is not proper. Thus the only subgroups of $(\mathbb{Z}_4,\oplus)$ are $\{\bar{0}\}$, S, and $\mathbb{Z}_4$ itself. This is indicated as follows:

$$(\mathbb{Z}_4,\oplus)$$
$$|$$
$$(S,\oplus)$$
$$|$$
$$(\{0\},\oplus)$$

This is called a *subgroup lattice*. The largest subgroup is at the top, the smallest is at the bottom, and the $\subset$ relationship is given by a vertical line or a series of such lines.

EXAMPLE D   Our second group of order 4, case II in the last section, has the following multiplication table:
$(V, *)$:

	e	a	b	c
e	e	a	b	c
a	a	e	c	b
b	b	c	e	a
c	c	b	a	e

This group is called *Klein's four group* (or *Viergruppe*), and we shall often denote it as V. It is easy to see that

$$H_1 = \{e,a\}$$
$$H_2 = \{e,b\}$$
$$H_3 = \{e,c\}$$

are subgroups. It is also easy to see that no three elements of V form a subgroup. Therefore $\{e\}$, H_1, H_2, H_3, and V are the only subgroups of V, so the subgroup lattice for V is as follows:

V is another group which is often very useful as an example or counterexample.

Exercises

1   Which of the following groups are subgroups of which other group on the list: $(\mathbb{Z},+)$, $(\mathbb{R}^+,\cdot)$, $(\mathbb{Z}_6,\oplus)$, $(\mathbb{Q}-(0),\cdot)$, $(\mathbb{R}-(0),\cdot)$, $(\mathbb{Z}_{12},\oplus)$, $(\bar{0},\bar{4},\bar{8}$ in $\mathbb{Z}_{12},\oplus)$? (*Warning:* $5+5+5+4$ equals 19 in $\mathbb{Z}$, 7 in $\mathbb{Z}_{12}$, and 1 in $\mathbb{Z}_6$, so $\oplus$ is not equal to $+$, nor do the various $\oplus$ equal each other.)

2   Show that, if G is a finite group, and x is an element of G, then $x^n = e$ for some positive integer n.

3   Show that, if H is a finite subset of a group $(G,*)$, and H is closed under $*$, then H is a subgroup of G.

4   Find all the subgroups of $(\mathbb{Z}_6,\oplus)$, and arrange them in a lattice. Do the same for $(\mathbb{Z}_8,\oplus)$.

5   Is the set $S = \{0,6,-6,12,-12,18,-18,...\}$ a subgroup of $(\mathbb{Z},+)$? Is $\mathbb{Z}^+ = \{1,2,3,...\}$ a subgroup of $(\mathbb{Z},+)$?

6   Find an example of a group $(G,*)$ and a subset H of G that is closed under $*$ but is not a subgroup. Find another subset J such that x in J implies x^{-1} is in J but J is not a subgroup.

7   If H and K are subgroups of G, then show that exactly one of the following is true:
(*a*) $H \cap K$ is a subgroup of G.
(*b*) $H \cup K$ is a subgroup of G.
(Try a few examples first, to eliminate the false assertion.)

Definition   If $(G,*)$ is a group and there is an element x in G such that $G = \{e,x,x^2,x^3,...,x^{-1},x^{-2},x^{-3},...\}$, then we say G is **cyclic** and that x is a **generator** for G. We denote the set $\{e,x,x^2,x^3,...,x^{-1},x^{-2},x^{-3},...\}$ as $\langle x \rangle$. Thus if G is cyclic with x as its generator, we can write this as $G = \langle x \rangle$.

Proposition The smallest subgroup of G containing x is $\langle x \rangle$. Also, $\langle x \rangle$ equals the intersection of all subgroups of G which contain x.

PROOF Obviously $\langle x \rangle$ is a subgroup of G since $\langle x \rangle =$ $\{e, x, x^2, \ldots, x^{-1}, x^{-2}, \ldots\}$ which provides e by definition. Also, $(x^n)^{-1} = x^{-n}$ is on the list, and $x^n x^m = x^{n+m}$, which verifies closure. Conversely, if $x \in H$ and H is a subgroup, then using closure repeatedly we have $x^2, x^3, x^4, \ldots$ all in H. Since we need an identity and inverses for H to be a subgroup, we have that x in H implies $\langle x \rangle \subseteq H$. Thus $\langle x \rangle \subseteq \cap H_i$, where the H_i are all the subgroups containing x. Since $\langle x \rangle$ is then one of the H_i, we have $\langle x \rangle = \cap H_i$.   ////

If H is a subgroup of G, and H is by itself cyclic, then H is naturally called a *cyclic subgroup* of G. A natural question, when investigating a group is, "What are the subgroups?" This is often very difficult, but the first step is just to choose an element and look at $\langle a \rangle = \{e, a, a^{-1}, a^2, a^{-2}, \ldots\}$. This method isn't deep but has already been used in Examples C and D and in Exercises 4 and 10. (It gets you ready for taking elements two at a time, which is much harder.)

Exercise

8 Find four proper subgroups of $(\mathbb{Q}^+, \cdot)$, including at least one which is not cyclic.

EXAMPLE E Our first example, I, of the two groups of order 4 gave the result $G = \{e, a, a^2, a^3\} = \langle a \rangle$, so G is cyclic. Incidentally, G also has a^3 as a generator. Klein's four group V has the four elements e, a, b, and c, and $\langle e \rangle = \{e\}$, $\langle a \rangle = \{e, a\}$, $\langle b \rangle = \{b, e\}$, and $\langle c \rangle = \{c, e\}$; so $V \neq \langle e \rangle$, $\langle a \rangle$, $\langle b \rangle$, or $\langle c \rangle$, and V is not cyclic. Using cyclic subgroups here has produced all three proper subgroups for us.

EXAMPLE F $(\mathbb{Z}_n, \oplus)$ is cyclic with $\bar{1}$ as its generator. Here we use additive rather than multiplicative notation, and we see that

$$\bar{1} = \bar{1}$$
$$\bar{2} = \bar{1} + \bar{1}$$
$$\bar{3} = \bar{1} + \bar{1} + \bar{1}$$
$$\cdots\cdots\cdots\cdots\cdots\cdots$$
$$\overline{n-1} = \bar{1} + \bar{1} + \cdots + \bar{1}$$
$$\bar{0} = \bar{n} = \bar{1} + \bar{1} + \cdots + \bar{1} + \bar{1}$$

Thus, $(\mathbb{Z}_n, \oplus) = \langle \bar{1} \rangle$

EXAMPLE G $(\mathbb{Q} - (0), \cdot)$, where $\mathbb{Q} - (0)$ denotes the nonzero rationals, is a noncyclic group. We show this by assuming it is cyclic and deriving a contradiction. If $(\mathbb{Q} - (0), \cdot)$ is cyclic, then $(\mathbb{Q} - (0), \cdot) = \langle a/b \rangle$. Thus any nonzero rational number is equal to $(a/b)^n$ for some n in $\mathbb{Z}$. So $2 = (a/b)^{n_1}$, $3 = (a/b)^{n_2}$, and $3^{n_1} = 2^{n_2}$. If n_1 or n_2 is negative, we multiply and end up with $3^m = 2^n$ or $3^l 2^k = 1$, with m, n, l, k in $\mathbb{Z}^+ \cup (0)$. Since no odd number is even, the first possibility leads to a contradiction. So, obviously, does the second.

Exercises

9   Every example of a cyclic group that we have examined has been abelian. Is every cyclic group abelian?

10   Show that $(\mathbb{Z}, +)$ is cyclic, and find two generators.

11   Show that, if G is a cyclic group with only one generator, then $|G| = 1$ or 2. (In fact, the same conclusion holds if we say G has an odd number of generators.)

12   What elements are in each of the following subgroups of $(\mathbb{Z}_6, \oplus)$: $\langle \bar{1} \rangle$, $\langle \bar{2} \rangle$, $\langle \bar{3} \rangle$, $\langle \bar{4} \rangle$, $\langle \bar{5} \rangle$?

13   If $(G, *)$ is given by the following multiplication table, is it cyclic? Is it abelian? Does it resemble any group we have looked at before? The operation here is multiplication modulo 8.

$*$	$\bar{1}$	$\bar{3}$	$\bar{5}$	$\bar{7}$
$\bar{1}$	$\bar{1}$	$\bar{3}$	$\bar{5}$	$\bar{7}$
$\bar{3}$	$\bar{3}$	$\bar{1}$	$\bar{7}$	$\bar{5}$
$\bar{5}$	$\bar{5}$	$\bar{7}$	$\bar{1}$	$\bar{3}$
$\bar{7}$	$\bar{7}$	$\bar{5}$	$\bar{3}$	$\bar{1}$

14   Show that the set $G_p = \{m/p^n \mid m, n \text{ are integers}\}$ for some fixed prime p is an abelian group under addition. Is G_p a cyclic group?

15   Show that the set of numbers $a + b\sqrt{2}$, where a and b are both in $\mathbb{Q}$ and not both zero, is a subgroup of $(\mathbb{R} - (0), \cdot)$.

16   If A is an abelian group and n is a positive integer, show that the elements x in A such that $x^n = e$ form a subgroup. [*Hint:* It is probably best to start by proving $(ab)^k = a^k b^k$ for all positive k.]

17   If a is an element of a finite group G, why is it correct to say $\langle a \rangle = \{a, a^2, a^3, \ldots, a^m, \ldots\}$? In other words, why are we guaranteed that the e, a^{-1}, a^{-2}, $\ldots$ will all show up on our list of positive powers?

*1.5   FUNCTIONS

We now come to another vital set of concepts. This material needs to be mastered by anyone with even a moderate interest in mathematics.

Definition (version 1)   A **function** f from a set S to a set T is a rule assigning each element of S to a unique element of T.

If S and T are sets, then $S \times T$ is the set of all ordered pairs (s,t) with $s \in S$, $t \in T$. Using this, we can give another definition which is better in the sense that the word "rule" isn't needed. This definition is derived from the way real-valued functions are usually graphed.

Definition (version 2)   A function f from S to T is a subset of $S \times T$ such that each $s \in S$ is the first component of exactly one element of the subset f. The t showing up in $(s,t) \in f$ is called $f(s) = t$. (And version 1 would say f is the rule taking s to t.)

EXAMPLE A   Using version 1, we would say $f(x) = x^2$ is a function from $\mathbb{R}$ to $\mathbb{R}$. Using version 2, we would say that $\{(x,x^2) \mid x \in \mathbb{R}\}$ is a function since it is a subset of $\mathbb{R} \times \mathbb{R}$ and each x in $\mathbb{R}$ is in exactly one pair (x,x^2).

EXAMPLE B $x^2 + y^2 = 25$ is *not* a function from $\mathbb{R}$ to $\mathbb{R}$ for several reasons. First, if $x < -5$ or $x > 5$, then x corresponds to no y. If we fix this up by asserting $x^2 + y^2 = 25$ is a function from $[5, -5]$ to $\mathbb{R}$, there is still trouble. For instance, $x = 3$ corresponds to both $y = 4$ and $y = -4$. If we eliminate one of these possibilities by also specifying $y \geq 0$, then we do have a function from $[5, -5]$ to $\mathbb{R}^+ \cup (0)$.

Exercises

1 If $S = (a,b,c)$ and $T = (1,2,3,4)$, which of the following are functions from S to T? Why?

Version 1:   (a) $f(a) = 1, f(b) = 2, f(c) = 4, f(b) = 3$
            (b) $f(a) = 4, f(b) = 3, f(c) = 2$
Version 2:   (c) $f = \{(a,1), (b,2), (c,4), (b,3)\}$
            (d) $f = \{(a,2),(b,2),(c,2)\}$
            (e) $f = \{(a,1),(c,3)\}$

2 Which of the following are functions from $\mathbb{R}$ to $\mathbb{R}$? Can the others be fixed up to be functions from S to $\mathbb{R}$, where S is a subset of $\mathbb{R}$?
(a) $f(x) = x^3$　　　　　(b) $f(x) = 1/(x - 3)$
(c) $f(x) = |x|$　　　　　(d) $x^2 + 8[f(x)]^2 = 9$
(e) $\{(x, \sin x) \mid x \in \mathbb{R}\}$　(f) $\{(x,y) \mid y^2 = x, y \text{ and } x \text{ in } \mathbb{R}\}$

Most of our experience with functions comes from the functions of high-school algebra and calculus. These are functions from $\mathbb{R}$ to $\mathbb{R}$ (as in Exercise 2) or, equivalently, subsets of $\mathbb{R} \times \mathbb{R}$. The main change in definition for this book is that we require a function to be defined everywhere on S. For instance,

$f(x) = (x^2 - 9)/(x - 3)$ is a fine calculus function, but we would require that $S = \mathbb{R} - \{3\}$ and not $S = \mathbb{R}$. In fact, in calculus, limits are introduced to fill in such missing points.

If $R \subseteq S \times T$ is any subset of $S \times T$, the *domain of* $R = \{s \mid (s,t) \in R\}$, and the *image of* $R = \{t \mid (s,t) \in R\}$. If R is also a function, then the domain of R has to be all of S. In calculus the word "range" might be used instead of "image."

A subset of $\mathbb{R} \times \mathbb{R}$ is a function if every vertical line in $\mathbb{R} \times \mathbb{R}$ crosses the function exactly once. Thus, if you have a reasonable graph you can tell if it represents a function.

EXAMPLE C   Let $S_1 = \{a,b,c,d\}$, $S_2 = \{1,2,3,4,5\}$, and $S_3 = \{x,y,z\}$. Then $f = \{(a,1), (b,1), (c,3), (d,4)\}$ is a function from S_1 to S_2. $g = \{(1,x), (2,y), (3,z), (3,x), (4,y), (5,z), (5,y)\}$ is not a function.

EXAMPLE D   If S is the set of all polynomials with real coefficients, then $D(p(x)) = p'(x) = dp(x)/dx$ is a function from S to S. $I(p(x)) = \int p(x)\, dx + C$ is not a function unless C is specified. If we now set $C = 0$, what can be said about $I(D(p(x)))$ and $D(I(p(x)))$?

Definition   A function f from S to T is **one-to-one** (or $1:1$) if $s_1 \neq s_2$ implies $f(s_1) \neq f(s_2)$.

Exercises

3   Express this definition both in terms of ordered pairs and as the contrapositive of the definition given.

4   Which of the functions in Exercises 1 and 2 are one-to-one?

5   † If f is a continuous function from $\mathbb{R}$ to $\mathbb{R}$, use the intermediate-value theorem from calculus to show that one-to-one is the same as monotonic. If f is differentiable, what is a convenient criterion for one-to-one? (Use the mean-value theorem.)

6   Which of the following functions from $\mathbb{R}$ to $\mathbb{R}$ are one-to-one?

(a) $f(x) = x^3$
(b) $f(x) = |x^3|$
(c) $f(x) = 3^x$
(d) $f(x) = 2[x] - x$, when $[x]$ is the largest integer less than or equal to x

Definition   If f is a function from S to T, then f is **onto** if for each t in T there is an s in S such that $f(s) = t$.

Intuitively, this means that we can use f to cover T by S. If T isn't covered, then we can define a new function from S to T', where T' is the part of T that is covered. One way to show that f is an onto function is to let t be any arbitrary element of T and to solve $f(\ ) = t$. The filled-in blank will almost always be expressed in terms of t. For instance, to show that $f(x) = 4 + x^3$ is onto (as a function from $\mathbb{R}$ to $\mathbb{R}$), we look at $f(\ ) = r$. We can fill in the blank with $(r - 4)^{1/3}$, since $f[(r - 4)^{1/3}] = [(r - 4)^{1/3}]^3 + 4 = r$. (If not obvious, the correct entry for the blank can be found by solving $x^3 + 4 = y$ for x.)

NOTATION $f: S \to T$ denotes that f is a function from S to T.

Exercises

7 Which of the functions in Exercises 1, 2, and 5 are onto? If $f: \mathbb{R} \to \mathbb{R}$ is a function and we know its graph, how can we tell whether f is onto?

8 How does the definition of onto translate into ordered pairs?

Proposition 1 If S and T are both sets containing exactly n elements, then a function from S to T is one-to-one if and only if it is onto.

PROOF We shall use the ordered-pairs definition of function so that $f = \{(s_1, t_1), (s_2, t_2), \ldots, (s_n, t_n)\}$. Then f is one-to-one if each t_i is distinct; if each of the n elements in T is used, then f is onto. ////

Exercise

9 If S is the set of integers, find a function $f: S \to S$ that is one-to-one but not onto, and a function $g: S \to S$ that is onto but not one-to-one.

As mentioned earlier, one of our main uses of functions is to show that the associative law holds in certain situations. Composition of functions is associative, as we now show. The next two sections will exploit this fact to provide two large important classes of nonabelian groups.

NOTATION If $f: S \to T$ and $g: T \to U$ are both functions, we can define a new function $g \circ f: S \to U$ by $(g \circ f)(s) = g(f(s))$. $g \circ f$ is called the *composition* of the functions f and g. Informally, this comes out as: take s from S to $f(s)$ in T and thence to $g(f(s))$ in U. If $f: S \to T$ and $g: T \to U$ are functions, then $g \circ f: S \to U$ is also a function, since each s in S goes to a unique $f(s)$ in T and then again to a unique $g(f(s))$ in U.

When are two functions equal? If f and g are both functions from S to T, then both are subsets of $S \times T$ and thus must contain the same elements. So $(x, f(x)) = (x, g(x))$ for each x in S and, stated in other terms, $f(x) = g(x)$ for all x in S. This is the crucial condition which, along with an observation that both f and g use the same S and T, gives us that two functions are equal.

We now prove that composition of functions is associative. Let $f: S \to T$, $g: T \to U$, and $h: U \to V$ be three functions. We want to show that $(h \circ g) \circ f =$

$h \circ (g \circ f)$. By the last paragraph we need only show $((h \circ g) \circ f)(x) = (h \circ (g \circ f))(x)$ for all x in S. This goes as follows:

$$((h \circ g) \circ f)(x) = (h \circ g)(f(x)) \qquad \text{all by the definition}$$
$$= h(g(f(x))) \qquad \text{of composition of functions}$$
$$= h((g \circ f)(x))$$
$$= (h \circ (g \circ f))(x)$$

Recapitulating, we have:

Theorem 1   If $f: S \to T$, $g: T \to U$, and $h: U \to V$ are functions, then $h \circ (g \circ f) = (h \circ g) \circ f$. Put briefly, composition of functions is associative.

When can a function $f: S \to T$ be reversed? If f is a subset of $S \times T$, then f^{-1} is a subset of $T \times S$ such that $(t,s) \in f^{-1}$ if and only if $(s,t) \in f$. So for any function f from S to T we have an f^{-1} which is read as "f inverse." It is a relation, but usually not a function.

EXAMPLE E $f(x) = x^2$ is a function which could be written as $\{(x,x^2) \mid x \in \mathbb{R}\} \subseteq \mathbb{R} \times \mathbb{R}$. Thus its inverse is $\{(x^2,x) \mid x \in \mathbb{R}\} \subseteq \mathbb{R} \times \mathbb{R}$. Since x^2 can never equal -5, no pair $(-5,x)$ is in f^{-1}. Moreover, both $(4,2)$ and $(4,-2)$ are in f^{-1}, so f^{-1} fails to be a function on two counts.

EXAMPLE F   Let $S = \{1,2,3\}$, $T = \{a,b,c,d\}$, and $f = \{(1,a), (2,b), (3,c)\}$. f^{-1} results from interchanging the pairs in f, so $f^{-1} = \{(a,1), (b,2), (c,3)\}$, and d is the first element of none of the three elements in f^{-1}. Thus f^{-1} is not a function. Is there any function f from this S to this T such that f^{-1} is a function?

Having seen two examples where f^{-1} is not a function, we now put conditions on f so that f^{-1} will be a function. $f^{-1} = \{(t,s) \mid (s,t) \in f\}$ will be a function from T to S if each t in T is in exactly one of the pairs in f^{-1}. Each t must show up at least once in a (t,s) in f^{-1}, so it also must be in some (s,t) in f. However, this condition is just that f is onto. If t shows up in two pairs (t,s_1) and (t,s_2) in f^{-1}, then (s_1,t) and (s_2,t) are both in f. To prevent this from happening we require that f be one-to-one, and this is precisely what is needed. Thus, if f is one-to-one and onto, each t in T will show up in exactly one ("at most one" and "at least one" equal "exactly one") pair $(t,s) \in f^{-1}$. And thus f^{-1} will be a function. If f^{-1} is a function, then f is indeed also one-to-one and onto. Stating this as a theorem, we have:

Theorem 2   If f is a function from S to T, then f^{-1} is a function from T to S if and only if f is both one-to-one and onto.

Exercises

10 If $f: S \rightarrow T$ is a function, show that $(f^{-1})^{-1} = f$.

11 Use Exercise 10 to show that if $f^{-1}: T \rightarrow S$ is a function then f^{-1} is one-to-one and onto.

12 Let $S = T = \mathbb{R}$. Find those linear functions $f(x) = mx + b$ such that $f = f^{-1}$. Graph several of those functions.

13 If f is a function from $\mathbb{R}$ to $\mathbb{R}$, show that the graph of f and the graph of f^{-1} are identical to each other after flipping about the $45°$ line through the origin. Try this with $f(x) = 10^x$ and $f^{-1}(x) = \log_{10} x$.

14 Graph each of the following conic sections. Which are functions from $\mathbb{R}$ to $\mathbb{R}$? Which have inverses that are functions? Which are equal to their inverses?

(a) $y^2 + x^2 = 4$ (b) $yx = 1$

(c) $y = x^2$ (d) $x = y^2$

(e) $y^2/4 + x^2/9 = 1$

15 † Where are inverse functions used in calculus?

We have been discussing inverses and hinting that groups come into this discussion later. This might lead one to suspect that identities had better be around somewhere. If A is any set, then $I_A: A \rightarrow A$ is the function such that $I_A(a) = a$ for all a in A. It is called the *identity function* on A.

If we are given a function $f: A \rightarrow B$, and $f^{-1}: B \rightarrow A$ turns out also to be a function, then $f^{-1} \circ f = I_A$. This is true since $(f^{-1} \circ f)(a) = f^{-1}(f(a)) = a = I_A(a)$ for all a in A. Then, since $(f^{-1})^{-1} = f$, we see that $(f \circ f^{-1})(b) = b = I_B(b)$ for all b in B, so $f \circ f^{-1} = I_B$.

As an example of this consider $A = \mathbb{R}$, $B = \mathbb{R}^+$, and $f(x) = 10^x$. Then f is one-to-one and onto and $f^{-1}(x) = \log_{10} x$. $(f^{-1} \circ f)(x) = f^{-1}(10^x) = \log_{10} 10^x = x = I(x)$ for any x in R. For any $x > 0$ we have

$$(f \circ f^{-1})(x) = f(\log_{10} x) \qquad \text{here we need } x > 0$$

$$= 10^{\log_{10} x}$$

$$= x$$

[Readers who have taken calculus might be more familiar with $f(x) = e^x$, $f^{-1}(x) = \ln x$, $e^{\ln x} = x$ for $x > 0$, and $\ln(e^x) = x$.]

Exercise

16 Let $(G, *)$ be a group, and let $S = T = G$. For any fixed element x in G, show that $f_x: G \rightarrow G$ is a one-to-one onto function, where $f_x(g) = g * x$.

1.6 2 × 2 MATRICES

In this section we consider 2×2 matrices with real-number entries. Various sets of 2×2 matrices will give us a vast number of examples of groups, both nonabelian and abelian. We then consider these matrices as functions and use the results of the last section.

Definition A 2×2 **matrix** over $\mathbb{R}$ is an ordered 4-tuple of real numbers (a,b,c,d). These four numbers are usually written

$$\begin{pmatrix} a & b \\ c & d \end{pmatrix}$$

Multiplication of 2×2 matrices is defined as follows:

$$\begin{pmatrix} a_1 & b_1 \\ c_1 & d_1 \end{pmatrix} \begin{pmatrix} a_2 & b_2 \\ c_2 & d_2 \end{pmatrix} = \begin{pmatrix} a_1 a_2 + b_1 c_2 & a_1 b_2 + b_1 d_2 \\ c_1 a_2 + d_1 c_2 & c_1 b_2 + d_1 d_2 \end{pmatrix}$$

For example,

$$\begin{pmatrix} 3 & \pi \\ 0 & 1 \end{pmatrix} \begin{pmatrix} -1 & 2 \\ 5 & 4 \end{pmatrix}$$

$$= \begin{pmatrix} 3(-1) + \pi 5 & 3 \cdot 2 + \pi 4 \\ 0(-1) + 1 \cdot 5 & 0 \cdot 2 + 1 \cdot 4 \end{pmatrix} = \begin{pmatrix} 5\pi - 3 & 6 + 4\pi \\ 5 & 4 \end{pmatrix}$$

and

$$\begin{pmatrix} 1 & 2 \\ 3 & 4 \end{pmatrix} \begin{pmatrix} 5 & 0 \\ 2 & 8 \end{pmatrix} = \begin{pmatrix} 1 \cdot 5 + 2 \cdot 2 & 1 \cdot 0 + 2 \cdot 8 \\ 3 \cdot 5 + 4 \cdot 2 & 3 \cdot 0 + 4 \cdot 4 \end{pmatrix} = \begin{pmatrix} 9 & 16 \\ 23 & 32 \end{pmatrix}$$

To compute the entry denoted by $*$ in the matrix

$$\begin{pmatrix} * & \\ & \end{pmatrix}$$

we use the row through $*$ of the left-hand matrix and the column through $*$ of the right-hand matrix; i.e., we look at

$$\begin{pmatrix} * & - \end{pmatrix} \begin{pmatrix} * \\ | \end{pmatrix}$$

and add the two products given by multiplying along the row and the column through $*$. Similarly,

$$\begin{pmatrix} - & * \end{pmatrix} \begin{pmatrix} * \\ | \end{pmatrix} = \begin{pmatrix} & * \end{pmatrix}$$

and so forth. (Any student who is at all unsure of how to multiply matrices will save some time by now working Exercise 1.)

Definition A 2×2 matrix $\begin{pmatrix} a & b \\ c & d \end{pmatrix}$ over $\mathbb{R}$ is **nonsingular** if $ad - bc \neq 0$.

Proposition 1 The set of 2×2 nonsingular matrices over $\mathbb{R}$ form a group under multiplication.

PROOF Our definition of multiplication gives us the fact that the product of 2×2 matrices is a 2×2 matrix. Thus we need only check that the product of nonsingular 2×2 matrices is nonsingular. We have

$$\begin{pmatrix} a_1 & b_1 \\ c_1 & d_1 \end{pmatrix}\begin{pmatrix} a_2 & b_2 \\ c_2 & d_2 \end{pmatrix} = \begin{pmatrix} a_1a_2 + b_1c_2 & a_1b_2 + b_1d_2 \\ c_1a_2 + d_1c_2 & c_1b_2 + d_1d_2 \end{pmatrix}$$

so if $a_1d_1 - b_1c_1$ and $a_2d_2 - b_2c_2$ are nonzero, we must show that

$$(a_1a_2 + b_1c_2)(c_1b_2 + d_1d_2) - (c_1a_2 + d_1c_2)(a_1b_2 + b_1d_2)$$

is also nonzero. Multiplying out, we get, after some cancellations, $a_1a_2 d_1d_2 + b_1b_2 c_1c_2 - a_2 b_1c_1d_2 - a_1b_2 c_2 d_1$ which factors to

$$(a_1d_1 - b_1c_1)(a_2 d_2 - b_2 c_2).$$

The product of nonzero terms is nonzero, so our result is nonzero.

Just multiplying out

$$\begin{pmatrix} a_1 & b_1 \\ c_1 & d_1 \end{pmatrix}\left[\begin{pmatrix} a_2 & b_2 \\ c_2 & d_2 \end{pmatrix}\begin{pmatrix} a_3 & b_3 \\ c_3 & d_3 \end{pmatrix}\right] = \left[\begin{pmatrix} a_1 & b_1 \\ c_1 & d_1 \end{pmatrix}\begin{pmatrix} a_2 & b_2 \\ c_2 & d_2 \end{pmatrix}\right]\begin{pmatrix} a_3 & b_3 \\ c_3 & d_3 \end{pmatrix}$$

proves associativity. This is Exercise 2.

It is easily seen that

$$\begin{pmatrix} 1 & 0 \\ 0 & 1 \end{pmatrix}\begin{pmatrix} a & b \\ c & d \end{pmatrix} = \begin{pmatrix} a & b \\ c & d \end{pmatrix}\begin{pmatrix} 1 & 0 \\ 0 & 1 \end{pmatrix} = \begin{pmatrix} a & b \\ c & d \end{pmatrix}$$

so $\begin{pmatrix} 1 & 0 \\ 0 & 1 \end{pmatrix}$ is the identity of this group.

The inverse of $\begin{pmatrix} a & b \\ c & d \end{pmatrix}$ is

$$\begin{pmatrix} \dfrac{d}{ad - bc} & \dfrac{-b}{ad - bc} \\ \dfrac{-c}{ad - bc} & \dfrac{a}{ad - bc} \end{pmatrix}$$

This is easily checked by multiplying out. Let $ad - bc = \Delta \neq 0$. Then

$$\begin{pmatrix} a & b \\ c & d \end{pmatrix}\begin{pmatrix} \dfrac{d}{\Delta} & \dfrac{-b}{\Delta} \\ \dfrac{-c}{\Delta} & \dfrac{a}{\Delta} \end{pmatrix} = \begin{pmatrix} \dfrac{ad - bc}{\Delta} & \dfrac{a(-b) + ba}{\Delta} \\ \dfrac{cd + d(-c)}{\Delta} & \dfrac{c(-b) + ad}{\Delta} \end{pmatrix} = \begin{pmatrix} 1 & 0 \\ 0 & 1 \end{pmatrix}$$

Similarly,

$$\begin{pmatrix} \dfrac{d}{\Delta} & \dfrac{-b}{\Delta} \\[2ex] \dfrac{-c}{\Delta} & \dfrac{a}{\Delta} \end{pmatrix} \begin{pmatrix} a & b \\ c & d \end{pmatrix} = \begin{pmatrix} 1 & 0 \\ 0 & 1 \end{pmatrix}$$

Since
$$\frac{d}{\Delta}\frac{a}{\Delta} - \frac{-b}{\Delta}\frac{-c}{\Delta} = \frac{ab - bc}{\Delta^2} = \frac{1}{ad - bc} \neq 0$$

the inverse of a 2×2 nonsingular matrix-is also nonsingular. Thus we have a group.   ////

Since

$$\begin{pmatrix} 1 & 2 \\ 0 & 3 \end{pmatrix} \begin{pmatrix} 0 & 2 \\ 1 & 0 \end{pmatrix} = \begin{pmatrix} 2 & 2 \\ 3 & 0 \end{pmatrix}$$

but

$$\begin{pmatrix} 0 & 2 \\ 1 & 0 \end{pmatrix} \begin{pmatrix} 1 & 2 \\ 0 & 3 \end{pmatrix} = \begin{pmatrix} 0 & 6 \\ 1 & 2 \end{pmatrix}$$

this group, usually denoted GL $(2,\mathbb{R})$, is nonabelian. GL $(2,\mathbb{R})$ is called the *general linear group* of degree 2 over $\mathbb{R}$.

The remainder of this section is optional, but it is recommended for any reader who wants to see the material on functions applied or who likes multiple proofs. We want to develop another proof that multiplication in GL $(2,\mathbb{R})$ is associative.

If $M = \begin{pmatrix} a & b \\ c & d \end{pmatrix}$ is a 2×2 matrix over $\mathbb{R}$, we want to make M into a function. This can be done by letting $S = T = \mathbb{R}^2 =$ the cartesian plane, and defining $(x,y)M = (x,y)\begin{pmatrix} a & b \\ c & d \end{pmatrix} = (ax + cy, bx + dy)$. [This brings up the peculiarity of writing $(x)f$ instead of $f(x)$, but this causes no difficulty if we let $(s)(f \circ g) = (((s)f)g).$] If $s = (x,y)$, then $sM = (x,y)\begin{pmatrix} a & b \\ c & d \end{pmatrix} = (ax + cy, bx + dy)$ also is in the cartesian plane and is uniquely determined if we know $a, b, c, d, x,$ and y. Thus M is a function.

Composition of functions corresponds to matrix multiplication. Let $s = (x,y)$, $M_1 = \begin{pmatrix} a_1 & b_1 \\ c_1 & d_1 \end{pmatrix}$, and $M_2 = \begin{pmatrix} a_2 & b_2 \\ c_2 & d_2 \end{pmatrix}$. Then

$$(s)(M_1 \circ M_2)$$

$$= (sM_1)M_2 = (a_1 x + c_1 y, b_1 x + d_1 y)\begin{pmatrix} a_2 & b_2 \\ c_2 & d_2 \end{pmatrix}$$

$$= ((a_1x + c_1y)a_2 + (b_1x + d_1y)c_2 , (a_1x + c_1y)b_2 + (b_1x + d_1y)d_2)$$
$$= ((a_1a_2 + b_1c_2)x + (c_1a_2 + c_1d_2)y, (a_1b_2 + b_1d_2)x + (c_1b_2 + d_1d_2)y)$$
$$= (x,y)\begin{pmatrix} a_1a_2 + b_1c_2 & a_1b_2 + b_1d_2 \\ c_1a_2 + d_1c_2 & c_1b_2 + d_1d_2 \end{pmatrix}$$
$$= (x,y)\begin{pmatrix} a_1 & b_1 \\ c_1 & d_1 \end{pmatrix}\begin{pmatrix} a_2 & b_2 \\ c_2 & d_2 \end{pmatrix}$$
$$= (x,y)(M_1 \circ M_2)$$

We now give a one-sentence proof that multiplication in $GL(2,\mathbb{R})$ is associative.

Matrix multiplication corresponds to composition of functions and matrices correspond to functions, so matrix multiplication is associative.

$(x,y)I = (x,y)\begin{pmatrix} 1 & 0 \\ 0 & 1 \end{pmatrix} = (x,y)$ for all (x,y) in $\mathbb{R}^2$ so I is the identity function. Since $M \circ M^{-1} = I = M^{-1} \circ M$, we have that M^{-1} is the inverse function for M.

We have seen that M^{-1} exists if $ad - bc \neq 0$ and $M = \begin{pmatrix} a & b \\ c & d \end{pmatrix}$. If $ad - bc = 0$, then both $(d,-b)M = (d,-b)\begin{pmatrix} a & b \\ c & d \end{pmatrix} = (0,0)$ and $(0,0)M = (0,0)$ [also, $(c,-a)M = (0,0)$]. Thus if $ad - bc = 0$, M is not one-to-one and M^{-1} doesn't exist.

Putting these remarks together, we see that M is nonsingular $\Leftrightarrow ad - bc \neq 0 \Leftrightarrow M^{-1}$ exists $\Leftrightarrow M$ is one-to-one and onto.

Since $(M^{-1})^{-1} = M$, we see that if M^{-1} exists, it in turn has an inverse, is one-to-one and onto, and thus must be nonsingular. If M_1 and M_2 are nonsingular, then both are one-to-one and onto. Then $M_1 \circ M_2$ is also one-to-one and onto (see Exercise 11 or the next section) and thus nonsingular.

All these remarks collectively give another proof that $GL(2,\mathbb{R})$ is a group under matrix multiplication. Summarizing this approach, which may be useful in other contexts, we have: Translate everything over to functions and use the facts there.

Exercises

1 Multiply the following matrices:

(a) $\begin{pmatrix} 1 & 0 \\ 0 & 0 \end{pmatrix}\begin{pmatrix} 3 & 5 \\ 2 & 9 \end{pmatrix}$ (b) $\begin{pmatrix} 0 & 0 \\ 0 & 1 \end{pmatrix}\begin{pmatrix} 3 & 5 \\ 2 & 9 \end{pmatrix}$

(c) $\begin{pmatrix} a & b \\ 0 & c \end{pmatrix}\begin{pmatrix} d & e \\ 0 & f \end{pmatrix}$ (d) $\begin{pmatrix} 1 & 2 \\ 3 & 0 \end{pmatrix}\begin{pmatrix} -2 & 5 \\ 7 & -1 \end{pmatrix}$

(e) $\begin{pmatrix} a & 0 \\ 0 & a \end{pmatrix}\begin{pmatrix} b & c \\ d & f \end{pmatrix}$ (f) $\begin{pmatrix} x & 0 \\ 0 & y \end{pmatrix}\begin{pmatrix} y & 0 \\ 0 & x \end{pmatrix}$

2 Show that

$$\left[\begin{pmatrix} a_1 & b_1 \\ c_1 & d_1 \end{pmatrix}\begin{pmatrix} a_2 & b_2 \\ c_2 & d_2 \end{pmatrix}\right]\begin{pmatrix} a_3 & b_3 \\ c_3 & d_3 \end{pmatrix} = \begin{pmatrix} a_1 & b_1 \\ c_1 & d_1 \end{pmatrix}\left[\begin{pmatrix} a_2 & b_2 \\ c_2 & d_2 \end{pmatrix}\begin{pmatrix} a_3 & b_3 \\ c_3 & d_3 \end{pmatrix}\right]$$

3 Show that the matrices of the form $\begin{pmatrix} a & b \\ 0 & d \end{pmatrix}$, where $ad \neq 0$, form a subgroup of GL $(2,\mathbb{R})$.

4 Show that the matrices $\begin{pmatrix} a & b \\ c & d \end{pmatrix}$, where $ad - bc = 1$, form a subgroup of GL $(2,\mathbb{R})$. This group is called SL $(2,\mathbb{R})$, the *special linear group* of degree 2 over the field $\mathbb{R}$.

5 Define addition of 2×2 matrices over $\mathbb{R}$ by

$$\begin{pmatrix} a_1 & b_1 \\ c_1 & d_1 \end{pmatrix} + \begin{pmatrix} a_2 & b_2 \\ c_2 & d_2 \end{pmatrix} = \begin{pmatrix} a_1 + a_2 & b_1 + b_2 \\ c_1 + c_2 & d_1 + d_2 \end{pmatrix}$$

Show that the set of all (nonsingular or otherwise) 2×2 matrices over $\mathbb{R}$ form an abelian group under this addition.

6 Do the following matrices form a subgroup of GL $(2,\mathbb{R})$?

$$A = \begin{pmatrix} 1 & 0 \\ 0 & 1 \end{pmatrix} \qquad B = \begin{pmatrix} 1 & 0 \\ 0 & -1 \end{pmatrix} \qquad C = \begin{pmatrix} -1 & 0 \\ 0 & -1 \end{pmatrix} \qquad D = \begin{pmatrix} -1 & 0 \\ 0 & 1 \end{pmatrix}$$

Write out a multiplication table.

7 What conditions on a, b, c, and d will guarantee that $\begin{pmatrix} 1 & 0 \\ 0 & 1 \end{pmatrix} = \begin{pmatrix} a & b \\ c & d \end{pmatrix}^2$?

8 If a, b, c, and d are integers, and $ad - bc \neq 0$, does this imply that $\begin{pmatrix} a & b \\ c & d \end{pmatrix}^{-1}$ has integral entries?

9 Show that the matrices $\begin{pmatrix} a & b \\ c & d \end{pmatrix}$, where $ad - bc = 1$ and a, b, c, and d are in $\mathbb{Z}$, form a group.

10 Show that in SL $(2,\mathbb{R})$ (see Exercise 3) there is a unique nonidentity element x such that $x^2 = \begin{pmatrix} 1 & 0 \\ 0 & 1 \end{pmatrix}$ (see Exercise 7).

11 (a) If $f: S \to T$ and $g: T \to U$ are both one-to-one, show that $g \circ f$ is one-to-one.

(b) What happens if we replace one-to-one with onto in part a?

1.7 PERMUTATIONS

Definition A function from a set A to A is a **permutation** on A if it is one-to-one and onto. The set of all permutations on A together with composition of functions as the operation is called the **symmetric group** on A and is denoted Sym (A). Any subgroup of Sym (A) is called a **permutation group** on A.

Our first proposition is merely a proof that the symmetric groups are groups. This, however, will provide us with an important class of examples. We shall later show that every group can be considered as a subgroup of a symmetric group, a fact which is of more philosophic than pragmatic interest. We start by proving a lemma about functions.

Lemma If $f: S \to T$ and $g: T \to U$ are functions, then:
(a) If f and g are one-to-one, then $g \circ f$ is one-to-one.
(b) If f and g are onto, then $g \circ f$ is onto.

PROOF (a) Let $(g \circ f)(s_1) = (g \circ f)(s_2)$. We want to show $s_1 = s_2$. Having nothing else available, we use the definition of composition of functions. $g(f(s_1)) = g(f(s_2))$. Since $f(s_1)$ and $f(s_2)$ are elements of T, and since g is one-to-one, we have $f(s_1) = f(s_2)$. In turn, f is one-to-one so we have $s_1 = s_2$, and $g \circ f$ is one-to-one.

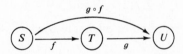

(b) We pick an arbitrary element u in U. Informally, we need to fill in the blank so that $(g \circ f)(\quad) = u$ is satisfied. We again work back to T and then to S. Since g is onto, there is a t in T such that $g(t) = u$. Since f is onto, there is an s in S such that $f(s) = t$. Thus, $(g \circ f)(s) = g(f(s)) = g(t) = u$. We can proceed in the same manner for any u in U, so $g \circ f$ is onto. ////

Proposition 1 If A is a set, then the set Sym (A) of all permutations on A is a group, where the group operation is composition of functions.

PROOF If f and g are both functions from A to A, then certainly $f \circ g$ is also a function from A to A. By the lemma, $f \circ g$ is one-to-one and onto, since both f and g are one-to-one and onto.

As we have shown in Sec. 1.5, composition of functions is associative.

If we define the function I_A by $I_A(a) = a$ for all a in A, I_A is easily seen to be a permutation. If f is any permutation of A, we have

$$(I_A \circ f)(a) = I_A(f(a)) = f(a) = f(I_A(a)) = (f \circ I_A)a$$

for any element a in A. Thus,

$$I_A \circ f = f = f \circ I_A$$

and I_A is an identity for Sym (A).

By the remarks of Sec. 1.5, if f is a one-to-one onto function, then f^{-1} is a function that is also one-to-one and onto. Since

$$(f \circ f^{-1})(a) = f(f^{-1}(a)) = a \qquad \text{by definition of } f^{-1}$$

and $\qquad (f^{-1} \circ f)(a) = f^{-1}(f(a)) = a \qquad \text{same reason}$

we have $\quad (f \circ f^{-1})(a) = (f^{-1} \circ f)(a) = a = I_A(a)$

This holds for any a in A, so

$$f \circ f^{-1} = f^{-1} \circ f = I_A$$

The f^{-1} turn out to be inverses not only as functions, but also as the inverses required by the group axioms, and Sym (A) is a group.   ////

If A is a finite set, we usually take $A = \{1,2,...,n\}$ and write Sym (n) for Sym (A). [$\mathscr{S}_n$ is used in many places instead of Sym (n).]

EXAMPLE A   Let us look at Sym (3), the group of all permutations of 3 elements. We can write a permutation of a finite set simply by telling where each element goes. If $f(1) = 2, f(2) = 3$, and $f(3) = 1$, we write this as $\begin{pmatrix} 1 & 2 & 3 \\ 2 & 3 & 1 \end{pmatrix}$. In general, if f is a permutation belonging to Sym (n), we write it as

$$\begin{pmatrix} 1 & 2 & \cdots & n \\ f(1) & f(2) & \cdots & f(n) \end{pmatrix}$$

It is trivial to compute the effect of the function. If g is $\begin{pmatrix} 1 & 2 & 3 & 4 \\ 2 & 1 & 4 & 3 \end{pmatrix}$, then

$$g(1) = \begin{pmatrix} 1 & 2 & 3 & 4 \\ 2 & 1 & 4 & 3 \end{pmatrix}(1) = 2$$

$$g(2) = \begin{pmatrix} 1 & 2 & 3 & 4 \\ 2 & 1 & 4 & 3 \end{pmatrix}(2) = 1$$

$$g(3) = \begin{pmatrix} 1 & 2 & 3 & 4 \\ 2 & 1 & 4 & 3 \end{pmatrix}(3) = 4$$

and $$g(4) = \begin{pmatrix} 1 & 2 & 3 & 4 \\ 2 & 1 & 4 & 3 \end{pmatrix}(4) = 3$$

Regarding permutations as functions also tells us how to multiply permutations. If in Sym (3) we have

$$f = \begin{pmatrix} 1 & 2 & 3 \\ 2 & 3 & 1 \end{pmatrix} \quad \text{and} \quad g = \begin{pmatrix} 1 & 2 & 3 \\ 3 & 2 & 1 \end{pmatrix}$$

then $$(f \circ g)(1) = \left(\begin{pmatrix} 1 & 2 & 3 \\ 2 & 3 & 1 \end{pmatrix} \circ \begin{pmatrix} 1 & 2 & 3 \\ 3 & 2 & 1 \end{pmatrix} \right)(1)$$

$$= \begin{pmatrix} 1 & 2 & 3 \\ 2 & 3 & 1 \end{pmatrix} \left(\begin{pmatrix} 1 & 2 & 3 \\ 3 & 2 & 1 \end{pmatrix}(1) \right)$$

$$= \begin{pmatrix} 1 & 2 & 3 \\ 2 & 3 & 1 \end{pmatrix}(3) = 1$$

Also $(f \circ g)(2) = \left(\begin{pmatrix} 1 & 2 & 3 \\ 2 & 3 & 1 \end{pmatrix} \circ \begin{pmatrix} 1 & 2 & 3 \\ 3 & 2 & 1 \end{pmatrix} \right)(2) = \begin{pmatrix} 1 & 2 & 3 \\ 2 & 3 & 1 \end{pmatrix}(2) = 3$

and $(f \circ g)(3) = \left(\begin{pmatrix} 1 & 2 & 3 \\ 2 & 3 & 1 \end{pmatrix} \begin{pmatrix} 1 & 2 & 3 \\ 3 & 2 & 1 \end{pmatrix} \right)(3) = \begin{pmatrix} 1 & 2 & 3 \\ 2 & 3 & 1 \end{pmatrix}(1) = 2$

Putting these three computations together, we see that

$$\begin{pmatrix} 1 & 2 & 3 \\ 2 & 3 & 1 \end{pmatrix}\begin{pmatrix} 1 & 2 & 3 \\ 3 & 2 & 1 \end{pmatrix} = \begin{pmatrix} 1 & 2 & 3 \\ 1 & 3 & 2 \end{pmatrix}$$

This procedure gives our multiplication procedure in Sym (n). (See Exercise 3.) Remember that we are dealing with functions, apply the function on the right first, then the one on the left, and run through the elements 1, 2, ..., n to determine the product permutation.

Before we write down a multiplication table and subgroup lattice for Sym (3), we give it a geometric interpretation. Assume we have a fixed hollow equilateral triangular frame with vertices labeled

Assume also that we have a triangle which fits into this frame and which can be taken out and fitted back in. Each such movement is called a symmetry. For instance, taking the triangle out and rotating it 120° clockwise will move the vertex at 1 to position 3, that at 3 to 2, and that at 2 to 1, so this rotation corresponds to $\begin{pmatrix} 1 & 2 & 3 \\ 3 & 1 & 2 \end{pmatrix}$. Let us see what happens if we take out the triangle, flip it about the line passing through 2 that bisects the triangle, and replace it in the frame. The vertices at 1 and 3 are interchanged, and the vertex at 2 is not moved. Therefore we can write this as $\begin{pmatrix} 1 & 2 & 3 \\ 3 & 2 & 1 \end{pmatrix}$. Multiplying as before, we see that

$$\begin{pmatrix} 1 & 2 & 3 \\ 3 & 1 & 2 \end{pmatrix}\begin{pmatrix} 1 & 2 & 3 \\ 3 & 2 & 1 \end{pmatrix} = \begin{pmatrix} 1 & 2 & 3 \\ 2 & 1 & 3 \end{pmatrix}$$

which corresponds to flipping the triangle about the bisector through 3. Thus, first rotating the triangle 120° clockwise and then flipping about is the same as just flipping the triangle about

If we let $a = \begin{pmatrix} 1 & 2 & 3 \\ 2 & 3 & 1 \end{pmatrix} = 240°$ rotation clockwise, and $b = \begin{pmatrix} 1 & 2 & 3 \\ 3 & 2 & 1 \end{pmatrix}$,

then $a^2 = 120°$ rotation clockwise $= \begin{pmatrix} 1 & 2 & 3 \\ 3 & 1 & 2 \end{pmatrix}$,

$$ab = \begin{pmatrix} 1 & 2 & 3 \\ 2 & 3 & 1 \end{pmatrix}\begin{pmatrix} 1 & 2 & 3 \\ 3 & 2 & 1 \end{pmatrix} = \begin{pmatrix} 1 & 2 & 3 \\ 1 & 3 & 2 \end{pmatrix},$$

and $a^2b = \begin{pmatrix} 1 & 2 & 3 \\ 3 & 1 & 2 \end{pmatrix}\begin{pmatrix} 1 & 2 & 3 \\ 3 & 2 & 1 \end{pmatrix} = \begin{pmatrix} 1 & 2 & 3 \\ 2 & 1 & 3 \end{pmatrix}$. We obtain the following multi-

plication table:

	I	a	a^2	b	ab	a^2b
I	I	a	a^2	b	ab	a^2b
a	a	a^2	I	ab	a^2b	b
a^2	a^2	I	a	a^2b	b	ab
b	b	a^2b	ab	I	a^2	a
ab	ab	b	a^2b	a	I	a^2
a^2b	a^2b	ab	b	a^2	a	I

Several items are worth noting. First,

$$ab = \begin{pmatrix} 1 & 2 & 3 \\ 2 & 3 & 1 \end{pmatrix}\begin{pmatrix} 1 & 2 & 3 \\ 3 & 2 & 1 \end{pmatrix} = \begin{pmatrix} 1 & 2 & 3 \\ 1 & 3 & 2 \end{pmatrix} \neq \begin{pmatrix} 1 & 2 & 3 \\ 2 & 1 & 3 \end{pmatrix}$$

$$= \begin{pmatrix} 1 & 2 & 3 \\ 3 & 2 & 1 \end{pmatrix}\begin{pmatrix} 1 & 2 & 3 \\ 2 & 3 & 1 \end{pmatrix} = ba$$

So Sym (3) is nonabelian. In fact, Sym (3) is the smallest nonabelian group, for, as we shall show, all groups of order 5 or less are abelian. The proper subgroups of Sym (3) are

$$H_1 = \langle b \rangle \qquad H_2 = \langle ab \rangle \qquad H_3 = \langle a^2b \rangle$$

each with two elements, and Alt (3) $= \langle a \rangle$ of order 3. So a subgroup lattice for Sym (3) is as follows:

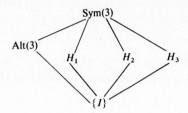

These subgroups have been produced by picking elements and looking at their powers, which form the cyclic subgroups. The notation Alt (3) will be explained later and has no special significance here. To show that these are the only subgroups, it suffices to examine a few cases, such as the following: If J is a subgroup containing e, b, and ab, then $ab \cdot b = a$ is in J as are $a \cdot a = a^2$ and $a^2 \cdot b = a^2b$, so $J = \text{Sym} (3)$.

Exercises

1   If $g \circ f: S \to U$ is onto, where $f: S \to T$ and $g: T \to U$ are functions, show that g is onto. Construct an example with g and $g \circ f$ onto but f not onto.

2   If $g \circ f: S \to U$ is one-to-one, where $f: S \to T$ and $g: T \to U$ are functions, show that f must be one-to-one but that (again produce an example) g need not be one-to-one.

3   Multiply the following permutations in Sym (5):

(a) $\begin{pmatrix} 1 & 2 & 3 & 4 & 5 \\ 3 & 2 & 4 & 1 & 5 \end{pmatrix}\begin{pmatrix} 1 & 2 & 3 & 4 & 5 \\ 2 & 3 & 4 & 1 & 5 \end{pmatrix}$

(b) $\begin{pmatrix} 1 & 2 & 3 & 4 & 5 \\ 2 & 1 & 3 & 4 & 5 \end{pmatrix}\begin{pmatrix} 1 & 2 & 3 & 4 & 5 \\ 1 & 3 & 2 & 4 & 5 \end{pmatrix}$

(c) $\begin{pmatrix} 1 & 2 & 3 & 4 & 5 \\ 2 & 3 & 4 & 5 & 1 \end{pmatrix}\begin{pmatrix} 1 & 2 & 3 & 4 & 5 \\ 2 & 1 & 3 & 4 & 5 \end{pmatrix}$

Proposition 2 $|\text{Sym} (n)| = n!$

PROOF   If $A = \{1,2,\dots,n\}$ and we look at all permutations of A, then 1 can be mapped to $1, 2, 3, \dots,$ or n. We have n choices of where to place 1. Once 1 is placed, we have $n - 1$ remaining choices of where to place 2. Then we have $n - 2$ choices for 3. Continuing, we find that we have $n(n - 1)(n - 2) \cdots (3)(2)(1) = n!$ ways to place all n elements, and thus $|\text{Sym} (n)| = n!$ ////

Exercise

4   †This last proof contains the word "continuing" and the three dots "$\cdots$," which disguise an induction proof. Give a rigorous proof by induction.

(It might be easier to show that if A and B both have k elements there are $k!$ functions from A onto B.)

EXAMPLE B Another important group is D_8, the dihedral group of order 8 or the group of symmetries of the square. Let us label the vertices of a square as follows:

Again consider this to be a fixed square frame to use in labeling the movements of a square fitted inside the frame. Then a 90° rotation counterclockwise is $\begin{pmatrix} 1 & 2 & 3 & 4 \\ 2 & 3 & 4 & 1 \end{pmatrix}$. For brevity let us set $a = \begin{pmatrix} 1 & 2 & 3 & 4 \\ 2 & 3 & 4 & 1 \end{pmatrix}$. Two 90° rotations performed successively make a 180° rotation, which is denoted $\begin{pmatrix} 1 & 2 & 3 & 4 \\ 3 & 4 & 1 & 2 \end{pmatrix}$. Checking, we see that

$$a^2 = \begin{pmatrix} 1 & 2 & 3 & 4 \\ 3 & 4 & 1 & 2 \end{pmatrix} = a \cdot a = \begin{pmatrix} 1 & 2 & 3 & 4 \\ 2 & 3 & 4 & 1 \end{pmatrix}\begin{pmatrix} 1 & 2 & 3 & 4 \\ 2 & 3 & 4 & 1 \end{pmatrix}$$

Similarly, $a^3 = \begin{pmatrix} 1 & 2 & 3 & 4 \\ 4 & 1 & 2 & 3 \end{pmatrix}$, and $a^4 = \begin{pmatrix} 1 & 2 & 3 & 4 \\ 1 & 2 & 3 & 4 \end{pmatrix} = I$.

Let b be the flip about the diagonal through the vertices at 1 and 3; then $b = \begin{pmatrix} 1 & 2 & 3 & 4 \\ 1 & 4 & 3 & 2 \end{pmatrix}$, and

$$ba = \begin{pmatrix} 1 & 2 & 3 & 4 \\ 4 & 3 & 2 & 1 \end{pmatrix} = \text{the ``vertical flip''}$$

$$ba^2 = \begin{pmatrix} 1 & 2 & 3 & 4 \\ 3 & 2 & 1 & 4 \end{pmatrix} = \text{``flip about the other diagonal''}$$

and $\quad ba^3 = \begin{pmatrix} 1 & 2 & 3 & 4 \\ 2 & 1 & 4 & 3 \end{pmatrix} = \text{``flip through the horizontal axis''}$

We note that $I = a^4 = 360°$ rotation, that $b^2 = I$, and that $ba^3 = ab$. Since $ba^3 \neq ba$, D_8 is not abelian. Just using $a^4 = b^2 = I$ and $ba^3 = ab$, we can write down a multiplication table:

	I	a	a^2	a^3	b	ba	ba^2	ba^3
I	I	a	a^2	a^3	b	ba	ba^2	ba^3
a	a	a^2	a^3	I	ba^3	b	ba	ba^2
a^2	a^2	a^3	I	a	ba^2	ba^3	b	ba
a^3	a^3	I	a	a^2	ba	ba^2	ba^3	b
b	b	ba	ba^2	ba^3	I	a	a^2	a^3
ba	ba	ba^2	ba^3	b	a^3	I	a	a^2
ba^2	ba^2	ba^3	b	ba	a^2	a^3	I	a
ba^3	ba^3	b	ba	ba^2	a	a^2	a^3	I

The following is a subgroup lattice for D_8 (we now use e instead of I just to make the terminology more uniform):

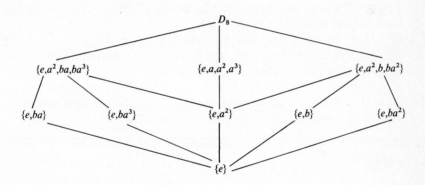

Exercises

5 Verify that $\{e,a^2,ba,ba^3\}$ is a subgroup of D_8.

6 Write down a multiplication table for a group, given only that $a^5 = e = b^2$ and $ab = ba^4$. This group is called D_{10}. What is a geometric interpretation of D_{10}?

7 Of the seven nonidentity elements of D_8, is there an element that commutes with every other element of D_8? Is there any such element in Sym (3)?

8 Show that, for any even number $2n$ greater than or equal to 6, there is at least one nonabelian group of order $2n$.

9 Compute a^2, ab, ba, and b^2, where a and b are permutations in Sym (3) as denoted on page 34.

10 Let $\rho = \begin{pmatrix} 1 & 2 & 3 & 4 & 5 & 6 \\ 4 & 2 & 1 & 3 & 6 & 5 \end{pmatrix}$ in Sym (6). What are the elements of $\langle \rho \rangle$?

11 If $a \in A$, let H be the subset of all f in Sym (A) such that $f(a) = a$. Show that H is a subgroup of Sym (A).

12 Let the 2×2 matrix $\begin{pmatrix} a & b \\ c & d \end{pmatrix}$ be the function from the real plane to the real plane that takes the point (x,y) to the point $(ax + cy, bx + dy)$. We write this as

$$(x,y)\begin{pmatrix} a & b \\ c & d \end{pmatrix} = (ax + cy, bx + dy)$$

(*a*) Compute $(1,0)\begin{pmatrix} 1 & 1 \\ 0 & 1 \end{pmatrix}$, $(5,6)\begin{pmatrix} 1 & 2 \\ 3 & 4 \end{pmatrix}$, and $(x,y)\begin{pmatrix} 4 & 4 \\ 0 & 0 \end{pmatrix}$.

(*b*) Find all points (x,y) such that $(x,y)\begin{pmatrix} 0 & 1 \\ 1 & 0 \end{pmatrix} = (x,y)$.

(*c*) Show that if $M = \begin{pmatrix} a & b \\ c & d \end{pmatrix}$ is nonsingular, M is a permutation of $\mathbb{R}^2$.

(*d*) Find all $M = \begin{pmatrix} a & b \\ c & d \end{pmatrix}$ such that $(1,0)\begin{pmatrix} a & b \\ c & d \end{pmatrix} = (1,0)$.

13 Consider the group of rigid motions of a cube. This corresponds to all motions of a cube that are possible by taking a cube out of a box, moving it about, and replacing it in the box.

(*a*) Show that this group has 24 elements.

(*b*) Let s be a side of the cube, and let H_s be the subgroup of all motions that take s to s. What is the order of H_s?

(*c*) Find two "geometric" subgroups other than those of part b.

14 Show that, if a is the only element of order 2 in a group G, then $xa = ax$ for all x in G. (*Hint*: Use Exercise 9 of Sec. 1.3.)

15 If $\alpha = \begin{pmatrix} 1 & 2 & 3 & 4 & 5 & 6 \\ 2 & 1 & 5 & 6 & 3 & 4 \end{pmatrix}$, what is α^{-1}? How can this be generalized?

*1.8   EQUIVALENCE RELATIONS

In this section we discuss equivalence relations. Equivalence relations generalize the idea of equality and are extremely useful in counting problems.

Definition   A **relation** R from S to T is a subset of $S \times T$, the set of ordered pairs (s,t) with $s \in S$ and $t \in T$. If $(s,t) \in R$, we often write this as sRt or $s \sim t$.

A function f from S to T is just a relation in which each s in S is in exactly one pair (s,t) in the relation. The unique t is often called $f(s)$.

Definition   A relation from S to S is called an **equivalence relation** on S if:

(*a*) $s \sim s$ for all s in S (reflexive law).

(*b*) If $s \sim t$, then $t \sim s$ (symmetric law).

(*c*) If $s \sim t$ and $t \sim v$, then $s \sim v$ (transitive law).

If $\sim$ is an equivalence relation, $s \sim t$ is read as "s is equivalent to t."

EXAMPLE A If S is any set and $\sim$ is $=$, then $\sim$ is an equivalence relation on S. Thus equality is an equivalence relation, and one way to view equivalence relations is as a generalization of equality.

EXAMPLE B Let S be the real numbers, and let $\sim$ be $\leq$. Then $s \sim s$ since $s \leq s$ for all s in $\mathbb{R}$. Also, if $s \leq t$ and $t \leq v$, then $s \leq v$, so the transitive law holds as well as the reflexive. But $s \leq t$ does not imply $t \leq s$, so $\leq$ is not an equivalence relation.

EXAMPLE C Let $S = \mathbb{Z}$, and let $a \sim b$ if $a - b$ is an even integer. We first note that this can be written as $a \sim b$ if and only if $a - b = 2k$, where $k \in \mathbb{Z}$.

   (a) Since $a - a = 0 = 2 \cdot 0$ for all a in $\mathbb{Z}$, we have $a \sim a$.
   (b) If $a - b = 2k$, then $b - a = -2k = 2(-k)$, so $a \sim b$ implies $b \sim a$.
   (c) If $a - b = 2k$ and $b - c = 2l$, adding gives us $(a - b) + (b - c) = a - c = 2(k + l)$, so $a \sim b$ and $b \sim c$ imply $a \sim c$.

Thus, this is an equivalence relation.

Exercise

1 Which of the following are equivalence relations?

   (a) $S = \mathbb{Z}$ and $a \sim b$ if $a - b = 5k$, where k is in Z.
   (b) $S = \mathbb{R}$ and $a \sim b$ if $a - b < 0$.
   (c) $S = $ the set of all functions from $\mathbb{R}$ to $\mathbb{R}$ (as in calculus), where $f \sim g$ if $f(x) - g(x) = c$ for some constant c.
   (d) $S = $ the plane, and $p \sim q$ with $p = (x_1, y_1)$, $q = (x_2, y_2)$ if $x_1^2 + y_1^2 = x_2^2 + y_2^2$.

Definition If $\sim$ is an equivalence relation on a set S, then for t in S, $\bar{t} = \{v \mid t \sim v\}$ is the **equivalence class** containing t.

EXAMPLE D $S = \mathbb{Z}$, $a \sim b$ if $a - b = 2k$, where $k \in Z$.

$$\bar{1} = \{n \mid 1 \sim n\} = \{n \mid 1 - n = 2k\} = \{n \mid n = 1 - 2k\}$$
$$= \text{the odd integers} = \bar{3} = \bar{5} = \cdots$$
$$\bar{0} = \{n \mid 0 \sim n\} = \{n \mid 0 - n = 2k\} = \{n \mid n = 2(-k)\}$$
$$= \text{the even integers} = \bar{2} = \bar{4} = -\bar{8} = \cdots$$

Here we note that by using $\sim$ we have divided the integers into two disjoint subsets. This is a typical use for equivalence relations.

   Lemma If S is a set, and $\sim$ is an equivalence relation on S, then $s \in \bar{t}$ implies $\bar{s} = \bar{t}$.

PROOF We shall show $\bar{s} \subseteq \bar{t}$ and $\bar{t} \subseteq \bar{s}$. If $x \in \bar{s}$, then $s \sim x$. Since $s \in \bar{t}$ we know $t \sim s$ and, using the transitive law, we have $t \sim x$, so $x \in \bar{t}$. Since x was any element of $\bar{s}$, we must have $\bar{s} \subseteq \bar{t}$. Conversely, let $y \in \bar{t}$. Then $t \sim y$ and we still have $t \sim s$. By the symmetric law we have $s \sim t$, and thus the transitive law gives $s \sim y$. Hence $y \in \bar{s}$ and again, as y is an arbitrary element of $\bar{t}$, we have $\bar{t} \subseteq \bar{s}$ and thus $\bar{t} = \bar{s}$. ////

Definition If S is a set, and $\{P_\alpha\}$ a collection of nonempty subsets of S, then $\{P_\alpha\}$ is a **partition** of S if

(a) $\bigcup_\alpha P_\alpha = S$

(b) $P_\alpha \cap P_\beta = \varnothing$ unless $\alpha = \beta$.

Stated less formally, this says that every element of S is in one of the P_α and in only one of the P_α.

Proposition 1 If S is a set, and $\sim$ is an equivalence relation on S, then the equivalence classes form a partition of S.

PROOF Since $t \in \bar{t}$ for all t in S, we see that every element in S is in at least one equivalence class. If we assume $t \in \bar{u} \cap \bar{v}$, then by the lemma $\bar{t} = \bar{u} = \bar{v}$, so t is in but one equivalence class (although this equivalence class may go by many names). So the equivalence classes form a partition. ////

Exercises

2 If S is a set partitioned by $\{P_a\}$, show that the following is an equivalence relation: $s \sim t$ if and only if s and t are in the same P_a.

3 In Exercise 1, what are the equivalence classes in those examples where $\sim$ turned out to be an equivalence relation?

The first application of these concepts will be in the second proof of Lagrange's theorem in the next section. In the exercises of the next section there will be another application in the proof of the class equation. Equivalence relations also arise naturally in the discussion of quotient fields, groups acting on sets, and isomorphisms, as will be seen later.

4 Equivalence classes are used extensively in geometry. We seldom are interested in two triangles being equal, but spend much time in establishing criteria for their being congruent (for example, the side-angle-side criterion).

(a) Let S be the set of all triangles in the plane, and let $\sim$ be defined by $T_1 \sim T_2$ if T_1 is congruent to T_2. Show that $\sim$ is an equivalence relation.

(b) Redo part a, replacing "congruent" with "similar."

(c) Let S be the set of all circles in the plane, and let $C_1 \sim C_2$ if C_1 and C_2 have the same center. Show that $\sim$ is an equivalence relation.

(d) Make up and verify a similar example.

1.9   LAGRANGE'S THEOREM

We now come to the first important theorem in group theory. If G is a finite group, and H a subgroup, we see by looking at examples that $|H|$ divides $|G|$. This is, in fact, always true, and we shall offer two proofs. The second uses equivalence relations and is a bit slicker than the first.

Definition   Let H and K be subsets of a group G. Then $HK = \{hk \mid h \in H, k \in K\}$. In the particular case that $K = \{g\}$, $H\{g\}$ is denoted Hg. In the even more particular case that H is a subgroup, Hg is called a **right coset** of H.

EXAMPLE A   Let $G = (\mathbb{R} - (0), \cdot)$, $H = \{1,2,5\}$, and $K = \{-1,3,\frac{3}{2}\}$. Then $HK = \{-1, -2, -5, 3, 6, 15, \frac{7}{2}, 7, \frac{3 \cdot 5}{2}\}$.

EXAMPLE B   Let $G = (\mathbb{R} - (0), \cdot)$, and let $H = K = \{2,4,6,8,...\}$. Then $HH = \{4,8,12,16,20,...\} \subsetneqq H$.

EXAMPLE C   If $G = \text{Sym}(3)$, $H = \{a,a^2\}$, and $K = \{e,b\}$, then $HK = \{a,a^2,ab,a^2b\}$.

Exercise

1   In the group D_8 find two subgroups H and K such that HK is not a subgroup. Can you find two distinct proper subgroups L and M such that LM is a subgroup?

Proposition 1   If H is a subgroup, then $HH = H$.

PROOF $HH = \{h_1 h_2 \mid h_1, h_2 \in H\} \subseteq H$ by the fact that H is closed. Since $H = \{h \mid h \in H\} = \{eh \mid h \in H\} = \{e\}H \subseteq HH$ we have the reverse inclusion, and thus $HH = H$.   ////

Proposition 2   If H is a subgroup of G, and Ha is any right coset of H, then there is a one-to-one function from H onto Ha.

PROOF   Define $f_a : H \to Ha$ by $f_a(h) = ha$. If $h_1 a$ is any member of Ha then $f_a(h_1) = h_1 a$, so f_a is onto. To show f_a is one-to-one, we assume $h_1 a = f_a(h_1) = f_a(h_2) = h_2 a$, and $h_1 = h_2$ by the right cancellation law. Thus f_a is one-to-one.   ////

Definition   If G is a finite group, then $|G|$, the number of elements in G, is called the **order** of G. If g is an element, then the **order of** g is the order of the subgroup $\langle g \rangle = \{e,g,g^2,...,g^{-1},g^{-2},...\}$. If $\langle g \rangle$ is finite, then

$$\langle g \rangle = \{g,g^2,...,g^k = e\},$$

so the order of g is the smallest positive integer k such that $g^k = e$. (See Exercise 17 in Sec. 1.4.)

Proposition 3 If H is a finite subgroup of a group G, then $|H| = |Ha|$.

PROOF Since there is a one-to-one onto map from H to Ha, they have the same number of elements. (In fact, with proper interpretation we can remove the word "finite," and Proposition 3 will still be true. Two infinite sets are said to *have the same cardinality* if there is a one-to-one function from one onto the other. Thus the proof of Proposition 2 shows that in all cases H and Ha have the same cardinality.) ////

EXAMPLE D Let H be the subgroup $\{1,-1\}$ of $(\mathbb{R} - (0), \cdot)$, and let $a = 5$. Then $Ha = \{5,-5\}$.

EXAMPLE E Let $H = 2Z$ be the subgroup of even integers of the group $(Z,+)$. Then $2Z + 5 = \{5,7,3,9,1,11,-1,1,...\} = 2Z + 1 =$ the odd integers. Here, since the group operation is addition, we write Ha as $H + a$.

EXAMPLE F In the group Sym (3), let $H = \{e,b\}$. Then $He = Hb = H$, $Ha = Hba = \{a,ba\}$, and $Ha^2 = Hba^2 = \{a^2,ba^2\}$. Note that the same coset can be labeled several ways.

Exercises

2 Find all the right cosets of $H = \{e,a,a^2,a^3\}$ in D_8. Do the same for $\{e,b\} = J$.
3 Find a group G with a subgroup H and an element g such that $Hg \neq gH$ (where $gH = \{g\}H$ is a *left coset* of H).
4 Check that all subgroups of D_8 have order 1, 2, 4, or 8, and that all subgroups of Sym (3) have order 1, 2, 3, or 6.

Lagrange's theorem If G is a finite group and H is a subgroup, then $|H|$ divides $|G|$. (Divides means divides evenly with no remainder.)

CLASSICAL PROOF Let $H = \{h_1,h_2,...,h_k\}$, and start writing the elements of G in the following manner:

$$h_1, h_2, \cdots, h_k$$
$$h_1a, h_2a, \cdots, h_ka$$
$$h_1b, h_2b, \cdots, h_kb$$

a is picked to be any element not in the first row, b to be any element not in the first two rows. Continue in this manner until all elements in G are listed.

$$
\begin{array}{ll}
h_1, \ h_2, \ ..., \ h_k & H \\
h_1a, h_2\,a, \ ..., \ h_k\,a & Ha \\
h_1b, h_2\,b, \ ..., \ h_k\,b & Hb \\
\cdots\cdots\cdots\cdots\cdots & \cdots \\
h_1x, h_2\,x, \ ..., \ h_k\,x & Hx
\end{array}
$$

Note that each row constitutes a right coset, so each row contains $|H|$ distinct elements by Proposition 2. What happens if two rows contain an element in common? Say Hy contains $h_i w$, where Hw is a row other than Hy. Then $h_i w = h_j y$ for some h_j in H. Then $y = h_j^{-1} h_i w \in Hw$. So y would have been included in an earlier row, and we would not have selected y to start a row. Thus no two rows have elements in common. If m is the number of distinct right cosets, we see that G can be divided into m rows of $|H|$ elements each, so $|G| = |H|m$, and $|H|$ divides $|G|$.

////

Corollary 1 If G is a finite group, H a subgroup, and m the number of distinct cosets of H, then $m|H| = |G|$. This number m is the **index** of H in G.

PROOF Use the same proof as the theorem. ////

Corollary 2 If g is an element of a finite group G, then the order of g divides $|G|$.

PROOF By definition, the order of g is $|\langle g \rangle|$, and $\langle g \rangle$ is a subgroup of G, so by Lagrange's theorem we have our result. ////

Corollary 3 If G is a group of prime order p, then G is cyclic.

PROOF Let g be any nonidentity element of G. Then $|\langle g \rangle| = 1$ or p, since $|G| = p$. Since e and g are in $\langle g \rangle$, we must have $|\langle g \rangle| = p = |G|$, so $G = \langle g \rangle$, and G is cyclic. ////

Exercises

5 Write down all multiplication tables for groups of order 5. But first spend 10 minutes trying to do this without Corollary 3.

6 If G is a group of order $2p$ where p is prime, show that every proper subgroup of G is cyclic.

7 Show that every proper subgroup of Sym (3) is cyclic.

8 Show that, if x has order n in G, and d is a positive integer dividing n, then G has an element of order d.

9 (From an old graduate record exam) If G is a group of order 60, can it have a subgroup of order 24?

Proposition 4 If G is a group and H is a subgroup of G, then $\sim$ is an equivalence relation on G, where we define $\sim$ by $x \sim y$ if $xy^{-1} \in H$. The equivalence classes for $\sim$ are the right cosets of H.

PROOF Since $e = xx^{-1}$ is in H, we see that $x \sim x$ for all x in G, and the reflexive property holds. If $x \sim y$, then $xy^{-1} \in H$ so $(xy^{-1})^{-1} =$

$(y^{-1})^{-1}x^{-1} = yx^{-1} \in H$, since H contains the inverse of any element in H. Thus $y \sim x$, and $\sim$ is symmetric. If $x \sim y$ and $y \sim z$, then xy^{-1} and yz^{-1} are in H, and since H is closed $(xy^{-1})(yz^{-1}) = xz^{-1}$ is in H. Hence $x \sim z$, and $\sim$ is transitive. Therefore $\sim$ is an equivalence relation.

Let us determine the equivalence class containing x. If $x \sim y$, then xy^{-1} is in H, so $(xy^{-1})^{-1} = yx^{-1}$ is in H and $yx^{-1} = h$ in H, so $y = hx$ for some h in H, so $y \in Hx$. Each of these steps is reversible, so $z \in Hx$ implies $x \sim z$, and the equivalence classes of $\sim$ are the right cosets of H.

////

SECOND PROOF OF LAGRANGE'S THEOREM   If G is the finite group and H is the subgroup, then we define $\sim$ on G by $x \sim y$ if $xy^{-1} \in H$. By Proposition 4, $\sim$ is an equivalence relation with the equivalence classes being the right cosets of H. Thus the right cosets partition G (see the last section). All the right cosets contain $|H|$ distinct elements by Proposition 3. So $|H|m = |G|$, where m is the number of distinct right cosets of H. Thus $|H|$ divides $|G|$.

////

It is interesting to note that Lagrange died in 1812, which was 20 years before Galois first defined group more or less as we know it, and 40 years before groups were discussed at all widely. In the last section of this chapter we discuss the theorem that Lagrange did prove.

The following set of exercises shows another powerful application of equivalence relations to finite groups. (This might be a good reading-period assignment or independent work project).

Exercises

10   †If G is a group, then $Z(G) = \{x \,|\, gx = xg \text{ for all } g \in G\} = $ the *center* of G. Find $Z(D_8)$, $Z((\mathbb{Z}, +))$, and $Z(\text{Sym}\,(3))$.

11   †If G is a group, define $\sim$ on G by $x \sim y$ if $y = g^{-1}xg$ for some g in G. Show that $\sim$ is an equivalence relation. Find the equivalence classes in Sym (3) and D_8.

12   †If $g \in G$, then $C_G(x) = \{g \,|\, xg = gx\} = $ the centralizer of x in G. Show that $C_G(x)$ is always a subgroup of G. Another way to put this is that $C_G(x)$ is the subgroup consisting of those elements that commute with x.

13   †Find $C_G(x)$ for several choices of x in each of Sym (3), D_8, any abelian group, and for $x = \begin{pmatrix} 1 & 1 \\ 0 & 1 \end{pmatrix}$ in $G = \text{GL}\,(2, \mathbb{R})$.

14   †If $x \in G$, $\bar{x}$ is the equivalence class of x under $\sim$, and G is finite, show $|G| = |\bar{x}| |C_G(x)|$.

15   †Show that $x \in Z(G)$ if and only if $\{x\} = \bar{x}$ if and only if $C_G(x) = G$.

16   †Use the fact that $\sim$ is an equivalence relation to show that G is the disjoint union of $Z(G)$ and those equivalence classes consisting of more than one element. Check this for Sym (3) and D_8.

17 †Show that $|G| = |Z(G)| + \sum_{i=1}^{k} [|G|/|C_G(x_i)|]$, where one x_i is chosen from each equivalence class containing more than one element.

18 †If p is a prime and $|G| = p^n$, show that $|Z(G)| > 1$. Does this check for D_8?

†Let us look at one example of cosets. Let $G = (\mathbb{R}^2, +)$, which we can view as the cartesian plane. Any line through the origin is a subgroup.

19 If $L = \{(x,mx), x \in R\}$ for some constant $m \in \mathbb{R}$, show that L is a subgroup of G.

For the sake of concreteness, let L be $\{(x,2x) | x \in \mathbb{R}\}$, which is just the points comprising the line $y = 2x$. Let us look at a right coset of L. If $a = (4,-1)$, then $L + a = \{(x,2x) | x \in \mathbb{R}\} + (4,-1) = \{(x+4,2x-1) | x \in \mathbb{R}\} = \{(x,2(x-4)-1) | x \in \mathbb{R}\} = \{(x,2x-9) | x \in \mathbb{R}\}$, which is the line $y = 2x - 9$, which is a line parallel to $y = 2x$.

20 If $L = \{(x,2x) | x \in \mathbb{R}\}$, show that any coset of L is a line parallel to $y = 2x$, and conversely.

21 If $L = \{(x,mx) | x \in \mathbb{R}\}$ and m is some constant in $\mathbb{R}$, show that the cosets of L are just the lines parallel to $y = mx$.

22 If $L = \{(0,x) | x \in \mathbb{R}\}$, show that the cosets of L are just the set of vertical lines.

Using the exercises and this picture of the cosets, we can offer a phony proof of Euclid's fifth postulate. Euclid's fifth postulate can be formulated as follows: Given a point and a line, there exists a line through the given point and parallel to the given line. For many years people tried to prove this postulate, using the earlier and more natural-sounding postulates; these efforts eventually led to the construction of noneuclidean geometries satisfying all but the fifth postulate. Thus the following proof must be invalid.

"PROOF" Let (p,q) be the given point, and L' the given line. L' is a right coset of $L = \{(x,mx) | x \in \mathbb{R}\}$, and m a constant in $\mathbb{R}$. $L'' = L + (p,q)$ is a right coset of L containing (p,q). Also, L'' is parallel to L', since two right cosets are disjoint or identical, as we proved in the proof of Lagrange's theorem. We recall that two lines in the plane are defined to be parallel if they are disjoint (i.e., do not intersect), and it costs nothing to say a line is parallel to itself. We skipped the case that L could equal $\{(0,x) | x \in \mathbb{R}\}$, but this case presents no difficulties.   ////

23 Why is this proof invalid?

1.10 ISOMORPHISMS

Having now considered groups for a while, we start thinking about "transportation" from one group to another. We want to find out about functions from one group to another that mesh with the two group operations.

Definition* A function ϕ from a group $(G, *)$ to a group $(H, \circ)$ is a **homomorphism** if $\phi(x * y) = \phi(x) \circ \phi(y)$ for all x and y in $(G, *)$. Without the operations written out explicitly, this becomes

$$\phi(xy) = \phi(x)\phi(y)$$

for all x and y in G. A **monomorphism** is a homomorphism that is one-to-one. A monomorphism from $(G, *)$ onto $(H, \circ)$ is an **isomorphism**. If such an isomorphism exists, then $(G, *)$ and $(H, \circ)$ are **isomorphic**.

EXAMPLE A Let ϕ be the function from $(\mathbb{R}, +)$ to $(\mathbb{R} - (0), \cdot)$ given by $\phi(x) = 3^x$. We see that

$$\phi(x + y) = 3^{x+y} = 3^x 3^y = \phi(x)\phi(y)$$

so ϕ is a homomorphism. Since $d\phi(x)/dx = \ln 3 \cdot 3^x > 0$ for all x, we see that $\phi(x)$ is strictly monotone increasing and therefore one-to-one. However, the range of ϕ is $\mathbb{R}^+$ and not $\mathbb{R} - (0)$, so ϕ is not onto. Thus ϕ is a monomorphism but not an isomorphism.

EXAMPLE A' $\phi(x) = 3^x$ is an isomorphism from $(\mathbb{R}, +)$ to $(\mathbb{R}^+, \cdot)$, since ϕ is still a monomorphism and now is onto. So we have shown $(\mathbb{R}, +)$ and $(\mathbb{R}^+, \cdot)$ to be isomorphic.

EXAMPLE B Any abelian group A can be mapped to itself by $\phi(a) = a^2$. Since $\phi(ab) = (ab)^2 = a^2 b^2 = \phi(a)\phi(b)$, ϕ is a homomorphism. Can the word abelian be omitted here?

EXAMPLE C Let ϕ be the map from $(Z, +)$ to $(\{1, -1\}, \cdot)$ sending all even integers to 1 and all odd integers to -1. One way to do this is to set $\phi(n) = (-1)^n$. Now

$$\phi(m + n) = (-1)^{m+n} = (-1)^m (-1)^n = \phi(m)\phi(n)$$

for all m and n in Z, so ϕ is a homomorphism. ϕ is not one-to-one and is not an isomorphism.

We note that if we have an isomorphism from G onto H, then the isomorphism respects the group structure and basically just relabels the elements. If there are many functions from G to H that are not isomorphisms, this does not prevent them from being isomorphic. It only takes one isomorphism to establish that G and H are isomorphic.

---

* This distinction between monomorphism and isomorphism is both clear and useful. It is currently used in ring theory but is slightly nonstandard elsewhere.

EXAMPLE D   The two groups of order 4 given as follows are not isomorphic:

V:	e	a	b	c
e	e	a	b	c
a	a	e	c	b
b	b	c	e	a
c	c	b	a	e

C:	e	x	x^2	x^3
e	e	x	x^2	x^3
x	x	x^2	x^3	e
x^2	x^2	x^3	e	x
x^3	x^3	e	x	x^2

Suppose ϕ were an isomorphism from C to V; then $\phi(x) = e, a, b,$ or c; $\phi(e)$ also equals one of these four. Since

$$\phi(x^2) = \phi(x)\phi(x) = e^2, a^2, b^2, \text{ or } c^2$$

in each case $\phi(x^2) = e$. Similarly,

$$\phi(e) = \phi(e^2) = \phi(e)\phi(e) = e$$

so $\phi(x^2) = \phi(e)$, and ϕ cannot be one-to-one.

Definition   Two sets S and T have the **same cardinality** (or are of the same cardinality) if there is a one-to-one and onto function from S to T.

In the finite case this just means that S and T have the same number of elements. In the infinite case the results are a bit more startling. $\mathbb{Z}^+$, $\mathbb{Z}$, $2\mathbb{Z}$, $\mathbb{Q}$, and $\mathbb{Q} - (0)$ all have the same cardinality, $\mathbb{R}$, $\mathbb{R} - (0)$, $\mathbb{R}^+$, $\mathbb{C}$ (the complex numbers), GL $(2,\mathbb{R})$, and the set of all continuous real-valued functions on $\mathbb{R}$ all have the same cardinality, although not the same cardinality as any of the sets listed in the previous sentence. Most of these facts are either easily established or found in one or both of the following books: Kamke, "Theory of Sets" [44], and Wilder, "Introduction to the Foundations of Mathematics" [90]. In Example A' we have shown that $\mathbb{R}$ and $\mathbb{R}^+$ have the same cardinality. The map $\phi: \mathbb{Z} \rightarrow 2\mathbb{Z}$ given by $\phi(n) = 2n$ is an isomorphism of $(\mathbb{Z}, +)$ to $(2\mathbb{Z}, +)$, so $\mathbb{Z}$ and $2\mathbb{Z}$ are of the same cardinality. The following map shows that $\mathbb{Z}^+$ and $\mathbb{Z}$ have the same cardinality:

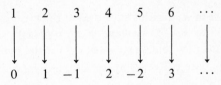

Exercises

1   Show that $2\mathbb{Z}$ and $5\mathbb{Z}$ have the same cardinality.
2   **Show that $\mathbb{Z}^+$ and $\mathbb{Q}$ have the same cardinality.
3   For any two sets S and T, let $S \sim T$ if S and T have the same cardinality. Show that $\sim$ is an equivalence relation.

Obviously if two groups have different cardinalities, then there is no way they can be isomorphic. Sym (3) cannot be isomorphic to D_8, and neither can be isomorphic to GL $(2,\mathbb{R})$.

We next prove a proposition about some of the properties that isomorphisms preserve. This theorem is designed to be useful in showing that two given groups are not isomorphic.

Proposition 1   If ϕ is an isomorphism from a group G onto a group H, then:

(a) $\phi(e_G) = e_H$, where e_G is the identity of G, and e_H the identity of H.

(b) $\phi(x^{-1}) = (\phi(x))^{-1}$ for all x in G.

(c) G and H are of the same cardinality.

(d) g and g' commute in G if and only if $\phi(g)$ and $\phi(g')$ commute in H.

(e) G is abelian if and only if H is abelian.

(f) $g = g'$ in G if and only if $(\phi(g))^k = \phi(g')$ in H.

(g) g and $\phi(g)$ have the same order.

(h) $x^k = g$ has just as many solutions in G as $x^k = \phi(g)$ has solutions in H. "Just as many" means two sets of solutions have the same cardinality.

APPLICATIONS OF PROPOSITION 1   As we have already mentioned, Sym (3) and D_8 cannot be isomorphic, since part c holds. Part e shows that $(\mathbb{R},+)$ and GL$(2,\mathbb{R})$ cannot be isomorphic; neither can $\mathbb{Z}_8$ and D_8. Applying part g to Example D, we note that a has order 4, and $\phi(a)$ has to have order 2 or 1, so ϕ cannot be an isomorphism. We could also use part h on Example D, noting that $x^2 = e$ has two solutions in C, e and a^2, whereas $x^2 = e$ has four solutions in V. $(\mathbb{Q} - (0),\cdot)$ is not isomorphic to $(\mathbb{Z},+)$ by part h, since $x^2 = 4$ has two solutions in $\mathbb{Q} - (0)$. But $\phi(4) = k$ is some integer k, and $2x = k$ has at most one solution in $(\mathbb{Z},+)$. (When we change the group operation here to addition, $x \cdot x = x^2$ becomes $x + x = 2x$.)

PROOF OF PROPOSITION 1   We start by proving $\phi(x^k) = (\phi(x))^k$ for all positive integers k.

For $k = 1$, this says $\phi(x) = \phi(x)$.

For $k = 2$, we have $\phi(x^2) = (\phi(x))^2$ since ϕ is a homomorphism. Assume now that $\phi(x^l) = (\phi(x))^l$ for the positive integer l. Then

$$\phi(x^{l+1}) = \phi(x^l x) = \phi(x^l)\phi(x)$$
$$= (\phi(x))^l \phi(x) \qquad \text{by assumption}$$
$$= (\phi(x))^{l+1}$$

which completes our proof by induction that $\phi(x^k) = (\phi(x))^k$ for all k in Z^+.

(a)   Since $e_G{}^2 = e_G$, we have

$$\phi(e_G) = \phi(e_G{}^2) = (\phi(e_G))^2$$

But we recall that, in a group, $x^2 = x$ implies $x = e$ (Exercise 1 of Sec. 1.2), so with $x = \phi(e_G)$ in the group H we have $e_H = \phi(e_G)$.

(b) $x \cdot x^{-1} = e_G$ implies $\phi(x \cdot x^{-1}) = \phi(e_G) = e_H = \phi(x)\phi(x^{-1})$. But $\phi(x)$ has a unique inverse, so $\phi(x^{-1}) = (\phi(x))^{-1}$. We can now extend $\phi(x^k) = \phi(x)^k$ to all integers k, since if $-k$ is negative then

$$\phi(x^{-k}) = \phi((x^{-1})^k) = (\phi(x^{-1}))^k = ((\phi(x))^{-1})^k = (\phi(x))^{-k}$$

(c)   Since the definition of cardinality just means the existence of a one-to-one and onto function, G and H have the same cardinality.

(d) $gg' = g'g$ implies $\phi(g)\phi(g') = \phi(gg') = \phi(g'g) = \phi(g')\phi(g)$, so g and g' commuting implies that $\phi(g)$ and $\phi(g')$ commute. If $\phi(g)\phi(g') = \phi(g')\phi(g)$, then $\phi(gg') = \phi(g'g)$; since ϕ is one-to-one, we have $gg' = g'g$.

(e)   If G is abelian, all g and g' commute, so all $\phi(g)$ and $\phi(g')$ commute; since ϕ is onto, this means all h and h' in H commute (see part d, above), so H is abelian. These steps can be reversed, so G is abelian if and only if H is abelian.

(f) $g^k = g'$ in G implies $(\phi(g))^k = \phi(g^k) = \phi(g')$ in H; applying ϕ^{-1} to $(\phi(g))^k = \phi(g^k) = \phi(g')$ gives $g^k = g'$.

(g)   If for any integer k we have $g^k = e_G$, then by part f above this is equivalent to $(\phi(g))^k = \phi(e_G) = e_H$.

(h)   If a satisfies $x^k = g$, then $a^k = g$; so, equivalently, $(\phi(a))^k = \phi(g)$, and $\phi(a)$ satisfies $x^k = \phi(g)$ in H. This also can be reversed. Thus ϕ takes the solutions of $x^k = g$ in G exactly onto the solutions of $x^k = \phi(g)$ in H, and ϕ is one-to-one.   ////

Exercises

4   Which of the following groups are isomorphic: $(\mathbb{Z}, +)$, $(2\mathbb{Z}, +)$, $(\mathbb{Z}_{20}, \oplus)$, $(\mathbb{Z}_8, \oplus)$, D_8, GL $(2, \mathbb{R})$, $(\mathbb{Q}, +)$, $(\mathbb{Q}^+, \cdot)$? (Apply your choice of part c or e first. After these two, part g is perhaps the most useful, often in the form of checking the number of square roots of the identity.)

5   Show that the elements $m + n\sqrt{2}$, where m and n are integers, form a group G under addition [actually this must be a subgroup of $(\mathbb{R}, +)$]. Show that the elements of the form $5^k 3^l$ under multiplication, where k and l are integers, form a subgroup H of $(\mathbb{R}, \cdot)$. Is H isomorphic to G?

6   Is $(\mathbb{R} - (0), \cdot)$ isomorphic to $(\mathbb{R}, +)$?

One use for equivalence relations is to partition a set so as to be able to count it. The other main use is to extend the concept of equality. We have seen how two groups which are isomorphic are very similar. The next proposition makes this more precise.

Proposition 2   Isomorphism of groups is an equivalence relation.

PROOF   If G is a group, then the identity map from G to G is obviously an onto isomorphism, so $G \sim G$ where $H \sim K$ means H is isomorphic to K. If $G \sim H$ and ϕ is the required isomorphism from G to H, then ϕ^{-1} is a one-to-one and onto map from H to G, as we have shown in Sec. 1.4. Since ϕ is onto for arbitrary elements h_1 and h_2 of H, we have unique g_1 and g_2 in G such that $\phi(g_1) = h_1$ and $\phi(g_2) = h_2$. Thus, $\phi(g_1)\phi(g_2) = \phi(g_1 g_2) = h_1 h_2$ so

$$\phi^{-1}(h_1 h_2) = \phi^{-1}(\phi(g_1 g_2)) = g_1 g_2 = \phi^{-1}(h_1)\phi^{-1}(h_2)$$

and ϕ^{-1} is a homomorphism and thus an isomorphism. Thus $H \sim G$ and $\sim$ is symmetric.

If $\phi: G \to H$ and $\psi: H \to K$ are isomorphisms, then both ϕ and ψ are one-to-one and onto, and so $\psi \circ \phi$ is one-to-one and onto (see Sec. 1.4). Since

$$
\begin{aligned}
(\psi \circ \phi)(g_1 g_2) &= \psi(\phi(g_1 g_2)) \\
&= \psi(\phi(g_1)\phi(g_2)) \qquad \text{since } \phi \text{ is a homomorphism} \\
&= \psi(\phi(g_1))\psi(\phi(g_2)) = (\psi \circ \phi)(g_1)(\psi \circ \phi)(g_2)
\end{aligned}
$$

$\psi \circ \phi$ is a homomorphism. Therefore $\sim$ is transitive and an equivalence relation.   ////

Exercises

7   If G and H are both cyclic groups of infinite order, show that G is isomorphic to H.

8   Can a group of order 30 be isomorphic to a subgroup of a group of order 72? Why?

9   Find a subgroup of GL $(2,\mathbb{R})$ isomorphic to $(\mathbb{R} - (0), \cdot)$. Find another isomorphic to $(\mathbb{R}, +)$. The exercises in Sec. 1.5 may be helpful here.

10   Let $(\mathbb{R}^2, +)$ be the group of all ordered pairs of real numbers, where $+$ is defined by

$$(x_1, y_1) + (x_2, y_2) = (x_1 + x_2, y_1 + y_2)$$

Show that $(\mathbb{R}^2, +)$ is an abelian group. Let $M = \begin{pmatrix} a & b \\ c & d \end{pmatrix}$ be a 2×2 matrix over $\mathbb{R}$, and let $G = (\mathbb{R}^2, +)$. If we let $(x,y)M = (x,y)\begin{pmatrix} a & b \\ c & d \end{pmatrix} = (ax + cy, bx + dy)$, show that $M: G \to G$ is a homomorphism. (Remember, the operation in G is $+$.) When is M an isomorphism?

The concept of isomorphism is very simple but vital. Instead of statements like " there are essentially two groups of order 4," we can now say more precisely " there are two nonisomorphic groups of order 4." In other words, if any new group of order 4 is produced, it must be isomorphic to V or C. In coming sections we shall classify all cyclic groups up to isomorphism. We shall discuss a similar classification for all finite abelian groups, as well as similar concepts for rings and vector spaces. The fundamental theorem of homomorphisms starts with a homomorphism and yields an isomorphism.

Thus isomorphism is a simple and important concept. It is also complicated. For instance, only in the last few years were the nonzero complex numbers shown to be isomorphic to the complex numbers of absolute value 1. See Clay [15].

*1.11 EUCLID'S ALGORITHM AND THE LINEAR PROPERTY

This section treats some elementary properties of the integers which turn out to be useful when we study groups. We start by recalling that if a is any integer, and b any positive integer, we can find unique integers c and r such that

$$a = b \cdot c + r \qquad 0 \le r < b$$

The greatest common divisor of two integers a and b is a number d such that d divides a, d divides b, and if any other integer e divides both a and b then e divides d. If we take d positive, then d is simply the largest number that divides a and b. For instance, the greatest common divisor of 100 and 175 is 25. -5 is also a common divisor of 100 and 175, and indeed -5 divides 25.

Definition If a and b are integers, then the **greatest common divisor** of a and b, written gcd (a,b), is the largest integer that divides both a and b.

One time-honored method of finding greatest common divisors is *Euclid's algorithm*. This method works without factoring a and b into primes and is even pretty efficient. We give an illustration first of how it works.

EXAMPLE A Find the greatest common divisor of 2,024 and 1,760. Using the division algorithm repeatedly, we have

$$2{,}024 = 1 \cdot 1{,}760 + 264 \qquad \text{(A)}$$
$$1{,}760 = 6 \cdot 264 + 176 \qquad \text{(B)}$$
$$264 = 1 \cdot 176 + 88 \qquad \text{(C)}$$
$$176 = 2 \cdot 88 + 0 \qquad \text{(D)}$$

and 88 is the gcd of 264 and 176. Thus, looking at (B), we see that 88 is the gcd of 264 and 1,760, and thus of 1,760 and 2,024. We also want to find integers a and b such that $88 = a \cdot 1{,}760 + b \cdot 2{,}024$. This is the linear property for the greatest common divisor of 1,760 and 2,024. Starting at (C), we have

$$88 = 264 - 1 \cdot 176$$

Using (B), we eliminate 176:

$$88 = 264 - (1{,}760 - 6 \cdot 264)$$
$$= 7 \cdot 264 - 1 \cdot 1{,}760$$

Using (A), we eliminate 264:

$$88 = 7(2{,}024 - 1 \cdot 1{,}760) - 1 \cdot 1{,}760$$
$$= 7(2{,}024) - 8(1{,}760)$$

Thus $a = -8$ and $b = 7$.

Another example of Euclid's algorithm is the following: What is the greatest common divisor of 7,440 and 41,261?

$$41{,}261 = 5 \cdot 7{,}440 + 4{,}061$$
$$7{,}440 = 1 \cdot 4{,}061 + 3{,}379$$
$$4{,}061 = 1 \cdot 3{,}379 + 682$$
$$3{,}379 = 4 \cdot 682 + 651$$
$$682 = 1 \cdot 651 + 31$$
$$651 = 21 \cdot 31 + 0$$

Therefore gcd $(7{,}440, 41{,}261) = 31$.

Again reversing the process, we have

$$31 = 682 - 651$$
$$= 682 - (3{,}379 - 4 \cdot 682)$$
$$= 5 \cdot 682 - 3379$$
$$= 5(4{,}061 - 3{,}379) - 3{,}379$$
$$= 5(4{,}061) - 6(3{,}379)$$
$$= 5(4{,}061) - 6(7{,}440 - 4{,}061)$$
$$= 11(4{,}061) - 6(7{,}440)$$
$$= 11(41{,}261 - 5 \cdot 7{,}440) - 6(7{,}440)$$
$$= 11(41{,}261) - 61(7{,}440)$$

If we now abstract this to the general situation of two integers a and b with, say, $b > 0$, we obtain

$$
\begin{aligned}
a &= d_1 b + r_1 & 0 \leq r_1 < b \\
b &= d_2 r_1 + r_2 & 0 \leq r_2 < r_1 \\
r_1 &= d_3 r_2 + r_3 & \\
&\cdots\cdots\cdots\cdots & \cdots\cdots\cdots \\
r_{n-2} &= d_n r_{n-1} + r_n & 0 \leq r_n < r_{n-1} \\
r_{n-1} &= d_{n+1} r_n + 0 &
\end{aligned}
$$

The existence of each step is justified by the division algorithm. How do we know that eventually the process stops? This follows from $b > r_1 > r_2 > \cdots > r_n \geq 0$, so there can be at most b steps. Why does Euclid's algorithm deliver the greatest common divisor? We first note that

$$\gcd (a,b) = \gcd (b,r_1)$$

This is a result of $a = d_1 b + r_1$, since any number that divides a and b must divide r_1, and any number that divides b and r_1 must divide a. Similarly, $\gcd (b,r_1) = \gcd (r_1,r_2)$, $\gcd (r_1,r_2) = \gcd (r_2,r_3)$, and $\gcd (r_{n-1},r_n) = \gcd (r_n,0) = r_n$. Therefore,

$$\gcd (a,b) = \gcd (b,r_1) = \gcd (r_1,r_2)$$
$$= \cdots = \gcd (r_{n-1},r_n) = r_n$$

and we have proved the following proposition:

Proposition 1   If a and b are integers with $b > 0$, then the greatest common divisor of a and b is r_n, where

$$a = bd_1 + r_1 \qquad 0 \leq r_1 < b$$
$$b = r_1 d_2 + r_2 \qquad 0 \leq r_2 < r_1$$
$$r_1 = r_2 d_3 + r_3$$
$$\cdots\cdots\cdots\cdots\cdots \qquad \cdots\cdots\cdots\cdots$$
$$r_{n-2} = r_{n-1} d_n + r_n \qquad 0 \leq r_n < r_{n-1}$$
$$r_{n-1} = r_n d_{n+1} + 0$$

where r_n is understood to be the last nonzero entry in the sequence b, $r_1, r_2, \ldots$.

Corollary   If r_n is the gcd of a and b, then r_n is the smallest positive integer that can be expressed as $na + mb$, where n and m are integers. This is the *linear property* for the greatest common divisor.

PROOF $r_n = na + mb$ by reversing the algorithm. If x is any other integer expressible as $x = ka + lb$ with k, l in $\mathbb{Z}$, then since r_n divides both a and b, it divides x.   ////

Exercises

1   Find the gcd of 800 and 122 by Euclid's algorithm, and express this gcd as $a \cdot 800 + b \cdot 122$ for some integers a and b. Do the same for 5,291 and 4,514.
2   Assume that every positive integer can be factored into a product of primes. Then another method for finding greatest common divisors can be found. Factor $a = 33,000$ and $b = 12,100$ into primes, and then find gcd (33,000, 12,100). How can this be generalized?

Two numbers a and b are called *relatively prime* if gcd $(a,b) = 1$. In many problems it becomes almost a reflex to rewrite "a and b are relatively prime" or "gcd $(a,b) = 1$" as "there exist n and m such that $na + mb = 1$." We shall see instances of this in the next section and throughout the text. For now we complete this section with several such exercises. The notation $b|a$ means that b divides a evenly, where both are integers. Equivalently, $b|a$ means there exists an integer m such that $a = bm$.

3   Show that gcd $(a, a + k)|k$.

4   Let $M = \begin{pmatrix} a & b \\ c & d \end{pmatrix}$, where $ad - bc = \pm 1$, and $a,b,c,d \in \mathbb{Z}$. Show that

$\text{gcd } (m,n) = \text{gcd } \left[(m,n)\begin{pmatrix} a & b \\ c & d \end{pmatrix} \right]$ where, as before,

$$(m,n)\begin{pmatrix} a & b \\ c & d \end{pmatrix} = (am + cn, bm + dn).$$

5   (a)   Find integers m and n such that $m/7 + n/15 = 1/105$.
   (b)   Find integers m and n such that $m/7 + n/15 = 23/105$.
   (c)   Find integers m and n such that $m/101 + n/17 = 25/1{,}717$.

6   (Euclid) If p is a prime and $p|mn$, show that $p|m$ or $p|n$. [*Hint*: Assume $p \nmid m$ so that gcd $(p,m) = 1$.]

1.12   CYCLIC GROUPS AND DIRECT PRODUCTS

In this section we do two things. First we classify all cyclic groups up to isomorphism. This is our first classification theorem; it is complete enough so that we lose interest in cyclic groups per se but are able to use cyclic groups easily as they enter into harder problems. Our second topic is a method for putting two groups together to get a new group. This will greatly extend our list of examples. We use the division algorithm as a key step in the proof of the following proposition about cyclic groups.

Proposition 1   Every subgroup of a cyclic group is cyclic.

PROOF   Let $G = \langle g \rangle$ be a cyclic group, and let H be a subgroup. If H is just the identity, then $H = \langle e \rangle$ and is cyclic, so we may assume H contains some element g^k other than the identity. H must also contain g^{-k}, so we may safely assume g^k is in H, where k is a positive integer. Let m be the smallest positive integer such that g^m is in H.

If we can now show $H = \langle g^m \rangle$, we will have shown H to be cyclic. Since $g^m \in H$ and H is closed, $g^{2m}, g^{3m}, \ldots. e, g^{-m}, g^{-2m}, \ldots$ are all in H, so $\langle g^m \rangle \subseteq H$. If g^n is in H, then by applying the division algorithm to n and m we have $n = cm + r$, for $0 \le r < m$. Therefore,

$$g^n = g^{cm+r} = (g^m)^c g^r \qquad 0 \le r < m$$

and
$$g^r = (g^m)^{-c} g^n$$

is in H since g^n and g^m are in H. But m was the smallest positive integer with g^m in H, so we have the following facts: $r = 0$, $n = cm$, $g^n = (g^m)^c$, and $g^n \in \langle g^m \rangle$. Thus $H \subseteq \langle g^m \rangle$, we have both inclusions, and $H = \langle g^m \rangle$. ////

To look at one example of this, we note that any subgroup of $(\mathbb{Z}, +)$ is $n\mathbb{Z}$ for some nonnegative integer n. These are all cyclic. Take $\mathbb{Z}_{20}$ as another example. $\mathbb{Z}_{20}$ has the following subgroups:

$$\bar{0} = \{\bar{0}\}$$
$$\bar{1} = \mathbb{Z}_{20}$$
$$\bar{2} = \bar{2}\mathbb{Z}_{20} = \{\bar{0}, \bar{2}, \bar{4}, \dots, \overline{18}\}$$
$$\bar{4} = \bar{4}\mathbb{Z}_{20} = \{\bar{0}, \bar{4}, \bar{8}, \overline{12}, \overline{16}\}$$
$$\bar{5} = \bar{5}\mathbb{Z}_{20} = \{\bar{0}, \bar{5}, \overline{10}, \overline{15}\}$$
$$\overline{10} = \overline{10}\mathbb{Z}_{20} = \{\bar{0}, \overline{10}\}$$

These are the only subgroups, and they are indeed all cyclic. The subgroup lattice is as follows:

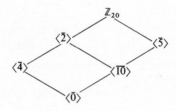

Exercises

1 Find the subgroup lattices for $\mathbb{Z}_{100}$ and $\mathbb{Z}_{101}$.
2 What is an example of a noncyclic abelian group all of whose proper subgroups are cyclic?
3 What is an example of a nonabelian group all of whose proper subgroups are cyclic?

Lemma If $G = \langle g \rangle$ is an infinite cyclic group, all the g^k are distinct, where k is in $\mathbb{Z}$.

PROOF Assume $g^k = g^l$, where, say, $k > l$. Then $g^{k-l} = e = g^m$ where $m = k - l$. For any g^n in G, we apply the division algorithm and obtain $n = cm + r$, for $0 \le r < m$. Thus $g^n = g^{cm+r} = (g^m)^c g^r = e^c g^r = g^r$, for $0 \le r < m$, so every g^n equals one of $e, g, g^2, \dots, g^{m-1}$; and G must be finite. ////

Proposition 2 All infinite cyclic groups are isomorphic.

PROOF Let $G = \langle g \rangle$ and $H = \langle h \rangle$ be any two infinite cyclic groups. Define $\phi: H \to G$ by $\phi(h^k) = g^k$ for all integers k. ϕ is a function and is onto, since for any g^m we have $\phi(h^m) = g^m$. Since

$$\phi(h^k h^l) = \phi(h^{k+l}) = g^{k+l} = g^k g^l = \phi(h^k)\phi(h^l)$$

ϕ is a homomorphism. We finish by showing that ϕ is one-to-one. If $\phi(h^m) = \phi(h^n)$, then $g^m = g^n$, and by the lemma m and n must be equal since G is infinite. Thus $h^m = h^n$, and ϕ is one-to-one.   ////

Corollary If G is an infinite cyclic group, then G is isomorphic to $(\mathbb{Z}, +)$.

Lemma If $G = \langle g \rangle$ is a cyclic group of order n then the elements $e, g, g^2, \ldots, g^{n-1}$ are all distinct.

PROOF Exercise (modify the proof of the last lemma).

Proposition 3 Any two finite cyclic groups of the same order are isomorphic.

PROOF Let $C = \langle c \rangle$ and $D = \langle d \rangle$ be two cyclic groups of order n. Then $\phi(c^k) = d^k$ for $0 \le k \le n - 1$ is an onto homomorphism as in the proof of Proposition 2. Since $|C| = |D|$, onto implies one-to-one, and the proposition is proved.   ////

Corollary If C is a cyclic group of order n, then C is isomorphic to $(\mathbb{Z}_n, \oplus)$.

EXAMPLE Let $G = \{\bar{1}, \bar{2}, \bar{3}, \bar{4}, \bar{5}, \bar{6}\}$ under multiplication modulo 7. After some trial and error we see the following:

$$\bar{3}^0 = \bar{1} \qquad \bar{3}^3 = \bar{6}$$
$$\bar{3}^1 = \bar{3} \qquad \bar{3}^4 = \bar{4}$$
$$\bar{3}^2 = \bar{2} \qquad \bar{3}^5 = \bar{5}$$

Thus $G = \langle \bar{3} \rangle$, G is cyclic of order 6, and so $G \simeq (\mathbb{Z}^6, \oplus)$.

Exercises

4 Find all the subgroups of G in the example above, using Proposition 1. Do the same for $H = \{\bar{1}, \bar{2}, \ldots, \overline{11}, \overline{12}\}$ under multiplication modulo 13.
5 If A is a finite abelian group of order n, and $\gcd(n, k) = 1$, show that the map $\phi: A \to A$ defined by $\phi(a) = a^k$ is an isomorphism. (*Hint*: There exist c

and d such that $nc + kd = 1$. So $a^1 = a^{nc+kd}$. This will help show that ϕ is onto and thus one-to-one.)

6   Show that every cyclic group is either finite or of the same cardinality as $\mathbb{Z}^+$.

Proposition 5 If $(G,*)$ and $(H,\circ)$ are groups, then $(G \times H, \#)$ is a group, where $G \times H = \{(g,h)\,|\,g \in G, h \in H\}$ and $(g_1,h_1) \# (g_2,h_2) = (g_1 * g_2, h_1 \circ h_2)$. This group is called the *direct product* of G and H. [Usually $\#$, $*$, and $\circ$ are suppressed, and this comes out $(g_1,h_1)(g_2,h_2) = (g_1 g_2, h_1 h_2)$.]

PROOF   By the definition of $\#$, it is closed. Since

$$((g_1,h_1) \# (g_2,h_2)) \# (g_3,h_3)$$
$$= (g_1 * g_2, h_1 \circ h_2) \# (g_3,h_3)$$
$$= ((g_1 * g_2) * g_3, (h_1 \circ h_2) \circ h_3) \qquad \text{since } * \text{ and } \circ \text{ are associative}$$
$$= (g_1 * (g_2 * g_3), h_1 \circ (h_2 \circ h_3))$$
$$= (g_1,h_1) \# (g_2 * g_3, h_2 \circ h_3)$$
$$= (g_1,h_1) \# ((g_2,h_2) \# (g_3,h_3))$$

we have that $\#$ is associative.

It is quickly checked that if e_G is the identity for G, and e_H is the identity for H, then (e_G,e_H) is the identity for $(G \times H, \#)$ and that $(g,h)^{-1} = (g^{-1},h^{-1})$. Thus, $(G \times H, \#)$ is indeed a group. This group is called the *direct product* of G and H.   ////

Proposition 6 If G and H are groups and $G \times H$ is their direct product, then

(a)   If G and H are finite, then $|G \times H| = |G|\,|H|$.
(b) G is isomorphic to the subset $\{(g,e_H)\,|\,g \in G\}$ of $G \times H$, and H is isomorphic to the subset $\{(e_G,h)\,|\,h \in H\}$.

PROOF   The set $G \times H = \{(g,h)\}$ must have $|G|\,|H|$ elements, so part a is trivial. $\phi(g) = (g,e_H)$ is obviously one-to-one and a homomorphism and so an isomorphism, and this establishes part b.   ////

EXAMPLE B   The most famous and most used example of direct products is just $(\mathbb{R},+) \times (\mathbb{R},+)$ or $\mathbb{R}^2$, the real plane. The elements are just the pairs (a,b) with a and b in $\mathbb{R}$. Since addition is used as the operation in each component, we use addition for the direct product. Thus, $(a,b) + (c,d) = (a + c, b + d)$.

EXAMPLE C   Sym $(3) \times (\mathbb{Z}_2, \oplus)$ might be pictured as

$$\bar{1} \quad \cdot \quad \cdot \quad \cdot \quad \cdot \quad \cdot \quad \cdot$$
$$\bar{0} \quad \cdot \quad \cdot \quad \cdot \quad \cdot \quad \cdot \quad \cdot$$
$$\quad\; I \quad a \quad a^2 \quad b \quad ab \quad a^2b$$

and, for instance, $(a,\bar{0})(ab,\bar{1}) = (a(ab), \bar{0} + \bar{1}) = (a^2b,\bar{1})$.

EXAMPLE D Klein's four group V is isomorphic to $(\mathbb{Z}_2, \oplus) \times (\mathbb{Z}_2, \oplus)$. Let $V = \{e, a, b, c\}$, and $\mathbb{Z}_2 \times \mathbb{Z}_2 = (\bar{0}, \bar{0})$, $(\bar{1}, \bar{0})$, $(\bar{0}, \bar{1})$, $(\bar{1}, \bar{1})$. Then $\phi(e) = (\bar{0}, \bar{0})$, $\phi(a) = (\bar{1}, \bar{0})$, $\phi(b) = (\bar{0}, \bar{1})$, and $\phi(c) = (\bar{1}, \bar{1})$ obviously define a one-to-one and onto function. Checking $\phi(x^2)$ and $\phi(xy)$ with $x \neq y$ will show that ϕ is a homomorphism and thus an isomorphism. Alternatively, $\mathbb{Z}_2 \times \mathbb{Z}_2$ has no element of order 4 and thus is not cyclic. Since there are but two nonisomorphic groups of order 4, we must have $\mathbb{Z}_2 \times \mathbb{Z}_2 \simeq V$.

Exercise

7 Show that if A and B are groups, then $A \times B$ is abelian if and only if A and B are both abelian.

We repeat a definition given in the exercises in Sec. 1.9.

Definition If G is a group, then the **center** of $G = \mathbf{Z}(G) = \{x \,|\, xg = gx,$ for all g in $G\}$ = the subgroup of G consisting of those elements which commute with all the elements in G.

EXAMPLE E Using the exercises of Sec. 1.9, recall that $\mathbf{Z}(\text{Sym}(3)) = I$ and $\mathbf{Z}(D_8) = \{e, a^2\}$. If A is abelian, then all elements commute with all other elements, so $\mathbf{Z}(A) = A$.

Exercises

8 Find the center of the following groups:
   (a) Klein's four group
   (b) $(\mathbb{Z}_2, \oplus) \times \text{Sym}(3)$
   (c)* Sym (4)
9 (A generalization of Exercise 7) Show $Z(A \times B) \simeq Z(A) \times Z(B)$.
10* Find two nonisomorphic subgroups of Sym (3) $\times$ Sym (3), both having order 18.

We have been using $\langle a \rangle$ to denote the subgroup $\{e, a, a^2, a^3, \ldots, a^{-1}, a^{-2}, \ldots\}$ which is also the smallest subgroup of G that contains a. If S is any subset of G, we can use $\langle S \rangle$ to denote the smallest subgroup of G containing S.

Definition and Proposition Suppose that G is a group and S is a non-empty subset of G. If $\langle S \rangle$ is the smallest subgroup of G containing S, then $S = \{s_1^{\alpha_1} s_2^{\alpha_2} \cdots s_k^{\alpha_k} \,|\, s_i \in S, \ \alpha_i = +1 \text{ or } -1\}$. This $\langle S \rangle$ is called the **subgroup generated by** S.

PROOF If $s \in S$, then $s^1 s^{-1} = e$ is in $\langle S \rangle$. If $s_1^{\alpha_1} s_2^{\alpha_2} \cdots s_k^{\alpha_k} \in \langle S \rangle$, then so is $(s_1^{\alpha_1} s_2^{\alpha_2} \cdots s_k^{\alpha_k})^{-1} = s_k^{-\alpha_k} s_{k-1}^{-\alpha_{k-1}} \cdots s_2^{-\alpha_2} s_1^{-\alpha_1}$. In addition, if

$s_1{}^{\alpha_1}s_2{}^{\alpha_2} \cdots s_k{}^{\alpha_k}$ and $t_1{}^{\beta_1}t_2{}^{\beta_2} \cdots t_l{}^{\beta_l}$ are in $\langle S \rangle$, we can relabel the $t_i{}^{\beta_i}$ as $s_{i+k}{}^{\alpha_{i+k}}$ Then

$$(s_1{}^{\alpha_1}s_2{}^{\alpha_2} \cdots s_k{}^{\alpha})(t_1{}^{\beta_1}t_2{}^{\beta_2} \cdots t_l{}^{\beta_l}) = s_1{}^{\alpha_1}s_2{}^{\alpha_2} \cdots s_{k+l}{}^{\alpha_{k+l}} \in \langle S \rangle. \qquad ////$$

As an example, we let $G = (\mathbb{R} - (0), \cdot)$ and $S = \{3, \frac{1}{5}\}$. Since G is abelian, we can collect terms and see that $\langle S \rangle = \{3^n(\frac{1}{5})^m \mid n, m \in \mathbb{Z}\} = \{3^n 5^k \mid n, k \in \mathbb{Z}\}$. If G is nonabelian, determining $\langle S \rangle$ is much harder. For instance, let G be GL $(2, \mathbb{R})$, and let $S = \left\{ \begin{pmatrix} 0 & -1 \\ 1 & 0 \end{pmatrix}, \begin{pmatrix} 0 & 1 \\ -1 & -1 \end{pmatrix} \right\}$. Thus $\begin{pmatrix} 0 & -1 \\ 1 & 0 \end{pmatrix}^2 = \begin{pmatrix} 1 & 0 \\ 0 & 1 \end{pmatrix} = \begin{pmatrix} 0 & 1 \\ -1 & -1 \end{pmatrix}^3$ is in $\langle S \rangle$, but so is $\begin{pmatrix} 0 & -1 \\ 1 & 0 \end{pmatrix}\begin{pmatrix} 0 & 1 \\ -1 & -1 \end{pmatrix} = \begin{pmatrix} 1 & 1 \\ 0 & 1 \end{pmatrix}$.

We now find that $\begin{pmatrix} 1 & 1 \\ 0 & 1 \end{pmatrix}^n = \begin{pmatrix} 1 & n \\ 0 & 1 \end{pmatrix}$ is in $\langle S \rangle$, which makes $\langle S \rangle$ infinite. We stop at this point, but it turns out that

$$\langle S \rangle = \left\{ \begin{pmatrix} a & b \\ c & d \end{pmatrix} a, b, c, d \in \mathbb{Z} \text{ and } ad - bc = 1 \right\},$$

which is called the *unimodular group*. See the books by Knopp [48] and Lehner [54].

Exercises

11   Show that $\langle S \rangle$ also equals the intersection of all those subgroups of G that contain S.

12   If $G = (\mathbb{R} - (0), \cdot)$ and $S = \{\frac{1}{2}, 3^2, 17\}$, find $\langle S \rangle$.

13   If $G = $ GL $(2, \mathbb{R})$ and $S = \left\{ \begin{pmatrix} 1 & 1 \\ 0 & 1 \end{pmatrix}, \begin{pmatrix} 1 & 0 \\ 0 & -1 \end{pmatrix} \right\}$, find $\langle S \rangle$.

Several more exercises of this type will be found at the end of Sec. 1.15. In Sec. 1.14 we use the concept of $\langle S \rangle$ to define commutator subgroups.

14   Suppose that $ab = ba$ for two elements a and b of a group G. Suppose also that the order of a is n and the order of b is m, where gcd $(m, n) = 1$. Show that $\langle a, b \rangle$ is a cyclic group of order nm with generator ab.

1.13   HOMOMORPHISMS AND NORMAL SUBGROUPS

In the study of homomorphisms from one group to another, a particular set of subgroups, the normal subgroups, turn out to be very important. We start this section by examining normal subgroups and some of their properties; then we examine homomorphisms. We conclude by examining the connection between the two and some of the applications of this connection.

The importance of normal subgroups was first noticed by Evariste Galois sometime before 1832. Galois is perhaps the most romantic figure in all of mathematics. He died in a duel at age 20, having written up most of his results the night before. We shall discuss more of his work in Chap. 10. For more details on his life and work, see Boyer [12] or Kline [47].

The idea of a normal subgroup is not difficult. A subgroup is normal if every right coset of the subgroup is also a left coset. It is remarkable that singling out this property leads to so much.

Definition   A subgroup N of a group G is **normal** in G if $x^{-1}Nx \subseteq N$ for all x in G. This is denoted $N \lhd G$.

REMARK $x^{-1}Nx$ is just the set $\{x^{-1}nx \mid n \in N\}$ which, as we have seen earlier, is a subgroup. We often verify that a subgroup N is normal by showing $x^{-1}nx$ is in N for all x in G and n in N. If N is normal and n_1 is in N, then $x^{-1}n_1x = n_2$ is also in N, though it is *not* necessary for n_1 to equal n_2. The next proposition gives some other conditions for normality.

Proposition 1   The following statements are equivalent for a subgroup N of a group G:

(a) N is normal in G.

(b) $x^{-1}Nx = N$ for all x in G.

(c) $xN = Nx$ for all x in G (that is, every right coset is a left coset, and vice versa).

PROOF   Part a implies part b: Since N is normal in G, we know $x^{-1}Nx \subseteq N$ for all x in G. Since x^{-1} is in G, we also have $(x^{-1})^{-1}Nx^{-1} \subseteq N$.

Then $(x^{-1})^{-1}Nx^{-1} = xNx^{-1} \subseteq N$, so $N = x^{-1}(xNx^{-1})x \subseteq x^{-1}Nx$. Thus we have $x^{-1}Nx = N$ for all x in G.

That part b implies part a is immediate.

Part b is equivalent to part c: Note that $x^{-1}Nx = N$ yields $x(x^{-1}Nx) = xN$ and $x(x^{-1}Nx) = x\{x^{-1}nx \mid n \in N\} = \{nx \mid n \in N\} = Nx = xN$. Similarly, $Nx = xN$, when multiplied on the left by x^{-1}, yields $x^{-1}Nx = N$.

/////

The next three propositions are criteria for determining when subgroups are normal.

Proposition 2   If A is an abelian group, then every subgroup of A is normal in A.

PROOF   Let N be a subgroup of A. Then for any x in A, $x^{-1}Nx = \{x^{-1}nx \mid n \in N\} = \{x^{-1}xn \mid n \in N\} = \{n \mid n \in N\} = N$, so N is normal in A.

/////

Proposition 2 gives us a large supply of examples of normal subgroups. It also says that we can turn our attention to nonabelian groups.

There are two common misunderstandings that occur at this point. First, it is very possible for a subgroup to be abelian but *not* be normal (see the following example). Second, if N is normal in G, it is usually *false* that $g^{-1}ng = n$ for n in N. Even if g is also in N, $g^{-1}ng = n$ usually is *false*.

Let us see which subgroups of Sym (3) are normal, where we may assume Sym (3) = $\{I,a,a^2,b,ab,a^2b\}$ and $a^3 = b^2 = I$, $a^2b = ba$. Sym (3) and $\{I\}$ have to be normal by a quick examination of x^{-1} Sym (3)x and $x^{-1}Ix$.

The other subgroups are Alt (3) = $\{I,a,a^2\}$, $H_1 = \{I,b\}$, $H_2 = \{I,ab\}$, and $H_3 = \{I,ab^2\}$. H_1 is not normal, since

$$a^{-1}H_1a = a^{-1}\{I,b\}a = \{a^{-1}Ia, a^{-1}ba\} = \{I,a^{-1}(ba)\}$$
$$= \{I,a^{-1}a^2b\} = \{I,ab\} = H_2 \nsubseteq H_1$$

Similarly, H_2 and H_3 are not normal in Sym (3). (Note that H_1 is abelian).

Alt (3) is normal, and to show this we need only examine g^{-1} Alt (3)g in the six cases $g = I$, a, a^2, b, ab, and a^2b. We can consider the first three cases simultaneously by taking $g = a^i$, for $i = 0$, 1, or 2. Then

$$g^{-1} \text{Alt (3)}g = a^{-i}\{I,a,a^2\}a^i = \{a^{-i}Ia^i, a^{-i}aa^i, a^{-i}a^2a^i\}$$
$$= \{I,a,a^2\} = \text{Alt (3)}$$

We also can group the last three cases by taking $g = a^ib$, for $i = 0$, 1, or 2. Here,

$$g^{-1} \text{Alt (3)}g = (a^ib)^{-1} \text{Alt (3)}a^ib = b^{-1}a^{-i} \text{Alt (3)}a^ib$$
$$= b^{-1} \text{Alt (3)}b = b^{-1}\{I,a,a^2\}b$$
$$= \{b^{-1}Ib, b^{-1}ab, b^{-1}a^2b\} = \{I,bab,ba^2b\}$$
$$= \{I, (a^2b)b, b(ba)\} = \{I,a^2,a\} = \text{Alt (3)}$$

These computations could be condensed a bit, but basically they illustrate the need for some further criteria to identify normal subgroups.

Proposition 3 If H is a subgroup of G such that H has only two right cosets (itself and one other), then H is normal in G. In the finite case this means that the order of H is one-half the order of G.

PROOF Any element in G is either in H or in $G - H$. If $x \in H$, then $xH = H = Hx$. If $x \notin H$, then xH is the set of elements $G - H$, since $H \cap Hx = \phi$. Thus, $xH = G - H = Hx$. Therefore, by part c of Proposition 1, H is normal in G. ////

An alternative phrasing says all subgroups of index 2 are normal.

Exercise

1 If x is a fixed element of the group G, then $f_x: G \to G$ defined by $f_x(g) = x^{-1}gx$ is an isomorphism from G to G.

Proposition 4 If H is a finite subgroup of G and is the only subgroup of order $|H|$, then $H \lhd G$. (With proper interpretation the word "finite" could be deleted.)

PROOF For any x in G we know by a previous exercise that $x^{-1}Hx$ is also a subgroup. $x^{-1}Hx = \{x^{-1}h_\alpha x \mid h_\alpha \in H\}$ and $H = \{h_\alpha \mid h_\alpha \in H\}$ obviously have the same cardinality, so $|x^{-1}Hx| = |H|$. By the uniqueness of H with respect to order, we have $x^{-1}Hx = H$. Since x was arbitrary, $H \lhd G$. ////

Proposition 5 If J is a subgroup of $Z(G)$, the center of a group G, then $J \lhd G$. In particular, $Z(G) \lhd G$.

PROOF Since every j in J commutes with all elements in G, we have the following for any x in G:

$$x^{-1}Jx = x^{-1}\{j \mid j \in J\}x = \{x^{-1}jx \mid j \in J\}$$
$$= \{x^{-1}xj \mid j \in J\} = \{j \mid j \in J\} = J$$

and $J \lhd G$. ////

Proposition 2 is a special case of this proposition.
If we now reexamine Sym (3) and its subgroups, we have

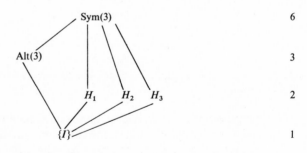

So Alt (3) $\lhd$ Sym (3) by either Proposition 3 or 4.

Exercises

2 Which of the subgroups of D_8 are normal? (*Hint*: We wrote out a subgroup lattice in Sec. 1.7.)
3 If $H \lhd G$ and $K \lhd H$, is $K \lhd G$?
4 If $N \lhd G$ and $|N| = 2$, show that $N \subseteq Z(G)$, the center of G.
5 If M and N are both normal subgroups of G, show that $M \cap N$ is also normal in G.

6 If M and N are normal subgroups of G and $M \cap N = \{e\}$, show that $mn = nm$ for all $m \in M$, $n \in N$.

7 Show that $\bigcap_{a \in G} C_G(a) = Z(G)$, where $C_G(a)$ consists of those elements in G commuting with a.

8 *Show that $|Z(G)| = |G|/2$ is impossible.

We now use a normal subgroup N of a group G to construct a smaller (and hopefully more manageable) group G/N which reflects the structure of G.

Proposition 6 A subgroup H of G is normal in G if and only if $xHyH = xyH$ for all x and y in G.

PROOF Since xH and yH are subsets of G,

$$xHyH = \{xhyh' \,|\, h \text{ and } h' \text{ in } H\}$$

Thus, if H is normal in G,

$$xHyH = x(yy^{-1})HyH = xy(y^{-1}Hy)H = xyHH = xyH$$

If H is not normal, then there is a y in G such that $y^{-1}Hy \neq H$, so $y^{-1}HyH \neq y^{-1}yH = H$, so $xHyH \neq xyH$ if we set $x = y^{-1}$. ////

Proposition 7 If $H \lhd G$, then G/H, the set of left cosets of H in G, can be made into a group by defining an operation $*$ on G/H by $xH * yH = xHyH$. Further, this holds only if $H \lhd G$.

PROOF $xH * yH = xHyH = xyH$ if and only if $H \lhd G$ by Proposition 6, so we have closure if and only if $H \lhd G$. Since

$$(xH * yH) * zH = (xy)zH = z(yz)H = xH * (yH * zH)$$

$(G/H, *)$ satisfies the associative law. $H = eH$ is the identity, since $eHxH = exH = xH = xHeH$. The inverse of xH is $x^{-1}H$, since $xH * x^{-1}H = xHx^{-1}H = xx^{-1}H = H = x^{-1}HxH$.

A quick glance would indicate that the proof is finished, but closer scrutiny brings up an unfinished detail. How do we know $*$ is a function? Specifically, if $x_1H = x_2H$ (which does *not* mean $x_1 = x_2$), do we have $x_1H * yH = x_2H * yH$? There is no trouble since

$$x_1H * yH = x_1HyH = x_2HyH = x_2H * yH$$

If $y_1H = y_2H$, we can similarly show that $x_1H * y_1H = x_2H * y_2H$, and $*$ is a well-defined function. ////

Definition If $N \lhd G$, then the group $(G/N, *)$ of left cosets of N with $xH * yH = xyN$ is called the **quotient group** or **factor group** of G over N. We usually omit the $*$ notation from this point on.

Definition and Proposition If N is a normal subgroup of G, and $\eta: G \to G/N$ is given by $\eta(g) = gN$, then η is called the **canonical homomorphism** from G to G/N.

PROOF (THAT η IS A HOMOMORPHISM AND IS ONTO)   First, $\eta(g)\eta(h) = gNhN = ghN = \eta(gh)$. Also, for any xN in G/N we have $\eta(x) = xN$.   ////

Thus, starting with any normal subgroup of a group, we obtain a homomorphism. This is the first half of the connection between normal subgroups and isomorphisms. After we prove the correspondence theorem, we shall be able to prove that any homomorphism gives us an accompanying normal subgroup. The correspondence theorem just collects some properties of groups preserved by homomorphisms.

Proposition 9 (the correspondence theorem) If G and $\bar{G}$ are groups and ϕ is a homomorphism from G onto $\bar{G}$, then
(a) $\phi(e_G) = e_{\bar{G}}$, where e_G is the identity of G, and $e_{\bar{G}}$ is that of $\bar{G}$.
(b) $(\phi(x))^{-1} = \phi(x^{-1})$.
(c) If H is a subgroup of G, then $\phi(H)$ is a subgroup of $\bar{G}$.
(d) If $\bar{H}$ is a subgroup of $\bar{G}$, then $\phi^{-1}(H)$ is a subgroup of G, where for any subset X of $\bar{G}$, $\phi^{-1}(X) = \{g \mid g \in G \text{ and } \phi(g) \text{ is in } X\}$.
(e) If $H \lhd G$, then $\phi(H) \lhd \bar{G}$.
(f) If $\bar{H} \lhd \bar{G}$, then $\phi^{-1}(\bar{H}) \lhd G$.

PROOF   We proved parts a and b when we proved Proposition 1 about isomorphisms in Sec. 1.6. However, the proofs are quite short, and we repeat them here.

(a)   Since $\phi(e_G) = \phi(e_G{}^2) = \phi(e_G)\phi(e_G) = (\phi(e_G))^2$, we have a solution to $x^2 = x$ in $\bar{G}$; by Exercise 1 of Sec. 1.1, we must have $x = e_{\bar{G}} = \phi(e_G)$.

(b)   Since $\phi(x)\phi(x^{-1}) = \phi(xx^{-1}) = \phi(e_G) = e_{\bar{G}}$, we have $\phi(x^{-1}) = (\phi(x))^{-1}$.

(c)   Let H be a subgroup of G. By part a, $e_{\bar{G}}$ is in $\phi(H)$, since $e_{\bar{G}} = \phi(e_G)$. By part b, if $\phi(h)$ is in $\phi(H)$ then so is $(\phi(h))^{-1}$, since $(\phi(h))^{-1} = \phi(h^{-1})$. Similarly, if $\phi(h_1)$ and $\phi(h_2)$ are in $\phi(H)$, so is $\phi(h_1)\phi(h_2) = \phi(h_1h_2)$, and thus $\phi(H)$ is closed and thus a subgroup.

(d)   Let $\bar{H}$ be a subgroup of $\bar{G}$. Then e_G is in $\phi^{-1}(H)$, since $\phi(e_G) = e_{\bar{G}} \in \bar{H}$. If x is in $\phi^{-1}(H)$, then $\phi(x) \in \bar{H}$ so $(\phi(x))^{-1} \in \bar{H}$; since $(\phi(x))^{-1} = \phi(x^{-1})$ we have x^{-1} in $\phi^{-1}(H)$. Thus it only remains to check closure. If g_1 and g_2 are in $\phi^{-1}(H)$, then $\phi(g_1)$ and $\phi(g_2)$ are in $\bar{H}$. Since $\bar{H}$ is closed we have $\phi(g_1)\phi(g_2) = \phi(g_1g_2)$ in $\bar{H}$, and g_1g_2 in $\phi^{-1}(H)$.

(e)   If H is a normal subgroup of G, then $\phi(H)$ is a subgroup of $\bar{G}$ by part c, so we need only show that $\phi(H)$ is normal in $\bar{G}$. If $\bar{g}$ is in $\bar{G}$ we know that $\bar{g} = \phi(g)$ for some g in G, since ϕ is onto. Therefore $\bar{g}^{-1}\phi(H)\bar{g} = (\phi(g))^{-1}\phi(H)\phi(g) = \phi(g^{-1})\phi(H)\phi(g) = \phi(g^{-1}Hg) = \phi(H)$ since $H \lhd G$, and thus $\phi(H) \lhd \bar{G}$.

(f) If $\overline{H}$ is normal in $\overline{G}$, then we want to show $g^{-1}(\phi^{-1}(\overline{H}))g \subseteq \phi^{-1}(\overline{H})$. Apply ϕ to $g^{-1}\phi^{-1}(\overline{H})g$. We get

$$\phi(g^{-1}\phi^{-1}(\overline{H})g) = \phi(g^{-1}) \cdot \overline{H}\phi(g) = \overline{H}$$

since $\overline{H} \lhd \overline{G}$ and thus $g^{-1}\phi^{-1}(\overline{H})g \subseteq \phi^{-1}(H)$ for all g in G. Thus, $\phi^{-1}(\overline{H})$ is normal in G. ////

REMARK We used the hypothesis "ϕ is onto" only in part e. In part f we only need $\overline{H} \lhd \phi(G)$.

Definition If $\phi: G \to \overline{G}$ is a homomorphism from G into $\overline{G}$, then ker $(\phi) = \{g \,|\, g \in G, \phi(g) = e_{\overline{G}}\}$ is the **kernel** of the homomorphism ϕ.

Corollary 1 If ϕ is a homomorphism from G into $\overline{G}$, then ker $(\phi) \lhd G$.

PROOF First note that ker $(\phi) = \phi^{-1}(e_{\overline{G}})$. Also note that $e_{\overline{G}} \lhd \overline{G}$, and then apply part f. ////

Thus for any homomorphism ϕ we obtain a normal subgroup; this is the second half of the connection between these two concepts.

Exercises

9 Which of the following functions ϕ are homomorphisms? Of the homomorphisms, which are isomorphisms?
   (a) $G = (\mathbb{R} - (0), \cdot) = \overline{G}$, $\phi(x) = 3x$
   (b) $G = (\mathbb{R}, +) = \overline{G}$, $\phi(x) = 3x$
   (c) $G = \mathrm{GL}\,(2, \mathbb{R})$, $\overline{G} = (\mathbb{R} - (0), \cdot)$, $\phi\begin{pmatrix} a & b \\ c & d \end{pmatrix} = ad - bc$
   (d) $(G, \cdot) = $ an abelian group $= (\overline{G}, \cdot)$, $\phi(x) = x^4$
   (e) $G = \left\{\begin{pmatrix} a & b \\ 0 & c \end{pmatrix}, ac \neq 0\right\}$ and $\overline{G} = (\mathbb{R} - (0), \cdot) \times (\mathbb{R} - (0), \cdot)$ and

$$\phi\begin{pmatrix} a & b \\ 0 & c \end{pmatrix} = (a, c)$$

10 For those parts of Exercise 9 which are homomorphisms, find ker (ϕ).

11 If A is an abelian group of order n, and ϕ is a function from A to A given by $\phi(a) = a^k$, where k is an integer, then show that
   (a) ϕ is a homomorphism.
   (b) ϕ is an isomorphism if gcd $(k, n) = 1$.

12 Let G be the group of all polynomials with real coefficients under addition, and let $\phi: G \to G$ be the derivative function [that is, $G = \{\sum_{n=0}^{i} a_i x^i \,|\, a_i \in \mathbb{R}\}$ and $\phi(p(x)) = p'(x) = dp(x)/dx = \sum_{n=0}^{i} a_{i-1} i x^{i-1}$].
   (a) Is ϕ a homomorphism?
   (b) Is ϕ one-to-one? Is it onto?

We now come to an important theorem which makes precise the notion of every homomorphism leading to a normal subgroup.

The fundamental theorem of homomorphisms If ϕ is a homomorphism from a group G onto a group $\bar{G}$, and $N = \ker(\phi)$, then G/N is isomorphic to $\bar{G}$. Further, there is a one-to-one correspondence between subgroups of $\bar{G}$ and those of G that contain N.

PROOF OF FIRST STATEMENT Since $\ker(\phi) = N \lhd G$, G/N exists. We have the following picture, where the map $G \to G/N$ is the canonical homomorphism:

We want to construct a map α from G/N to $\bar{G}$

which has some relationship to ϕ, and then show that α is an isomorphism.

If Ng is a typical element of G/N, what is a logical choice for $\alpha(Ng)$? We need an element in $\bar{G}$ related to both ϕ and Ng, so we try $\alpha(Ng) = \phi(g)$. We first must show that while $Ng = Nh$ does not imply that $g = h$, this ambiguity does not make α multiple valued.

$$Ng = Nh$$

if and only if

$$Ngh^{-1} = N$$

if and only if

$$gh^{-1} \in N = \ker(\phi)$$

if and only if

$$\phi(g)(\phi(h))^{-1} = \phi(g)\phi(h^{-1}) = \phi(gh^{-1}) = e_{\bar{G}}$$

if and only if

$$\alpha(Ng) = \phi(g) = \phi(h) = \alpha(Nh)$$

Thus α is a function. Since every step is reversible, reading from bottom to top yields a proof that α is one-to-one.

If $\bar{g}$ is any element of $\bar{G}$, we note that ϕ is onto so $\bar{g} = \phi(g)$ for some g in G. Thus, $\alpha(Ng) = \phi(g) = \bar{g}$, and α is an onto function.

Since N is normal in G, we have $Ngh = NgNh$. Thus, $\alpha(Ng)\alpha(Nh) = \phi(g)\phi(h) = \phi(gh) = \alpha(Ngh) = \alpha(NgNh)$, and α is a homomorphism and thus an isomorphism.   ////

To prove the second statement, we start with a lemma.

Lemma   If $\phi: G \to \bar{G}$ is a homomorphism and $H \supseteq N = \ker(\phi)$ is a subgroup of G, then $\phi^{-1}(\phi(H)) = H$.

PROOF OF LEMMA   If $x \in \phi^{-1}(\phi(H))$, then $\phi(x) \in \phi(H)$ so $\phi(x) = \phi(h)$ for some $h \in H$. Thus, $\phi(x)\phi(h)^{-1} = \phi(xh^{-1}) = e_{\bar{G}}$, and $xh^{-1} \in \ker(\phi) = N$. Thus $x \in Nh \subseteq Hh = H$, and $\phi^{-1}(\phi(H)) \subseteq H$. By definition of ϕ^{-1} we must have $\phi^{-1}(\phi(H)) \supseteq H$.   ////

PROOF OF SECOND STATEMENT   Returning to the proof of the second statement, we see that if H and K are two distinct subgroups of G both containing N, then $\phi(H)$ and $\phi(K)$ are distinct in $\bar{G}$. This follows from the lemma since if $\phi(H) = \phi(K)$, then $\phi^{-1}(\phi(H)) = \phi^{-1}(\phi(K)) = H = K$. Thus ϕ, considered as a function from the subgroups of G containing N to subgroups of $\bar{G}$, is one-to-one. It is also onto, because if $\bar{H}$ is any subgroup of $\bar{G}$, then $\phi^{-1}(\bar{H})$ is a subgroup of G containing N, and $\phi(\phi^{-1}(\bar{H})) = \bar{H}$.   ////

As an illustration of the second part of this theorem, let $\phi: D_8 \to V$ be given by $\phi(a) = \bar{a}$, $\phi(b) = \bar{b}$, where $a^4 = b^2 = e$, and $b^{-1}ab = a^{-1}$ in D_8 and $\bar{a}^2 = \bar{b}^2 = (\overline{ab})^2 = \bar{e}$ in V. Then $N = \ker(\phi) = \{e, a^2\}$ in D_8. Recalling the subgroup lattices involved, we have

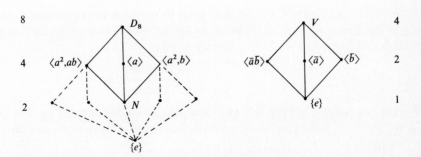

Thus, the subgroup lattice above N corresponds with the subgroup lattice of V.

Exercises

13   Show that if $\phi: G \to A$ is a homomorphism from a group G to an abelian group A, and $N = \ker(\phi)$, then any subgroup of G containing N is normal in G. (Imitate the proof that $N \lhd G$, and use Proposition 2.)

14   (a)   Find the lattice of all subgroups of $\mathbb{Z}_4 \times \mathbb{Z}_2$.

(b)   Every group of order 4 is isomorphic to V or $\mathbb{Z}_4$. Use this to identify $\mathbb{Z}_4 \times \mathbb{Z}_2 / N$, where $N = \{(\bar{0},\bar{0}), (\bar{0},\bar{1})\}$. What if $N = \{(\bar{0},\bar{0}), (\bar{2},\bar{0})\}$ or $\{(0,0), (\bar{2},\bar{1})\}$?

(c)   If ϕ is a homomorphism from $\mathbb{Z}_4 \times \mathbb{Z}_2$ onto $\mathbb{Z}_4$, what are the possibilities for $\ker(\phi)$?

EXAMPLE A   If $G = (\mathbb{R} - (0), \cdot)$ and $\phi(x) = x^2$, then it is easy to see that ϕ is a homomorphism from G onto the range of ϕ, which is the positive reals. The fundamental theorem applies to an onto homomorphism, and we compute the kernel:

$$\begin{aligned} \ker(\phi) &= \{x \,|\, \phi(x) = 1\} \\ &= \{x \,|\, x^2 = 1\} \\ &= \{+1, -1\} \end{aligned}$$

Therefore, $(\mathbb{R} - (0), \cdot)/\{+1, -1\} \simeq (\mathbb{R}^+, \cdot)$. This could be roughly interpreted as saying that nonzero reals are the same as the positive reals if we disregard signs.

EXAMPLE B   By Exercise 9c we see that $\phi: \mathrm{GL}\,(2,\mathbb{R}) \to (\mathbb{R} - (0), \cdot)$, given by $\begin{pmatrix} a & b \\ c & d \end{pmatrix} = ad - bc$, is a homomorphism. If r is any nonzero real number, then $\phi \begin{pmatrix} r & 0 \\ 0 & 1 \end{pmatrix} = r$ so ϕ is onto, and we can apply the fundamental theorem:

$$\ker(\phi) = \left\{ \begin{pmatrix} a & b \\ c & d \end{pmatrix} \,\bigg|\, \phi\begin{pmatrix} a & b \\ c & d \end{pmatrix} = 1 \right\} = \left\{ \begin{pmatrix} a & b \\ c & d \end{pmatrix} \,\bigg|\, ad - bc = 1 \right\}$$

This normal subgroup of $\mathrm{GL}\,(2,\mathbb{R})$ is of great importance in mathematics, and even in theoretical physics. It is called the *special linear group* of degree 2 over the reals and is denoted $\mathrm{SL}\,(2,\mathbb{R})$. Therefore we have

$$\frac{\mathrm{GL}\,(2,\mathbb{R})}{\mathrm{SL}\,(2,\mathbb{R})} \simeq (\mathbb{R} - (0), \cdot)$$

15   If m is a positive integer, and any other integer n is written $n = cm + r$, for $0 \le r \le m - 1$, then define ϕ by $\phi(n) = \bar{r}$ so that ϕ is a function from $(Z, +)$ to $(Z_m, \oplus)$.

(a)   Show that ϕ is onto and a homomorphism.

(b)   What is $\ker(\phi)$?

(c)   What does the fundamental theorem say?

16 If $G = \mathbb{R}^3$ and $\bar{G} = \mathbb{R}^2$, both under addition, and $\phi(a,b,c) = (a,b)$, is ϕ a homomorphism? What is ker (ϕ)? What does the fundamental theorem say? This example resembles the situation of taking a photograph, since ϕ takes 3-space to 2-space. Some information is lost (the kernel, which is one-dimensional), but a photograph can still give very valuable information about 3-space.

17 If H and K are subgroups of G with K normal in G, show that

(*a*) HK is a subgroup of G, and $(H \cap K) \triangleleft H$. (We wish to show $HK/K \simeq H/H \cap K$, and to do this we want to use the fundamental theorem.)

(*b*) $\phi(hk) = (H \cap K)h$ is a homomorphism from HK to $H/H \cap K$.

(*c*) ϕ is onto $H/H \cap K$.

(*d*) ker (ϕ) is K.

(*e*) ϕ is not multiple-valued [if $h_1 k_1 = h_2 k_2$, show that $(H \cap K)h_1 = (H \cap K)h_2$].

(*f*) $HK/K \simeq H/H \cap K$.

This is often called the *second isomorphism theorem*.

The next set of exercises is designed to show that much of high-school mathematics is related to homomorphisms.

18 (The trigonometry homomorphism) Let $\theta: (\mathbb{R}, +) \to GL\,(2, \mathbb{R})$ be given by $\theta(x) = \begin{pmatrix} \cos x & \sin x \\ -\sin x & \cos x \end{pmatrix}$. Show that θ is a homomorphism into $(GL\,(2, \mathbb{R}), \cdot)$. [An alternative statement here would be that $\theta(r) = \cos r + i \sin r$ is a homomorphism from $(\mathbb{R}, +)$ to $(\mathbb{C} - (0), \cdot)$.] What happens if we expand

$$\begin{pmatrix} \cos x & \sin x \\ -\sin x & \cos x \end{pmatrix}^2 = \begin{pmatrix} \cos 2x & \sin 2x \\ -\sin 2x & \cos 2x \end{pmatrix}$$

19 (The logarithm homomorphism) Show that $\alpha(\mathbb{R} - (0), \cdot) \to (\mathbb{R}, +)$, given by $\alpha(x) = \log_{10} (x)$, is an isomorphism.

20 (The absolute-value homomorphism) Show that $\beta: (\mathbb{R} - (0), \cdot) \to (\mathbb{R}^+, \cdot)$ is a homomorphism, where $\beta(x) = |x|$. What is ker (β)?

21 (The "casting out 9s" homomorphism) We have shown that $\theta(_B) = \bar{r}$, where $n = c \cdot 9 + r$, for $0 \le r \le 8$, is a homomorphism from $(\mathbb{Z}, +)$ to $(\mathbb{Z}_9, \oplus)$. First prove that $\overline{10} = \bar{1}$, $\overline{100} = \bar{1}$, ..., $\overline{10^n} = \bar{1}$ for all positive integers n. Now note that a number in decimal expansion can be written $a_k a_{k-1} \cdots a_1 a_0 = a_k 10^k + a_{k-1} 10^{k-1} + \cdots + a_1 10 + a_0$. For instance, 723 $= 7 \cdot 10^2 + 2 \cdot 10 + 3$. Using this, we show that

$$\overline{a_k a_{k-1} a_{k-2} \cdots a_1 a_0} = \overline{a_k} + \overline{a_{k-1}} + \overline{a_{k-2}} + \cdots + \overline{a_1} + \overline{a_0}$$

(a) Prove that

$$9 \mid a_k a_{k-1} \cdots a_1 a_0 \Leftrightarrow \overline{a_k a_{k-1} \cdots a_1 a_0} \in \ker(\theta)$$

$$\Leftrightarrow \overline{a_k a_{k-1} \cdots a_1 a_0} = \overline{a_k} + \overline{a_{k-1}} + \cdots + \overline{a_1} + \overline{a_0} = \overline{0}.$$

Use this to see which of the following are divisible by 9: 1,003,847, 73,127,894, 987,654,321.

(b) Since θ is a homomorphism, we can use casting out 9s to check addition or subtraction. If $a_k a_{k-1} \cdots a_1 a_0 + b_l b_{l-1} \cdots b_1 b_0 = c_m c_{m-1} \cdots c_1 c_0$, show that

$$\overline{a_k} + \overline{a_{k-1}} + \cdots + \overline{a_1} + \overline{a_0} + \overline{b_l} + \overline{b_{l-1}} + \cdots + \overline{b_1} + \overline{b_0}$$

$$= \overline{c_m} + \overline{c_{m-1}} + \cdots \overline{c_1} + \overline{c_0}$$

For instance, $723 + 131 = 864$ must be wrong since $\overline{8} = \overline{17} = \overline{7} + \overline{2} + \overline{3} + \overline{1} + \overline{3} + \overline{1} = \overline{8} + \overline{6} + \overline{4} = \overline{18} = \overline{0}$, a contradiction. (This also works when we check multiplication, since θ is also a ring homomorphism.)

(c) What is $\ker(\theta)$? How does $\ker(\theta)$ relate to those errors which this method will not detect?

Exercise 22 provides a convenient criterion for a homomorphism to be one-to-one. Exercise 23 is used once in Chap. 5 to prove a key theorem on solvable groups which, in turn, is used in Chap. 10 to show the existence of polynomials not solvable by radicals.

22 Let ϕ be a homomorphism from G to $\overline{G}$. Show that ϕ is one-to-one (i.e., a monomorphism) if and only if $\ker(\phi) = e_G$.

23 (The first isomorphism theorem) Let H and K be normal subgroups of G with $H \triangleleft K$. We want to show $(G/H)/(K/H) \simeq G/K$, and one way to proceed is as follows: Define

$$\phi: \frac{G}{H} \to \frac{G}{K} \quad \text{by} \quad \phi(gH) = gK$$

(a) Show that ϕ is well-defined, onto, and a homomorphism.
(b) Show that $\ker(\phi) = K/H$.
(c) Apply the fundamental homomorphism theorem to ϕ.

24 Apply Exercise 23 to $G = \mathbb{Z}$, $K = 25\mathbb{Z}$, and $H = 100\mathbb{Z}$.

†1.14 THE COMMUTATOR SUBGROUP AND A UNIVERSAL MAPPING PROPERTY

(This section is of less vital importance than preceding sections. It might be advisable to omit it on a first reading.)

A group G is abelian if $ab = ba$ for all a and b in G. The equation $ab = ba$ is equivalent to $a^{-1}b^{-1}ab = e$ and leads us to examine elements of the form $x^{-1}y^{-1}xy$. If $S = \{x^{-1}y^{-1}xy \,|\, x,y \in G\}$ then $\langle S \rangle$, the subgroup generated by S, is called G', the *commutator subgroup* of G. In many examples $S = \langle S \rangle$, but there are groups where $S \subsetneqq \langle S \rangle = G'$. A typical element of S is $c_1 c_2 \cdots c_k$, where each $c_i = a_i^{-1} b_i^{-1} a_i b_i$ is a commutator. We have shown that if A is abelian then $A' = e$; so in a sense the bigger G' is, the "less abelian" G is.

EXAMPLE A If D_{2n} is the dihedral group of order $2n$ given by $a^n = b^2 = e$ and $b^{-1}ab = a^{-1}$, then

$$a^{-i}a^{-j}a^i a^j = e$$
$$a^{-1}b^{-1}ab = a^{-2}$$
$$b^{-1}a^j b = (a^{-1})^j = a^{-j}$$
$$a^{-j}b^{-1}a^j b = a^{-2j}$$
$$b^{-1}(ba^j)^{-1}bba^j = b^{-1}a^{-j}b^{-1}bba^j = a^{2j}$$

and finally

$$b^{-1}b^{-1}bb = e$$

These computations give that $S = \{a^{-2j}\}$, so

$$\langle S \rangle = \{e, a^2, a^4, \dots, a^{-2}, a^{-4}, \dots\} = \langle a^2 \rangle = D'_{2n}$$

If n is even, then $G' = \{e, a^2, a^4, \dots, a^{n-2}\}$ is a normal subgroup of order $n/2$. If $n = 2k + 1$ is odd, then $G' = \{e, a^2, a^4, \dots, a^{2k}, a, a^3, \dots, a^{2k-1}\}$ is a normal subgroup of order n. Before doing more examples, it becomes convenient to prove a proposition. This proposition often is useful for finding an upper bound on the possibilities for G'; computations similar to those above give us a lower bound.

Exercises

1 Show that G' is a normal subgroup of G.

2 What is G' where $G = \left\{ \begin{pmatrix} 1 & a \\ 0 & b \end{pmatrix} \,\middle|\, a,b \in \mathbb{R}, b \neq 0 \right\}$?

Proposition 1 If G is a group and G' is its commutator subgroup, then

(a) G' is a normal subgroup of G.
(b) G/G' is abelian.
(c) If H is a normal subgroup such that G/H is abelian, then $H \supseteq G'$.

PROOF (a) Let $\Pi_i \, x_i^{-1} y_i^{-1} x_i y_i = c$ be an arbitrary element of $\langle S \rangle$, and let g be any element of G. Since $g^{-1}cg = c(c^{-1}g^{-1}cg) \in \langle S \rangle$, $\langle S \rangle$ must be normal in G.

(*b*)   Let xG' and yG' be two typical elements of G/G'. It will suffice to show $xG'yG' = yG'xG'$. Since $x^{-1}y^{-1}xy \in G'$, we have $x^{-1}y^{-1}xyG' = G'$ or $xyG' = yxG'$. Thus,

$$xG'yG' = xyG' = yxG' = yG'xG'$$

(*c*)   We want to show $x^{-1}y^{-1}xy \in H$ for all x and y in G. This will show $S \subseteq H$ which, since H is a subgroup, implies $\langle S \rangle \subseteq H$. Since $xyH = xHyH = yHxH = yxH$ we have $x^{-1}y^{-1}xyH = H$ or $x^{-1}y^{-1}xy \in H$.

$/\!/\!/\!/$

Exercise

3 Show that if G is a group and H is a subgroup such that $H \supseteq G'$, then $H \lhd G$. (Use the correspondence theorem on $\eta: G \rightarrow G/G'$.)

Having this proposition, we can see what it says in some known cases. Part c tells us again that $A' = e$ for an abelian group A. If we wanted to compute D'_{2m}, m odd, we could note that $a^{-1}b^{-1}ab = a^2$ and $\langle a^2 \rangle = \langle a \rangle$, and thus $D'_{2m} \supseteq \langle a \rangle$; however since $D'_{2m}/\langle a \rangle$ is of order 2, it is cyclic and abelian. Using part c, we see that $\langle a \rangle \supseteq D'_{2m}$ so we again have $D'_{2m} = \langle a \rangle$, but this time with less computation. Similarly, if m is even, $D'_{2m}/\langle a^2 \rangle$ is of order 4 and thus abelian.

EXAMPLE B   Let $G = \text{GL}(2, \mathbb{R})$. We want to know what G' is, and straight-forward computation of $M_1^{-1}M_2^{-1}M_1M_2$ is not too illuminating. Remember from the last section that $\det: G \rightarrow (\mathbb{R} - (0), \cdot)$ is an onto homomorphism, and $(\mathbb{R} - (0), \cdot)$ is abelian. The basic homomorphism theorem states that

$$\frac{G}{\ker(\det)} = \frac{G}{\text{SL}(2, \mathbb{R})} \simeq (\mathbb{R} - (0), \cdot)$$

and thus we have that $\text{SL}(2, \mathbb{R}) \supseteq G'$. Going the other way, we note that $\begin{pmatrix} r & 0 \\ 0 & 1 \end{pmatrix}^{-1}\begin{pmatrix} 1 & 1 \\ 0 & 1 \end{pmatrix}^{-1}\begin{pmatrix} r & 0 \\ 0 & 1 \end{pmatrix}\begin{pmatrix} 1 & 1 \\ 0 & 1 \end{pmatrix} = \begin{pmatrix} 1 & 1-r^{-1} \\ 0 & 1 \end{pmatrix} \in G'$, so $\begin{pmatrix} 1 & x \\ 0 & 1 \end{pmatrix} \in G'$ for all $x \neq 1$. Since

$$\begin{pmatrix} 1 & \frac{1}{2} \\ 0 & 1 \end{pmatrix}\begin{pmatrix} 1 & \frac{1}{2} \\ 0 & 1 \end{pmatrix} = \begin{pmatrix} 1 & 1 \\ 0 & 1 \end{pmatrix}$$

we can say all $\begin{pmatrix} 1 & x \\ 0 & 1 \end{pmatrix} \in G'$. Similarly, all $\begin{pmatrix} 1 & 0 \\ y & 1 \end{pmatrix} \in G'$. Also,

$$\begin{pmatrix} 0 & 1 \\ 1 & 0 \end{pmatrix}^{-1}\begin{pmatrix} x & 0 \\ 0 & y \end{pmatrix}^{-1}\begin{pmatrix} 0 & 1 \\ 1 & 0 \end{pmatrix}\begin{pmatrix} x & 0 \\ 0 & y \end{pmatrix} = \begin{pmatrix} xy^{-1} & 0 \\ 0 & yx^{-1} \end{pmatrix} = \begin{pmatrix} z & 0 \\ 0 & z^{-1} \end{pmatrix} \in G'$$

if $z \neq 0$. Summing up, we have

$$\begin{pmatrix} 1 & x \\ 0 & 1 \end{pmatrix}, \begin{pmatrix} 1 & 0 \\ y & 1 \end{pmatrix}, \text{ and } \begin{pmatrix} z & 0 \\ 0 & z^{-1} \end{pmatrix} \in G'$$

for all x, y in $\mathbb{R}$ and z in $\mathbb{R} - (0)$. Now

$$\begin{pmatrix} 1 & 0 \\ y & 1 \end{pmatrix}\begin{pmatrix} z & 0 \\ 0 & z^{-1} \end{pmatrix}\begin{pmatrix} 1 & x \\ 0 & 1 \end{pmatrix} = \begin{pmatrix} z & zy \\ zy & zxy + z^{-1} \end{pmatrix}$$

Let $z = a$, $zx = b$ or $x = b/a$, $y = c/a$. Then if $ad - bc = 1$ we shall have $zxy + z^{-1} = bc/a + 1/a = (bc + 1)/a = ad/a = d$. Thus,

$$\begin{pmatrix} 1 & 0 \\ ca^{-1} & 1 \end{pmatrix}\begin{pmatrix} a & 0 \\ 0 & a^{-1} \end{pmatrix}\begin{pmatrix} 1 & ba^{-1} \\ 0 & 1 \end{pmatrix} = \begin{pmatrix} a & b \\ c & d \end{pmatrix}$$

if $ad - bc = 1$. Thus $G' = \mathrm{SL}\,(2, \mathbb{R})$. This is incomplete (since $a \neq 0$) and un-motivated, but a reasonably important result. The following exercise will remedy the incompleteness. Motivation comes from linear algebra, where the theory of elementary row operations via elementary matrices is discussed. In fact it is true that $\mathrm{GL}\,(n, \mathbb{R})' = \mathrm{SL}\,(n, \mathbb{R})$, which will follow from this same theory. It is even true that $\mathrm{SL}\,(n, \mathbb{R})' = \mathrm{SL}\,(n, \mathbb{R})$ for all $n \geq 2$.

Exercises

4   If $a = 0$, modify the above computations to show that $\begin{pmatrix} 0 & b \\ c & d \end{pmatrix} \in G'$ when $bc = -1$.

5   *Show that if $\mathrm{SL}\,(2, \mathbb{R}) = H$ then $H' = H$.

6   If $J = \left\{ \begin{pmatrix} a & b \\ 0 & c \end{pmatrix} \middle| ac \neq 0 \right\}$, find J'.

7   Let $\mathcal{Q}_{4n}$ be the group of order $4n$ given by $a^i b^j$, where $i = 0, 1, \ldots, 2n$, and $j = 0, 1$. Also, $a^{2n} = e = b^4$, $b^2 = a^n$, and $b^{-1}ab = a^{-1}$. Find $\mathcal{Q}'_{4n}$.

8   Let $G = \{(a,b) \mid a, b \in Z\}$, and $*$ be given by $(a,b) * (c,d) = (a + c(-1)^b, b + d)$. We can show easily that G is a nonabelian group. What is G'?

9   (Out of place) Show that $\mathrm{Sym}\,(n)' = \mathrm{Alt}\,(n)$.

10 Find the commutator subgroup of $\mathrm{Sym}\,(3)$ and $\mathrm{Sym}\,(4)$.

11 Show that if $G \simeq H \times K$, then $G' \simeq H' \times K'$.

If we translate Proposition 1 into homomorphisms we have an example of a universal mapping property.

Proposition 1′   If G is a group, G' its commutator subgroup, and $G_{ab} = G/G'$, then

(a) G' is normal in G, so G_{ab} is a group.

(b) G_{ab} is abelian.

(c)   If h is a homomorphism from G onto an abelian A, then $h = g \circ f$, where $G \xrightarrow{f} G_{ab} \xrightarrow{g} A$, f and g are homomorphisms, and $f(x) = xG'$ is the canonical map of G to G/G'.

PROOF   To prove part c, let $K = \ker(h)$, where h is onto A. Then, by the basic homomorphism theorem, $G/K \simeq A$. So by Proposition 1c, $K \supseteq G'$. Let g take xG' to xK. g is easily a well-defined homomorphism, and $(g \circ f)(x) = g(f(x)) = g(xG') = xK = h(x)$, so $f \circ g = h$.   ////

G_{ab} is sometimes called G *abelianized*. The situation of Proposition $1'c$ can be diagramed as

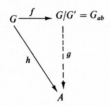

where the dashed line means that if h exists then g exists. We also say h "factors through" G_{ab}, since $g \circ f = h$.

EXAMPLE C   If $G = \mathrm{GL}(2,\mathbb{R})$, let $h\begin{pmatrix} a & b \\ c & d \end{pmatrix} = 1$ if $ad - bc > 0$ and $h\begin{pmatrix} a & b \\ c & d \end{pmatrix} = -1$ if $ad - bc < 0$. Then h is a homomorphism, and $\{\pm 1\}$ under multiplication is an abelian group. Using Example B we know $G \xrightarrow{f} G/G'$ is just $G \xrightarrow{\det} G/G' = \mathrm{GL}(2,\mathbb{R})/\mathrm{SL}(2,\mathbb{R}) \simeq (\mathbb{R} - (0), \cdot)$. Thus, to construct g we just need the appropriate homomorphism from $(\mathbb{R} - (0), \cdot)$ to $(\pm 1, \cdot)$, and

$$g(r) = \begin{cases} +1 & \text{if } r > 0 \\ -1 & \text{if } r < 0 \end{cases}$$

does the job.

Exercises

12   If G is a group such that $G = G'$ and $K \lhd G$, show that $(G/K)' = G/K$. In particular, this applies to $\mathrm{SL}(2,\mathbb{R})$ by Exercise 5. (*Hint:* Use the fact that $G_{ab} = \{e\}$.)

13   If N is a normal subgroup of G and $N \cap G' = \{e\}$, show that $N \subseteq Z(G)$.

We have not yet defined a universal mapping property. In rough terms it goes as follows: We want to examine all functions of some type from (or to) a set S. The type of function involved allows us to find another set S^* and a function $f: S \to S^*$ which could be called the universal map. Then, if $\alpha: S \to T$ is any function of the desired type, we can find $\beta: S^* \to T$ such that $\alpha = \beta \circ f$.

Diagrammatically, this becomes

where the dashed arrow means that the correct β can be found after the rest of the diagram is set up.

The first example of a universal mapping property that students encounter is "least common multiples," which usually show up in the teaching of addition of fractions. To set this up as above, let S be the set of all ordered pairs of positive integers. We shall say that $\alpha: S \to \mathbb{Z}^+$ is "desirable" if $\alpha(n_1,n_2) = m$ means m is a multiple of both n_1 and n_2. We also set $S^* = \mathbb{Z}^+$ and set $f(n_1,n_2) = \text{lcm}(n_1,n_2) = $ the least common multiple. Now the existence of a lcm means all "desirable" functions can be factored through f. The function β would just multiply integers by the integer $m/\text{lcm}(n_1,n_2)$. Thus the plebian fact that the least common multiple divides all other multiples can be translated into a universal mapping property.

REMARKS   (1) Reversing all the arrows gives a second kind of universal mapping property. (2) More details are given in the books by Freyd [24] and Mitchell [62], or any other text on category theory.

†1.15 ODDS AND ENDS

We have seen how to construct a new group G as a direct product of two other groups A and B. We wish to reverse this process, but first we prepare via the following exercise.

Exercise

1   Show $A \simeq A \times \{e_B\}$ and $B \simeq \{e_A\} \times B$. Also, show that both $A \times \{e_B\}$ and $\{e_A\} \times B$ are normal subgroups of $A \times B$. By identifying A and B with their isomorphic copies, we can thus say A and B are normal subgroups of $A \times B$.

Observing Exercise 1 closely, we proceed to the following proposition:

Proposition 1   If G is a group with two normal subgroups M and N such that $M \cap N = \{e\}$ and $MN = G$, then $G \simeq M \times N$.

PROOF   Every element g in G can be written $g = mn$, since $G = MN$. We basically want to use the function $\theta(g) = (m,n)$ and to show that θ is the desired isomorphism. This is pretty straightforward, except that $g = mn = m_1 n_1$ might be possible with $m_1 \neq m$. This would give us $\theta(g) = (m_1, n_1)$, and θ would not be a function. We show that this cannot happen.

$$mn = m_1 n_1 \Leftrightarrow m_1{}^{-1} m = n_1 n^{-1} \in M \cap N = \{e\}$$
$$\Leftrightarrow m_1{}^{-1} m = n_1 n^{-1} = e$$
$$\Leftrightarrow m_1 = m \text{ and } n_1 = n$$

So θ is a well-defined function from G to $M \times N$. θ is onto since $\theta(mn) = (m,n)$ for any $m \in M$, $n \in N$. If $\theta(m_1 n_1) = \theta(m_2 n_2)$, then $(m_1, n_1) = (m_2, n_2)$, so $m_1 = m_2$, $n_1 = n_2$, and $m_1 n_1 = m_2 n_2$. We now need only show θ is a homomorphism. Let $g = m_1 n_1$ and $h = m_2 n_2$ be any two elements. Then

$$\begin{aligned}
\theta(gh) &= \theta(m_1 n_1 m_2 n_2) \\
&= \theta(m_1 m_2 n_1 n_2) \quad \text{by Exercise 6 of Sec. 1.13} \\
&= (m_1 m_2, n_1 n_2) \\
&= (m_1, n_1)(m_2, n_2) \\
&= \theta(m_1 n_1)\theta(m_2 n_2) = \theta(g)\theta(h)
\end{aligned}$$

and this finishes the proof.   ////

Let us apply the proposition to some groups where we have a good supply of normal subgroups.

EXAMPLE A   Klein's four group $V = \{e,a,b,c\}$ is abelian, so all subgroups of V are normal. $M = \{e,a\}$ and $N = \{e,b\}$ are both cyclic of order 2; they intersect in $\{e\}$, and $V = MN$. Thus $V \simeq M \times N \simeq \mathbb{Z}_2 \times \mathbb{Z}_2$. We have seen this isomorphism before in the section on isomorphisms. Similarly, the next example is familiar.

EXAMPLE B   In $(\mathbb{R} - (0), \cdot)$ we note that both $\mathbb{R}^+$ and $\{\pm 1\}$ are normal subgroups. They intersect in $\{1\}$. The law of trichotomy says every nonzero real number is positive or negative, so $(\mathbb{R} - (0), \cdot) = \mathbb{R}^+ \times \{\pm 1\}$.

Exercise

2   Which of the following groups can be written as a direct sum of proper subgroups: $(\mathbb{Z}_6, \oplus)$, $(\mathbb{Z}_7, \oplus)$, $(\mathbb{Z}_9, \oplus)$, $(\mathbb{Z}_{12}, \oplus)$? Can you conjecture a proposition covering $(\mathbb{Z}_m, \oplus)$?

EXAMPLE C   Let D_{12}, the dihedral group of order 12, be given by $a^6 = b^2 = e$ and $b^{-1}ab = a^{-1}$. Then $A = \{e,a^3\} = Z(D_{12})$, and

$$B = \langle a^2, b \rangle = \{e, a^2, a^4, b, a^2 b, a^4 b\} \simeq D_6$$

A and B are normal in D_{12}. (Why?) $A \cap B = \{e\}$, and $D_{12} = AB$. Thus $D_{12} \simeq A \times B \simeq \mathbb{Z}_2 \times D_6$.

Exercises

3 Show that $D_{4m} \simeq \mathbb{Z}_2 \times D_{2m}$ for all odd integers $m \geq 3$.
4 Show that $(\mathbb{Q}, +)$ cannot be written as $A \times B$ (or $A + B$) unless A or B equals (0). The same is true for $(\mathbb{Z}, +)$.

Definition If $G = A \times B$, then the maps $\pi_1(a,b) = a$ and $\pi_2(a,b) = b$ are called the **projection maps** from G to A and B.

Exercises

5 If $G = A \times B$, what are ker (π_1) and ker (π_2)? If N is a normal subgroup of G, show that $\pi_1(N) \lhd A$ and $\pi_2(N) \lhd B$.
6 Suppose we have two homomorphisms $f: A \to B$ and $g: B \to C$, and $g \circ f$ is an isomorphism from A onto C. Show the following:
(a) f is one-to-one.
(b) g is onto.
(c) *If, in addition, $f(A) \lhd B$, show that $B \simeq f(A) \times$ ker (g).

The last exercises show that direct products can be viewed in terms of homomorphisms. The project of putting all concepts in terms of homomorphisms is useful for solving problems and is also interesting in itself.

7 If $G = A \times B$, and π_1 and π_2 are the projection homomorphisms of G onto A and B, show that a function h from a group J to G is a homomorphism if and only if $\pi_1 \circ h: J \to A$ and $\pi_2 \circ h: J \to B$ are homomorphisms. It might be useful to prove first that if $h_1: J \to A$ and $h_2: J \to B$ are homomorphisms then there is a homomorphism $h: H \to A \times B$ so that $\pi_1 \circ h = h_1$ and $\pi_2 \circ h = h_2$.

Again using solid arrows for the maps that exist by hypothesis, and dashed arrows for those implied, we have

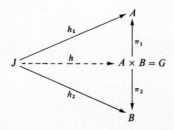

Thus, direct product is another example of a universal mapping property. See any text on category theory or homological algebra for much more on this type of property.

At this point we introduce a new notation for permutations. Let's start with an example:

$$\alpha = \begin{pmatrix} 1 & 2 & 3 & 4 & 5 & 6 & 7 \\ 7 & 1 & 2 & 6 & 5 & 4 & 3 \end{pmatrix}$$

Since the top row is almost wasted, we suspect there must be a more efficient way to write α without losing any information. Since α takes 1 to 7, we could write $\alpha = (17 \cdots) \cdots$. Instead of now seeing where 2 goes, let us note that α takes 7 to 3:

$$\alpha = (173 \cdots) \cdots$$

Next, α takes 3 to 2, and 2 back to 1. So we can write $\alpha = (1732) \cdots$ with the understanding that α takes each entry to the following entry, except that the last entry in parenthesis is taken back to the first entry in parenthesis.

This is all well, except that we have ignored 4, 5, and 6. We repeat the process starting with 4, and find that

$$\alpha = (1732)(46) \cdots$$

Now the only entry we have left out is 5. Since α takes 5 to 5, the complete situation is described by

$$\alpha = (1732)(46)(5)$$

This is called the *cycle notation* for α.

We can shorten this a little more by dropping (5) and having an understanding that α takes each omitted entry to itself.

Exercises

8 Rewrite the following permutations in the cycle notation:

(a) $\begin{pmatrix} 1 & 2 & 3 & 4 & 5 \\ 2 & 3 & 4 & 5 & 1 \end{pmatrix}$ (b) $\begin{pmatrix} 1 & 2 & 3 & 4 & 5 & 6 \\ 2 & 1 & 4 & 3 & 6 & 5 \end{pmatrix}$

(c) $\begin{pmatrix} 1 & 2 & 3 & 4 & 5 & 6 & 7 \\ 2 & 3 & 1 & 4 & 5 & 7 & 6 \end{pmatrix}$

9 The following permutations in Sym (8) are written in the cycle notation. Rewrite them in the old notation.
(a) (13572468)
(b) (123)(47)
(c) (18)(27)(36)(45)
(d) What is $(13572468)^{-1}$?

This new method of writing permutations is not only convenient, but leads to some important new concepts.

Definitions A permutation of the form $(a_1, a_2, \ldots, a_k)$ is called a **cycle** or a k-**cycle**. A 2-cycle is called a **transposition**. Two cycles, $\alpha = (a_1, a_2, \ldots, a_n)$ and $\beta = (b_1, b_2, \ldots, b_m)$, are **disjoint** if $\{a_1, a_2, \ldots, a_n\} \cap \{b_1, b_2, \ldots, b_m\} = \varnothing$. A set Γ of cycles is disjoint if any two distinct cycles in Γ are disjoint.

The new insight here is that a permutation like $\alpha = (135)(2468)$ can be also viewed as a product of two permutations in Sym (8). That is, $\alpha = \alpha_1 \alpha_2$, where $\alpha_1 = (135)$ (α_1 fixes 2, 4, 6, 7, and 8) and $\alpha_2 = (2468)$.

Exercises

10 If α and β are disjoint cycles, show that $\alpha\beta = \beta\alpha$.

11 (15)(14)(13)(12) as written represents the composite of four functions. What is the product of these four functions? Show that any cycle can be written as the product of transpositions.

12 We started this topic by showing $\begin{pmatrix} 1 & 2 & 3 & 4 & 5 & 6 & 7 \\ 7 & 1 & 2 & 6 & 5 & 4 & 3 \end{pmatrix}$ can be written as the product of disjoint cycles. Imitate this to prove that any permutation of a finite set can be written as a product of disjoint cycles.

Definition Any permutation can be written as a product of transpositions, as we see by using Exercises 11 and 12. If a permutation is the product of an even number of transpositions, then it is called an **even permutation**. Otherwise it is called an **odd permutation**. The set of all even permutations is called the **alternating group**.

13 Which of the following permutations are even?
  (a) (15)(14)(13)(12)
  (b) $(ab)(ab)(ab)$
  (c) $\begin{pmatrix} 1 & 2 & 3 & 4 & 5 & 6 & 7 \\ 7 & 1 & 2 & 6 & 5 & 4 & 3 \end{pmatrix}$
  (d) (123)(426)(132)

14 (12)(13)(14)(25)(45)(14) = (25)(23) Is this permutation even or odd? What do you think about this situation?

15 Show that an even permutation times an even permutation is even. Check the other three possible cases.

It is a fact that any permutation on a finite set is even or odd but not both. This is unfortunately harder to prove than it would seem to be. A brief proof is included in the appendix at the end of this chapter. A charming, short, but incomplete proof can be found in the June, 1969, issue of the *American Mathematical Monthly*. Using this fact and the last exercise, it is easy to show that the alternating group is a subgroup of the symmetric group. If

$$\alpha = (a_1 a_2)(b_1 b_2) \cdots (x_1 x_2)$$

is even, so is $(x_1x_2)\cdots(b_1b_2)(a_1a_2) = \alpha^{-1}$. $I = (ab)(ab)$ is also even, and closure for the alternating group follows from the last exercise.

16   Show that Alt (n), the alternating group on n letters, has order $n!/2$ for all $n \geq 2$.

17   What is a homomorphism ϕ from Sym (n) onto $(\mathbb{Z}_2, \oplus)$ such that ker $(\phi) =$ Alt (n)?

18   Give two proofs that Alt $(n) \lhd$ Sym (n).

Another fact that we shall use but not prove is that Alt (n) is a simple group for all n except $n = 4$. A **simple** group is one with no proper normal subgroups. Thus Alt (n), $n \neq 4$, has no proper normal subgroups. Proofs can be found in Herstein [38], Paley and Weichsel [68], and van der Waerden [85].

19   Let ϕ: Alt $(n) \to G$ be a homomorphism, where $n \neq 4$. Show that either ϕ is one-to-one or $\phi(\text{Alt }(n)) = \{e_G\}$.

20   *Write out the elements of Alt (4), and find all subgroups. In particular, show that, despite the fact that $6 \,|\, |\text{Alt (4)}|$, Alt (4) has no subgroup of order 6. This is the smallest counterexample showing that Lagrange's theorem does not have a complete converse.

21   Show that Alt (5) has no subgroup of order 30.

22   Find $\langle S \rangle$ in Sym (6), where
(a) $S = \{(123),(456)\}$ (b) $S = \{(12),(13),(14),(15),(16)\}$

23   Show that Alt $(4) \simeq T$, where T is the group of rotations of a regular tetrahedron.

24   If G is any permutation group on n letters, show that either half or all the elements of G are even.

25   (a)   Find the orders of the following permutations: (123)(456), (12)(3456), $(abc)(de)$, $(abcd)(efgh)(jkl)$.
(b)   Make up a theorem about the order of a product of disjoint cycles.

26   Consider the common puzzle where a number above, below, or either side of the empty space can be slid into the empty space.

1	2	3	4
5	6	7	8
9	10	11	12
13	14	15	

(If you've never seen this puzzle, skip the problem.) Show that the set of reachable positions form a group. Show that this group is isomorphic to Alt (15) if we put the blank square in the bottom right corner.

APPENDIX

Proposition   If α is a permutation of a finite set $S = \{1,2,\ldots,n\}$, then α is not both even and odd.

PROOF   We introduce (from algebraic number theory) a function
$$f = (r_1 - r_2)(r_1 - r_3) \cdots (r_1 - r_n)(r_2 - r_n) \cdots (r_{n-1} - r_n) = \prod_{i<j} (r_i - r_j),$$
where we can take the r_i to be any n distinct real numbers.

Define $\alpha(f) = \prod_{i<j} (r_{\alpha(i)} - r_{\alpha(j)})$, and observe that $\alpha(f)$ is either f of $-f$. We insert a lemma.   ////

Lemma   If α is a transposition, $\alpha(f) = -f$.

PROOF OF LEMMA   If $\alpha = (st)$ with $s < t$, then α changes the sign of $(r_s - r_t)$. If q is less than s, α interchanges $r_q - r_s$ and $r_q - r_t$; and similarly if q is greater than t. When $s < q < t$, then $r_s - r_q$ and $r_q - r_t$ are interchanged and both change signs. Thus, the total effect of α is to change the sign of $r_s - r_t$, and $\alpha(f) = -f$.   ////

Returning to the main proof, we see that if α is written as a product of n transpositions, then $\alpha(f) = (-1)^n f$. This must be constant and independent of the mode of representation. Thus, if α is also a product of m transpositions, then $\alpha(f) = (-1)^m f = (-1)^n f$. Thus m and n must be both even or both odd, and α is thus either even or odd, but not both.   ////

†1.16   SOME HISTORICAL REMARKS

A fact that has always fascinated me is that Lagrange died before groups were invented. In this section we discuss the theorem that Lagrange actually did prove in his 1770-1 paper, "Reflexions sur la resolution álgebrique des équations" [51].

The theorem states that the number of different formal values which are obtained by permuting the n elements of a given function of n variables in every possible manner is a divisor of $n!$. What does this mean? We give an informal explanation.

Let $f(x_1,x_2,x_3) = 5x_1 + 5x_2 + x_3{}^3$. If we pick, say, $x_1 = 2$, $x_2 = 3$, and $x_3 = 4$, then $f(2,3,4) = 89$. If we interchange x_1 and x_2, we see $f(3,2,4) = 89$. Trying the other permutations, we have $f(4,2,3) = f(2,4,3) = 57$ and $f(4,3,2) = f(3,4,2) = 43$. Thus, permuting these three values in all 3! possible ways, we get three different values. If we had picked, say, $x_1 = x_2 = x_3 = \frac{4}{3}$, we of course would have obtained only one value. Such duplication can be avoided, and usually simple trial and error suffice.

Definition   If $f(x_1,\ldots,x_n)$ is a function of n variables, then the n-tuple $(c_1,c_2,\ldots,c_n)$ is called **generic** if the $n!$ permutations of the c_i yield the largest number of different values of f possible. This largest number will be called the **number of formal values.**

Exercises

1 If $f(x_1,x_2,x_3,x_4,x_5) = x_1 + x_2 + x_3 + 3x_4{}^2 + 3x_5{}^2$, find a 5-tuple such that permuting the entries of the 5-tuple will yield 10 values for f.

2 If $f(x_1,x_2,x_3,x_4) = x_1{}^3 + x_2{}^3 + 3x_3 + x_4{}^2$, what is the number of formal values f can take on?

Now let us give a proof of Lagrange's theorem. Let $f(x_1,x_2,...,x_n)$ be our function, and $(c_1,c_2,...,c_n)$ a generic point. We let Sym (n) act on the symbols $\{c_1,c_2,...,c_n\}$. If $f(c_1,c_2,...,c_n) = w$, then let

$$H = \{\sigma \mid \sigma \in \text{Sym } (n) \text{ and } f(c_1,c_2,...,c_n) = f(\sigma(c_1), \ldots, \sigma(c_n))\}.$$

H is just those permutations in Sym (n) that take w to w. H is a subgroup (Exercise 3). If v is another formal value of f, then $v = f(\alpha(c_1), \alpha(c_2), \ldots, \alpha(c_n))$, and the left coset αH consists of all elements in Sym (n) yielding the value v. Thus $|H| \cdot (\text{number of formal values}) = |\text{Sym } (n)| = n!$. In particular, the number of formal values divides $n!$.

Exercise

3 Show that, in the above proof, H is indeed a subgroup of Sym (n).

This theorem is interesting also in that it relates classical algebra with modern algebra. It led to many similar but more refined results about formal values, and to much research about Sym (n) and its subgroups. For instance, let us consider two results about Sym (n).

1 Sym (n) contains no subgroup of order 15 for $n = 5$, 6, or 7, even though 15 divides 5!, 6!, and 7!.

2 If $n \geq 5$, then Sym (n) has no subgroup of index 3, 4, $\ldots$, $n - 1$.

Result 2 is discussed in Chap. 4, and we will present result 1 in the exercises. But for the moment let us assume their validity and translate them.

2′ No function of n variables can have 3, 4, $\ldots$, $n - 1$ formal values if $n \geq 5$ (due to Abbati for the case $n = 5$).

1′ A function of five variables cannot have eight formal values. A function of six variables cannot take on 48 formal values. Similarly, a function of seven variables cannot take on 336 formal values. Note that $48 = 6!/15$.

Throughout, we are assuming that no nonsense is taking place, like first letting $f(x_1,x_2,x_3) = x_1{}^2 + 2x_2 + 0 \cdot x_3$ and then calling it a function of three variables.

The next three exercises show that Sym (5), Sym (6), and Sym (7) contain no subgroup of order 15.

Exercises

4 (a) If G is nonabelian, show that $G/Z(G)$ is not cyclic.

(b) Show that every group of order p^2 is abelian (use Exercise 18 in Sec. 1.9).

(c) If G is nonabelian and has order pq, where p and q are distinct primes, then $Z(G) = \{e\}$.

(d) Use the class equation to show that every group of order 15 is abelian.

(e) Show that no abelian group of order 15 can have more than two elements of order 3 or more than four elements of order 5. Thus show that all groups of order 15 are cyclic.

5 (See Exercise 8c in Sec. 1.12.) The smallest n such that Sym (n) contains an element of order pq is $n = p + q$, where p and q are distinct primes. Prove this. In particular, Sym (8) is the smallest symmetric group containing an element of order 15.

6 Show that neither Sym (5), Sym (6), nor Sym (7) contains a subgroup of order 15.

7 Show that all groups of order 33 are cyclic, thus that Sym (14) is the smallest symmetric group containing a subgroup of order 33, and then that functions of 11 variables cannot take on $10 \cdot 9 \cdot 8 \cdot 7 \cdot 6 \cdot 5 \cdot 4 \cdot 2$ formal values.

8 (a) Show that the function

$$f(x_1, x_2, x_3, x_4) = [(x_1 - x_3)/(x_1 - x_4)](x_2 - x_4)/(x_2 - x_3)$$

takes on six formal values. This is the cross-ratio function from geometry and complex analysis.

(b) Find all x such that, for $(x, 0, 1, \infty)$, f takes on fewer than six formal values.

These problems have been designed to show where some of the original interest developed for the symmetric group and its subgroups.

Cauchy developed such concepts as cycle and transposition. Other mathematicians who worked in this area were Galois, Netto, Cayley, and Jordan. Galois realized the importance of normal subgroups and solvable groups. He also proved that all groups of order less than 60 are solvable. We shall discuss solvable groups in Chap. 5, and Galois theory in Chap. 10. The next stage was the realization that many of the concepts developed from permutation groups and other definite groups generalized to abstract groups. Cayley's theorem must have been reassuring to those who were working with abstract groups, since it meant any result would have some interpretation for permutation groups. The next important result for abstract groups was the Sylow theory, which appeared in 1870. The Sylow theorems are treated in Chap. 4, as is Cayley's theorem.

One other place where finite groups show up in fourth-grade arithmetic is given in the following exercises. These examples have been known a long time and were perhaps influential in the development of number theory and abstract group theory.

9 (a) Show that the remainders obtained while dividing 7 into 1 are 3, 2, 6, 4, 5, and 1.

(b) Compute $\langle \bar{3} \rangle$ in the group $(\{\bar{1}, \bar{2}, \bar{3}, \dots, \bar{6}\}, \odot_7) = G$.

(c) Show that G is cyclic.

10 (a) Find the remainders obtained by dividing 13 into 1, and show that they form a subgroup of $G = (\{\bar{1}, \bar{2}, \dots, \overline{12}\}, \odot_{13})$.

(b) Show that the remainders obtained while dividing 13 into 2 form a coset of this subgroup.

(c) Is G cyclic?

Chapter 7 on number theory develops these concepts further.

11 Where have you heard of Cauchy previously? Cayley? Jordan?

2.1 DEFINITIONS AND EXAMPLES

In this chapter we discuss ring theory, and this section is devoted to the preliminary definitions and examples. The definitions of ring and field seem to be due to Dedekind, who proposed them in the middle of the nineteenth century. The concept of an ideal arose at about the same time. In trying to prove Fermat's conjecture that $x^n + y^n = z^n$ has no nontrivial solutions in integers when n exceeds 2, Dedekind and Kummer saw the need to look at the integers with various roots of unity adjoined. These enlarged sets retained many of the properties of the integers, and rings were defined to be those sets with such properties. Basically, a ring is a set together with two operations, which we call $+$ and $\cdot$, such that $\alpha + \beta$, $\alpha \cdot \beta$, and $-\alpha$ are in the ring whenever α and β are. Although $+$ and $\cdot$ are very suggestive, and although many of the examples lie inside the complex numbers, these are just two abstract binary operations satisfying only those properties required by the definition. It would be more precise, though more cumbersome, to use $\circ$ and $*$.

Definition A **ring** $(R, +, \cdot)$ is a set R together with two binary operations $+$ and $\cdot$ such that the following are true:

(a) R together with $+$ is an abelian group.

(b) $\cdot$ is closed and associative. That is, $x \cdot y$ is in R, and $x \cdot (y \cdot z) = (x \cdot y) \cdot z$ for all x, y, and z in R.

(c) $x \cdot (y + z) = x \cdot y + x \cdot z$ and $(x + y) \cdot z = x \cdot z + y \cdot z$ for all x, y, and z in R.

The additive identity is denoted by 0.

As we shall see, rings appear naturally in the study of number theory, linear algebra, analysis, and set theory, as well as in abstract algebra.

EXAMPLE A $(\mathbb{Z}, +, \cdot)$, the integers under ordinary addition and multiplication, is a ring. We have already seen that $(\mathbb{Z}, +)$ is an abelian group. $\mathbb{Z}$ is obviously closed under multiplication, and the other properties are true for the integers.

EXAMPLE B Let $(A, +)$ be any abelian group, and define $\cdot$ on A by $a \cdot b = 0$ for all a and b in A. Then $(A, +, \cdot)$ is easily checked to be a ring, called the **trivial ring on** A.

EXAMPLE C The even integers with the usual multiplication and addition form a ring. The product of two even integers is even, and the other properties obviously hold, so this is easily checked.

Before proceeding with more examples, we introduce some more definitions. Rings are more manageable if their multiplication satisfies more of the group properties.

Definition A ring is a **commutative ring** if $a \cdot b = b \cdot a$ for all a, b in R.

Definition A ring with **unity** is a ring R with an element 1 such that $1 \cdot a = a \cdot 1 = a$ for all a in R.

Definition A commutative ring R with unity satisfying the cancellation laws for $(R - (0). \cdot)$ is called an **integral domain.**

Definition A ring R such that $(R - (0), \cdot)$ is a group is a **division ring.**

Definition A commutative division ring is a **field.**

Example A, B, and C are all commutative; Example A has 1 as its unity. Example A is also an integral domain.

EXAMPLE D Let $R = \{0, x, y, z\}$, and define $+$ and $\cdot$ with the following tables:

+	0	x	y	z
0	0	x	y	z
x	x	0	z	y
y	y	z	0	x
z	z	y	x	0

and

·	0	x	y	z
0	0	0	0	0
x	0	x	y	z
y	0	x	y	z
z	0	0	0	0

$(R,+)$ is isomorphic to $\mathbb{Z}_2 \times \mathbb{Z}_2$. By considering a few cases, the multiplication can be shown to be closed and associative. By inspection R has no unity, since x and y work on the left but not on the right. Since $x \cdot z = y \cdot z$, R does not satisfy the cancellation laws.

EXAMPLE E $(\mathbb{Z}_n, +, \cdot)$ is the ring of integers, where both $+$ and $\cdot$ are computed modulo n. That is, if $x + y = an + r$, where $0 \leq r < n$ in $\mathbb{Z}$, then $\bar{x} + \bar{y} = \bar{r}$ in $\mathbb{Z}_n$. Similarly, $x \cdot y = an + r$ in $\mathbb{Z}$ is equivalent to $\bar{x} \cdot \bar{y} = \bar{r}$ in $\mathbb{Z}_n$. $(\mathbb{Z}_n, +, \cdot)$ is commutative and has $\bar{1}$ as a unity. If n is composite so that $n = km$, for $k, m > 1$, then $\bar{k}\bar{m} = \bar{0} = \bar{k}\bar{0}$ in $\mathbb{Z}_n$; thus, the cancellation law does not hold in $\mathbb{Z}_n$ unless n is prime. This particular example is of great importance in number theory, as will be discussed in Chap. 7.

EXAMPLE F Let $R = \{\bar{0}, \bar{2}, \bar{4}\}$ with modulo-6 addition and multiplication. This is a commutative ring with 4 as the unity.

EXAMPLE G $(\mathbb{Q}, +, \cdot)$ is a field [as is $(\mathbb{R}, +, \cdot)$]. This is easily shown, since we know that $(\mathbb{Q}, +)$ and $(\mathbb{Q} - (0), \cdot)$ are commutative groups and that the distributive law holds for $+$ and $\cdot$ in $\mathbb{Q}$.

Before continuing this list of rings, we prove a few elementary propositions.

Proposition 1 If $(R, +, \cdot)$ is a ring, then

(a) $a \cdot 0 = 0 = 0 \cdot a$ for all a in R.
(b) $(-a) \cdot b = a \cdot (-b) = -(a \cdot b)$ for all a, b in R.
(c) $(-a) \cdot (-b) = a \cdot b$.

PROOF Since $(R, +)$ is an abelian group, we can use the additive properties without special mention.

(a) $a + 0 = a$

so
$$a \cdot a = a(a + 0)$$
$$a \cdot a = a \cdot a + a \cdot 0 \qquad \text{distributive law}$$
$$a \cdot a + 0 = a \cdot a + a \cdot 0$$
$$0 = a \cdot 0 \qquad \text{cancellation for } (R, +)$$

Similarly, $0 \cdot a = 0$.

(b)
$$a + (-a) = 0$$

so
$$[a + (-a)]b = 0 \cdot b = 0 \qquad \text{by part } a$$
$$= a \cdot b + (-a) \cdot b \qquad \text{distributive law}$$

Thus $-(a \cdot b) = (-a) \cdot b$ by the uniqueness of the inverse element in $(R, +)$. The proof that $-(a \cdot b) = a \cdot (-b)$ is similar.

(c)
$$(-a) \cdot (-b) = (-a)(-b) + a \cdot 0$$
$$= (-a)(-b) + a[(-b) + b]$$
$$= (-a)(-b) + a(-b) + a \cdot b$$
$$= [(-a) + (a)](-b) + a \cdot b$$
$$= 0(-b) + a \cdot b$$
$$= 0 + a \cdot b = a \cdot b \qquad \text{////}$$

REMARK   Part (c) provides an actual proof of one of the gospel truths of high-school mathematics; minus times minus equals plus. Instead of memorizing these proofs, the reader should just observe that they all follow easily from the distributive laws. Since the distributive laws connect the addition with the multiplication, and this proposition tells how some of the additive concepts mesh with some of the multiplicative concepts, it is no surprise that the distributive law is the key to the proofs.

Definition   In a ring R with a multiplicative unity 1, an element a is called a **unit** or **invertible** if there is an element b in R such that $ab = ba = 1$. Such an element b is then often labeled a^{-1}.

Proposition 2   If R is a ring with unity 1, then $U(R) =$ *the unit group of* R is the group of all invertible elements in R.

PROOF   To verify that $U(R)$ is a group, we need only check that $U(R)$ is closed under multiplication and the taking of inverses. We have the associative law since we are in a ring, and 1 obviously is in $U(R)$.

If x, y are in $U(R)$ then x^{-1}, y^{-1} are in R, and $(x \cdot y)^{-1} = y^{-1} \cdot x^{-1}$ is also in R, so $x \cdot y$ is in $U(R)$. Also, if x is in $U(R)$ then x^{-1} is in R. Since $(x^{-1})^{-1} = x$ is in R, x^{-1} itself is in $U(R)$. Therefore $U(R)$ is a group. ////

EXAMPLE H $(\mathbb{Z},+,\cdot)$ has only two invertible elements, $+1$ and -1, so $U(\mathbb{Z}) = \{1,-1\}$, which is isomorphic to $\mathbb{Z}_2$, the cyclic group of order 2.

EXAMPLE I $(\mathbb{Z}_{14},+,\cdot)$ has 14 elements. Solving $xy = 1$ in $\mathbb{Z}_{14}$ is possible only if both x and y are not divisible by 2 or 7, since 2 and 7 do not divide 1. Looking at the remaining elements, 1, 3, 5, 9, 11, and 13, we see that

$$1 = 1^2 = 3 \cdot 5 = 9 \cdot 11 = 13^2$$

in $\mathbb{Z}_{14}$, so

$$
\begin{array}{ccc}
1^{-1} = 1 & 3^{-1} = 5 & 5^{-1} = 3 \\
9^{-1} = 11 & 11^{-1} = 9 & 13^{-1} = 13
\end{array}
$$

Since

$$
\begin{array}{ccc}
3 = 3^1 & 9 = 3^2 & 13 = 3^3 \\
11 = 3^4 & 5 = 3^5 & 1 = 3^6
\end{array}
$$

we see that $U(\mathbb{Z}_{14})$ has six elements, is abelian, and, being generated by 3, is cyclic.

REMARK If D is a division ring, then every nonzero element has an inverse, so $U(D) = (D - \{0\}, \cdot)$. This example includes all fields.

Most of the examples we have discussed have been commutative rings. Section 2.3 discusses an important class of noncommutative rings.

EXAMPLE J Let β be the set of subsets of a set S. We define addition and multiplication in β as follows:

$$X + Y = (X \cup Y)/(X \cap Y) = \text{all elements in } X \cup Y \text{ but not in } X \cap Y$$

This is called the *symmetric difference* of X and Y.

$$X \cdot Y = X \cap Y$$

β, which is called a *Boolean ring*, is important in set theory. It is also important in truth-table logic and in the electrical theory of switching circuits. A statement is either true or false, a switch is either on or off, and an element is either contained or not contained in a given subset of a set. By properly setting up the equivalence of these three statements, we can convert many problems in electrical circuits to calculations in β. Before doing an example we verify some of the ring axioms for β. Since both $X + Y = (X \cup Y)/(X \cap Y) = (Y \cup X)/(Y \cap X) = Y + X$ and $X \cdot Y = X \cap Y = Y \cap X = Y \cdot X$, we have that β is commutative for both $+$ and $\cdot$. Closure is obvious since if X and Y are both subsets, so are $X \cap Y$ and $(X \cup Y)/(X \cap Y)$.

The associative laws are verified as follows:

$$(X \cdot Y) \cdot Z = (X \cap Y) \cap Z = X \cap Y \cap Z = X \cap (Y \cap Z) = X \cdot (Y \cdot Z)$$
$$
\begin{aligned}
(X + Y) + Z &= ((X \cup Y)/(X \cap Y)) + Z \\
&= ((X \cup Y)/(X \cap Y)) \cup Z)/((X \cup Y)/(X \cap Y)) \cap Z) \\
&= (((X \cup Y) \cup Z)/(X \cap Y))/((X \cup Y)) \cap Z)) \cup (X \cap Y \cap Z) \\
&= ((X \cup Y \cup Z)/((X \cap Y) \cup (X \cap Z) \cup (Y \cap Z))) \cup (X \cap Y \cap Z) \\
&= X + (Y + Z)
\end{aligned}
$$

since the factors are symmetric.

The following Venn diagrams may explain these computations:

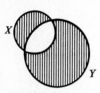

The shaded area is
$X + Y$.

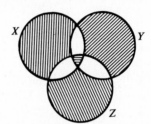

The shaded area represents
$(X + Y) + Z = X + (Y + Z) = X + Y + Z.$

Now $X + \varnothing = \varnothing + X = X$ and $X \cdot S = S \cdot X = X$ for all subsets X of S, so the additive identity is $\varnothing$; the multiplicative unity is the whole set S. The additive inverse of X is X itself, since

$$X + X = (X \cup X)/(X \cap X) = X/X = \varnothing$$

Similarly, we note that $X \cdot X = X \cap X = X$ for all X in B. The verification of the distributive law is similar to that of associativity.

Putting this together, we have the following:

Proposition The Boolean ring β is a commutative ring with unity and satisfies $X \cdot X = X$ for all X in β.

We shall prove a supplement to this proposition in Chap. 8, where the Wedderburn theorems are discussed.

When convenient, we now drop the $\cdot$ for multiplication so that $XY = X \cdot Y$, $X^2 = X \cdot X$, and so forth.

Exercises

1 In $\mathbb{Z}_{12}$, find $9 + 4$, $9 \cdot 4$, and $(7 + 3) \cdot 2$. If the time now is nine o'clock, what time will it be in four hours?

2   If R is a commutative ring, show that the binomial theorem holds. That is, show that if a and b are in R, and n is a positive integer, then $(a + b)^n = \sum_{i=0}^{n} \binom{n}{i} a^{n-i} b^i$. What is $(x + y)^2$ in Example D? What does this imply? The symbol $\binom{n}{k}$ denotes $n!/[k!(n - k)!]$, and we set $0! = 1$.

Definition   An element a in a ring is called **nilpotent** if $a^k = 0$ for some positive integer.

3   What are the nilpotent elements in $\mathbb{Z}_{10}$, in $\mathbb{Z}_{20}$, in β (Example J), and in Example D.
4   What is the group of units of β (Example J)?

Definition   A **zero divisor** in a ring R is a nonzero element a such that there is a nonzero element b such that $ab = 0$ or $ba = 0$. If $ab = 0$, where both a and b are nonzero, then a is a **left zero divisor** while b is a **right zero divisor**.

5   Find all the zero divisors in $\mathbb{Z}_{10}$, in $\mathbb{Z}_{20}$, in β, and in Example D.
6   If R is a commutative ring with unity, show that R has no zero divisors if and only if R is an integral domain.

Definition   An element x is **idempotent** if $x^2 = x$.

7   Show that a division ring has only two idempotent elements. Can this be generalized?
8   We define a ring $R = \mathbb{Q} \times \mathbb{Q} = \{(r,s), r \in \mathbb{Q}, s \in \mathbb{Q}\}$ with

$$(r_1,s_1) + (r_2,s_2) = (r_1 + r_2, s_1 + s_2)$$

and

$$(r_1,s_1) \cdot (r_2,s_2) = (r_1,r_2,s_1 s_2)$$

Show that R has four idempotent elements, no nonzero nilpotent elements, and many zero divisors. Is R a field?

Exercises 9 to 12 refer to the Boolean ring β discussed in Example J:

9   Show that if A, B, and C are subsets of S and $A \supseteq B$, then $(A/B) \cap C = (A \cap C)/(B \cap C)$, so that $(A/B) \cdot C = (A \cdot C)/(B \cdot C)$.
10   If in β we define $X + Y = X \cup Y$ instead of $X + Y = (X \cup Y)/(X \cap Y)$, do we get a ring? Why?
11   Using the new definition of $+$ in Exercise 10, we can set up the following equivalence: X in β corresponds to a switch X° denoted

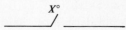

which is on if a is in X, or off if a is not in X. Show that $X \cdot Y$ corresponds to

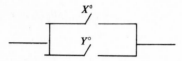

and that $X + Y$ corresponds to

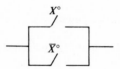

12 Again using $+$ as in Exercise 10, show the following:

(a) $(X + Y)Z = XZ + YZ$

(b) $(X + Y) \cdot X = X$

(c) If $\overline{X}$ denotes S/X, show that $X + \overline{X} = S$ and $X \cdot \overline{X} = \varnothing$.

(d) If $\overline{X}^\circ$ is a switch that is off when X° is on, and on when X° is off, show that

is always off, and

is always on. Thus, the latter can be replaced, probably at lower cost, by a straight piece of wire.

(e) Show, using part b, that

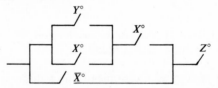

can be replaced by a straight piece of wire with only one switch.

*†2.2 TWO IMPORTANT EXAMPLES

The two examples discussed in this section are the continuous functions on $[0,1]$ and the complex numbers.

The remaining text does not depend on the ring of continuous functions, so in that sense the material is optional. The complex numbers are used freely throughout the remainder of the text. However, if you are already acquainted with complex numbers you may want to go over this material quite quickly.

EXAMPLE A The set of real-valued functions continuous on the closed interval [0,1] form a ring. Some elements of this ring are $x^2 + 3$, $|x - \frac{1}{2}|$, $\cos x$, e^{5x+2} and $\tan(x/2)$.

In this example we consider the two functions $f(x) = \sin^2 x$ and $g(x) = 1 - \cos^2 x$ to be identical, since $f(x) = g(x)$ for all x between 0 and 1. If we are dealing strictly with polynomials, we sometimes use a different definition. Here, though, we say f and g are the same function if $f(x) = g(x)$ for all x with $0 \leq x \leq 1$.

To prove $C[0,1]$ is a ring, we need some theorems from calculus and some of the ring properties of the real numbers. Our definition of $+$ and $\cdot$ is pointwise; that is, we define $f + g$ by

$$(f + g)(x) = f(x) + g(x)$$

and $f \cdot g$ by

$$(f \cdot g)(x) = f(x)g(x)$$

Since $f(x)$ and $g(x)$ are real numbers, $f(x) + g(x)$ and $f(x)g(x)$ are uniquely defined, and $f + g$ and fg are functions. That they are also continuous is the content of the two following standard theorems of calculus, namely, the sum of two continuous functions is continuous, and the product of two continuous functions is continuous.

The associative, commutative, and distributive laws follow from the same laws in $\mathbb{R}$. That is, $(fg)h = f(gh)$, $(f + g) + h = f + (g + h)$, $fg = gf$, $f + g = g + f$, and $f(g + h) = fg + fh$ because for any fixed x in $[0, 1]$ we have

$$(f(x)g(x))h(x) = f(x)(g(x)h(x))$$
$$(f(x) + g(x)) + h(x) = f(x) + (g(x) + h(x))$$
$$f(x)g(x) = g(x)f(x)$$
$$f(x) + g(x) = g(x) + f(x)$$
$$f(x)(g(x) + h(x)) = f(x)g(x) + f(x)h(x)$$

since $f(x)$, $g(x)$, and $h(x)$ all belong to the ring $\mathbb{R}$.

Considering 0 as the constant function 0, we see that 0 is the additive identity for $C[0,1]$, since $0 + f(x) = f(x)$ for all x in $[0,1]$. Similarly, the constant function 1 is the multiplicative identity for $C[0,1]$. The additive inverse for f is the function $-f$ defined by $(-f)(x) = -(f(x))$ which works since $f(x) - f(x) = 0$ for all x. Thus we have proved $C[0,1]$ is a commutative ring with unity.

What advantages are there in this interplay between calculus and modern algebra? The first is notational. Saying "$C[0,1]$ is a commutative ring with unity" is much quicker than saying "the set of continuous functions has the properties that the sum or product of two such functions is also in the set and has lots of other properties." There is some fascination in being able to sum up several days of hard work with delta-epsilon proofs in one short sentence.

The other advantage is that it suggests other natural questions to ask about $C[0,1]$. How near is $C[0,1]$ to being a field? Since $f(x) = x - \frac{1}{2}$ is 0 at $x = \frac{1}{2}$, $C[0,1]$ cannot be a field since $f(x)$ cannot have an inverse. The following exercises extend these ideas.

Exercises

1 *Show that $C[0,1]$ has zero divisors and thus is not an integral domain.

2 *Which of the following functions in $C[0,1]$ have multiplicative inverses: $x^2 - 2$, x^2, $1/(x^2 + 2)$, e^x, $\log_e x$?

3 **If $f(x_0) > 0$, $x_0 \in [0,1]$, and f is continuous, show that there is an $\varepsilon > 0$ such that $f(x) > 0$ for all x in $(x_0 - \varepsilon, x_0 + \varepsilon)$.

4 On the open interval $(0,1)$ find an example of a continuous function f such that $f(x) > 0$ for all x, but $f(x)$ takes on values arbitrarily near 0.

5 *Characterize functions on $C[0,1]$ which have multiplicative inverses.

6 Let $C(-\infty,\infty)$ be the set of all functions continuous on the whole real line. Using pointwise addition and multiplication, show that $C(-\infty,\infty)$ is a commutative ring with 1.

7 Let $\mathbb{R}^{\mathbb{R}}$ be the set of all functions from the reals to the reals. Show that $\mathbb{R}^{\mathbb{R}}$ is a commutative ring with 1, again using pointwise addition and multiplication.

8 *Let $I(-\infty,\infty)$ be those real-valued functions f defined on the whole real line such that $\int_{-\infty}^{\infty} f(x)\, dx$ is finite. Find some elements of $I(-\infty,\infty)$, and show that this ring has no multiplicative identity. Again use pointwise operations.

9 *If S is a set and R is a ring, show that R^S, the set of all functions from S to R, together with pointwise addition and multiplication, forms a ring.

We shall ask more questions about $C[0,1]$ after we have introduced subrings and ideals.

EXAMPLE B   In this example we discuss the field of complex numbers. Since the square of any real number is either zero or positive, it is impossible to find a real number x such that $x^2 = -1$. If we define a new number i such that $i^2 = -1$, we can overcome this difficulty. If we solve $x^2 + x + 1$ using the quadratic formula, we run into

$$x = \frac{-1 \pm \sqrt{-3}}{2} = \frac{-1}{2} \pm \frac{\sqrt{3}\sqrt{-1}}{2} = \frac{-1}{2} \pm \frac{\sqrt{3}i}{2}$$

Thus we need formulas for adding and multiplying quantities of the form $a + bi$, where a and b are real numbers. After that we need to eliminate vague expressions like "quantities of the form."

We define addition and multiplication by operating pretty much as though the $a + bi$ were real numbers, except that we simplify by using $i^2 = -1$ when possible. Thus

$$(a + bi) + (c + di) = (a + c) + (b + d)i \qquad a, b, c, d \text{ in } \mathbb{R}$$

and

$$(a + bi)(c + di) = ac + bci + adi + bdi^2 = (ac - bd) + (bc + ad)i$$

We graph the complex number $a + bi$ by assigning it the point (a,b); this gives us the rigorous way to treat the complex numbers. We merely replace $a + bi$ with the ordered pair (a,b), and use addition and multiplication as above. One thing that this achieves immediately is that we can tell that $a + bi = c + di$ if and only if $a = c$ and $b = d$. This is true since we know that, for ordered pairs, $(a,b) = (c,d)$ if and only if $a = c$ and $b = d$.

Theorem   Let $\mathbb{C}$ be the set of all ordered pairs (a,b) of real numbers, where the following addition and multiplication are defined:

$$(a,b) + (c,d) = (a + c, b + d)$$

$$(a,b) \cdot (c,d) = (ac - bd, ad + bc)$$

Then $\mathbb{C}$ forms a field.

PROOF $(\mathbb{C}, +)$ is the same as $(\mathbb{R}^2, +)$ in Example B of Sec. 1.12. This group is easily shown to be an abelian group with identity $(0,0)$ and $-(a,b) = (-a,-b)$.

Our definition of multiplication is obviously closed. The associative law follows from

$$[(a,b)(c,d)](e,f)$$

$$= (ac - bd, bc + ad)(e,f)$$

$$= ((ac - bd)e - (bc + ad)f, (bc + ad)e + (ac - bd)f)$$

$$= (a(ce - df) - b(de + cf), b(ce - df) + a(de + cf))$$

$$= (a,b)(ce - df, de + cf)$$

$$= (a,b)[(c,d)(e,f)]$$

(1,0), the complex number that corresponds to 1 or $1 + 0i$, is the multiplicative identity, since

$$(a,b)(1,0) = (a \cdot 1 - b \cdot 0, 1 \cdot b + 0 \cdot a) = (a,b)$$
$$= (1,0)(a,b)$$

To find multiplicative inverses we define the **conjugate** of (a,b) to be $\overline{(a,b)} = (a,-b)$ and note that

$$(a,b)\overline{(a,b)} = (a,b)(a,-b) = (a \cdot a - (b)(-b), a(-b) + ab)$$
$$= (a^2 + b^2, 0)$$

Thus, if $(a,b) \neq (0,0)$, the additive identity, we have

$$(a,b)\left(\frac{a}{a^2 + b^2}, \frac{-b}{a^2 + b^2}\right) = \left(\left(\frac{a^2 + b^2}{a^2 + b^2}\right), 0\right) = (1,0)$$

and

$$(a,b)^{-1} = \left(\frac{a}{a^2 + b^2}, \frac{-b}{a^2 + b^2}\right)$$

$$(a,b)(c,d) = (ac - bd, ad + bc) = (ca - db, da + cb) = (c,d)(b,a)$$

so multiplication is commutative. The distributive law is equally automatic:

$$(a,b)[(c,d) + (e,f)] = (a,b)(c + e, d + f)$$
$$= (a(c + e) - b(d + f), b(c + e) + a(d + f))$$
$$= (ac - bd + ae - bf, bc + ad + be + af)$$
$$= (ac - bd, bc + ad) + (ae - bf, be + af)$$
$$= (a,b)(c,d) + (a,b)(e,f)$$

Thus $\mathbb{C}$ is a field.   ////

In terms of $a + bi$ we would have

$$(a + bi)^{-1} = \frac{a}{a^2 + b^2} - \frac{b}{a^2 + b^2} i$$

If r is a real number, we define $r(a,b) = (ra,rb)$ or, informally, $r(a + bi) = ra + rbi$.

We can map the real numbers into the complex numbers by mapping r to $(r,0)$. Graphically the complex numbers are the same as the real plane, with the real numbers as the x axis. In Fig. 1 we have illustrated some of the numbers associated with $z = (3,4)$. Note that i corresponds to $(0,1)$, since $i = 0 + 1i$.

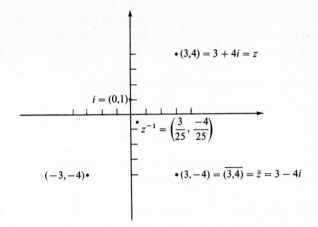

We also define $|z| = \sqrt{z\bar{z}} = \sqrt{(a,b)\overline{(a,b)}} = \sqrt{(a + bi)(a - bi)} = \sqrt{a^2 + b^2}$, where $z = a + bi$. The positive square root of a positive real number (like $a^2 + b^2$) is well-defined, so $|z|$ is well-defined.

Having now set up the field of complex numbers, we define $a + bi = (a,b)$, where a and b are real, and use whichever notation is convenient. In fact, we often denote a typical element of $\mathbb{C}$ as z. We now define $\bar{z} = a - bi$ and $|z| = \sqrt{z\bar{z}} = \sqrt{a^2 + b^2}$, where $z = a + bi$. If $z = 3 + 3i$, we see that $\bar{z} = 3 - 3i$ and $|z| = \sqrt{3^2 + 3^2} = 3\sqrt{2}$ or, considering $|z|$ as the complex number $(|z|,0)$, we have the following diagram:

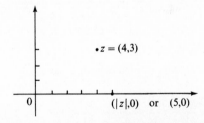

Using the pythagorean theorem, we see that the distance from $z = a + bi$ to the origin 0 is $|z|$.

Exercises

10 Write a multiplication table for the elements 1, i, -1, and $-i$, and show that they form a group.

11 If M is a subgroup of $(\mathbb{C} - (0), \cdot)$, show that if $2 + 2i$ is in M, then M must be infinite.

12 Let $(\mathbb{C}, +)$ be the additive group for $\mathbb{C}$. Show that the mapping ϕ such that $\phi(z) = \bar{z}$ is an automorphism of $(\mathbb{C}, +)$.

13 Let $(\mathbb{C}-\{0\}, \cdot)$ be the multiplicative group of complex numbers, and show that the mapping θ with $\theta(z) = |z|$ is a homomorphism from $(\mathbb{C}-\{0\}, \cdot)$ onto the multiplicative group of positive real numbers. What is ker θ?

What are some uses for the complex numbers?

1   $x^2 + 1$ can be solved in $\mathbb{C}$ but not in $\mathbb{R}$.
2   Every polynomial with real coefficients can be solved in $\mathbb{C}$.
3   Every polynomial with complex coefficients can be solved in $\mathbb{C}$.

Of these, item 1 is obvious, item 3 is a major theorem, and item 3 includes item 2 as a special case. In fact, item 3 is the fundamental theorem of algebra and is proved using a little heavy analysis. A standard proof follows easily from Liouville's theorem and is one of the gems almost always included in a first course in complex analysis. A very pretty proof using a little calculus and a very clever induction due to Gauss can be found in van der Waerden [85]. Gauss gave five proofs of this theorem.

Another nice trick available with complex numbers is the condensation of some trig formulas. We recall some power series from calculus:

$$e^y = 1 + y + \frac{y^2}{2!} + \frac{y^3}{3!} + \cdots = 1 + \sum_{j=1}^{\infty} \frac{y^j}{j!}$$

$$\sin y = y - \frac{y^3}{3!} + \frac{y^5}{5!} - \frac{y^7}{7!} + \cdots$$

$$\cos y = 1 - \frac{y^2}{2!} + \frac{y^4}{4!} - \frac{y^6}{6!} + \cdots$$

Let $y = ix$, and we have

$$e^{ix} = 1 + ix + \frac{(ix)^2}{2!} + \frac{(ix)^3}{3!} + \frac{(ix)^4}{4!} + \frac{(ix)^5}{5!} + \cdots$$

$$= 1 + ix - \frac{x^2}{2!} - i\frac{x^3}{3!} + \frac{x^4}{4!} + i\frac{x^5}{5!} + \cdots$$

$$= \left(1 - \frac{x^2}{2!} + \frac{x^4}{4!} \cdots\right) + i\left(x - \frac{x^3}{3!} + \frac{x^5}{5!} \cdots\right) = \cos x + i \sin x$$

This "proof" is made rigorous in a course in complex analysis, where the proper convergence theorems are proved.

If n is an integer we thus have DeMoivre's theorem (A):

$$e^{inx} = (e^{ix})^n \qquad \text{where } x \text{ is a real number}$$

or, more dramatically, DeMoivre's theorem (B):

$$\cos(nx) + i \sin(nx) = (\cos x + i \sin x)^n$$

For $n = 3$, for instance, we get the interesting fact that $\cos 3x + i \sin 3x = (\cos x)^3 + 3i \cos^2 x \sin x - 3 \cos x \sin^2 x - i \sin^3 x$. This can only happen if the real and imaginary parts are equal, so

$$\cos 3x = \cos^3 x - 3 \cos x \sin^2 x$$

or

$$\cos 3x = \cos^3 x - 3 \cos x (1 - \cos^2 x)$$
$$= 4 \cos^3 x - 3 \cos x$$

and

$$\sin 3x = 3 \cos^2 x \sin x - \sin^3 x$$
$$= 3 \sin x - 4 \sin^3 x$$

The interest here is not the trig formulas themselves, but the fact that DeMoivre's theorem is easy to remember and such trig formulas are easy to forget.

Exercises

14 Derive formulas for $\cos 2x$ and $\sin 2x$.

15 What is $\cos 5x$ in terms of $\sin x$ and $\cos x$?

16 Let $G = a + bi$, for $a, b \in \mathbb{Z}$. Show that G is a commutative ring with unity. This ring is called the ring of *Gaussian integers*. What is $U(G)$, the group of units of G?

17 Find both solutions of $z^2 = 4i$. Then find all the solutions of $z^3 = 1 + i$.

18 Let $(\mathbb{R}^2, +)$ be the set of ordered pairs of real numbers with $(a,b) + (c,d) = (a + c, b + d)$ and $(a,b)(c,d) = (ac,bd)$. Show that this more natural looking multiplication on $\mathbb{R} \times \mathbb{R}$ yields a commutative ring but not a field.

Those elements on the unit circle, the z with $z = 1$, can be represented as $(\cos \theta, \sin \theta)$, where θ is the angle between the x axis and the line from 0 to z. For example,

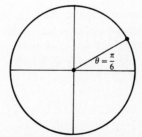

$$z = \frac{\sqrt{3}}{2} + \frac{1}{2} i = \left(\frac{\sqrt{3}}{2}, \frac{1}{2}\right) = \left(\cos \frac{\pi}{6}, \sin \frac{\pi}{6}\right) = \cos \frac{\pi}{6} + i \sin \frac{\pi}{6} = e^{(\pi/6)i}$$

Similarly, we can write any element of $\mathbb{C}$ other than 0 as $re^{i\theta}$, where $r = |z| = a^2 + b^2$, and $\tan \theta = b/a$ (or, more geometrically, θ) is the angle between the x axis and the line from 0 to z. (This is not quite complete, but if $a = 0$ then $\theta = \pi/2$ or $3\pi/2$.)

In this form, multiplication of complex numbers can be looked at geometrically: $r_1 e^{\theta_1 i} \cdot r_2 e^{\theta_2 i} = r_1 r_2 e^{(\theta_1 + \theta_2) i}$, so the distances from 0 multiply, and the angles add.

Using this fact and DeMoivre's theorem, we can find all the nth roots of unity. We note that $e^{i\theta} = e^{i(\theta + 2\pi)} = e^{i(\theta + 4\pi)} = e^{i(\theta + 2n\pi)}$, for $n \in \mathbb{Z}$. In particular, $1 = e^{2\pi i 0} = e^{2\pi i} = e^{4\pi i} = e^{2\pi n i}$, for $n \in \mathbb{Z}$. We use this to solve the equation $z^n = 1$, where n is some fixed positive integer.

$$z^n = 1 = e^0 = e^{2\pi i} = e^{4\pi i} = \cdots = e^{(n-1)2\pi i}$$

Thus $1 = e^0$, $e^{2\pi i/n}$, $e^{4\pi i/n}$, ..., $e^{(n-1)2\pi i/n}$ all are solutions of $z^n = 1$. For $n = 5$, the five solutions appear as follows:

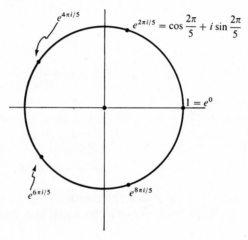

The origin of the term cyclic group might come from the fact that these five neatly arranged roots of unity do form a cyclic group under multiplication. Any solution z_0 of $z^n = 1$ must satisfy $|z_0| = 1$ and thus $z_0 = e^{i\theta}$. Now $z_0{}^n = e^{i\theta n} = e^{2\pi i k} = 1$, so $\theta = 2\pi i k/n$ for some integer k, which can be assumed to be one of $0, 1, 2, \ldots, n - 1$.

Exercises

19 (a) Show that the solutions of $z^n = 1$ (n a positive integer) form an abelian group under multiplication.

(b) Is this group isomorphic to any group we have seen before?

20 If $S = 1 + z_0 + z_0{}^2 + \cdots + z_0{}^n$ and $z_0 \neq 1$, show that $S = (1 - z_0{}^{n+1})/(1 - z_0)$.

21 Find and graph all the solutions of $z^4 + z^3 + z^2 + z + 1 = 0$ and $z^{n-1} + z^{n-2} + z^{n-3} + \cdots + z + 1$. Factor $z^4 + z^3 + z^2 + z + 1$ into two quadratic factors with real coefficients.

22 *For any real number θ, show that

$$\frac{1}{2} + \cos \theta + \cos 2\theta + \cdots + \cos n\theta = \frac{\sin (n + \frac{1}{2})\theta}{\sin \frac{1}{2}\theta}$$

(*Hint:* Let $z = \cos \theta + i \sin \theta$, and look at $1 + z + z^2 + \cdots + z^n$. You may need the half-angle formulas from trigonometry. This is a crucial formula for proving some convergence theorems in Fourier series.)

23 Explain the following fallacy: Let m and n be any two integers. Then $e^{2\pi i m} = e^{2\pi i n} = 1$. Raising both sides to the ith power, we have $e^{-2\pi m} = e^{-2\pi n}$. Since e^x is a one-to-one function from the real numbers to the positive reals, we obtain $-2\pi m = -2\pi n$ and thus $m = n$ for any two integers m and n.

24 *Let n be an integer with $n \geq 2$. If $\varepsilon = e^{2\pi i/n}$, show that $1 + \varepsilon + \varepsilon^2 + \cdots + \varepsilon^{n-1} = 0$.

25 (a) Prove both geometrically and algebraically that $|z_1| + |z_2| \geq |z_1 + z_2|$.

(b) Show that $|z_1| + |z_2| = |z_1 + z_2|$ if and only if $\theta_1 = \theta_2$, where $z_1 = r_1 e^{i\theta_1}$ and $z_2 = r_2 e^{i\theta_2}$.

26 Let $\theta(a + bi) = \begin{pmatrix} a & b \\ -b & a \end{pmatrix}$. Show θ is an isomorphism both from $(\mathbb{C}, +)$ into $(M_2(\mathbb{R}), +)$ and from $(\mathbb{C} - (0), \cdot)$ into $\text{GL}(2, \mathbb{R})$. Here $M_2(\mathbb{R})$ denotes the additive group of all 2×2 matrices over $\mathbb{R}$.

27 Show by factoring over $\mathbb{C}$ that every polynomial with real coefficients factors into a product of linear and quadratic factors with coefficients in $\mathbb{R}$.

2.3 MATRIX RINGS

In this section we want to expand our list of examples of rings; in particular, we want to discuss some noncommutative rings. The first example is $M_2(\mathbb{R})$, the set of all (not necessarily nonsingular) 2×2 matrices with real entries. $+$ and $\cdot$ are defined as before

$$\begin{pmatrix} a & b \\ c & d \end{pmatrix} + \begin{pmatrix} e & f \\ g & h \end{pmatrix} = \begin{pmatrix} a + e & b + f \\ c + g & d + h \end{pmatrix}$$

and

$$\begin{pmatrix} a & b \\ c & d \end{pmatrix}\begin{pmatrix} e & f \\ g & h \end{pmatrix} = \begin{pmatrix} ae + bg & af + bh \\ ce + dg & cf + dh \end{pmatrix}$$

We saw, in a previous example, that $(M_2(\mathbb{R}), +)$ is an abelian group with $\begin{pmatrix} 0 & 0 \\ 0 & 0 \end{pmatrix}$ as identity and with $\begin{pmatrix} -a & -b \\ -c & -d \end{pmatrix}$ as the inverse for $\begin{pmatrix} a & b \\ c & d \end{pmatrix}$.

Multiplication is associative, closed, and noncommutative, as we have also seen previously. To finish verifying that we have a ring, we must verify the distribution law:

$$\begin{pmatrix} a_1 & b_1 \\ c_1 & d_1 \end{pmatrix}\left[\begin{pmatrix} a_2 & b_2 \\ c_2 & d_2 \end{pmatrix} + \begin{pmatrix} a_3 & b_3 \\ c_3 & d_3 \end{pmatrix}\right]$$

$$= \begin{pmatrix} a_1 & b_1 \\ c_1 & d_1 \end{pmatrix}\begin{pmatrix} a_2 + a_3 & b_2 + b_3 \\ c_2 + c_3 & d_2 + d_3 \end{pmatrix}$$

$$= \begin{pmatrix} a_1(a_2 + a_3) + b_1(c_2 + c_3) & a_1(b_2 + b_3) + b_1(d_2 + d_3) \\ c_1(a_2 + a_3) + d_1(c_2 + c_3) & c_1(b_2 + b_3) + d_1(d_2 + d_3) \end{pmatrix}$$

$$= \begin{pmatrix} a_1 a_2 + b_1 c_2 & a_1 b_2 + b_1 d_2 \\ c_1 a_2 + d_1 c_2 & c_1 b_2 + d_1 d_2 \end{pmatrix} + \begin{pmatrix} a_1 a_3 + b_1 c_3 & a_1 b_3 + b_1 d_3 \\ c_1 a_3 + d_1 c_3 & c_1 b_3 + d_1 d_3 \end{pmatrix}$$

$$= \begin{pmatrix} a_1 & b_1 \\ c_1 & d_1 \end{pmatrix}\begin{pmatrix} a_2 & b_2 \\ c_2 & d_2 \end{pmatrix} + \begin{pmatrix} a_1 & b_1 \\ c_1 & d_1 \end{pmatrix}\begin{pmatrix} a_3 & b_3 \\ c_3 & d_3 \end{pmatrix}$$

The right distributive law is proved in a similar manner. Thus we have an example of a noncommutative ring. It does have a multiplicative identity, $\begin{pmatrix} 1 & 0 \\ 0 & 1 \end{pmatrix}$.

Since $\begin{pmatrix} 0 & 0 \\ a & 0 \end{pmatrix}\begin{pmatrix} 0 & 0 \\ b & 0 \end{pmatrix} = \begin{pmatrix} 0 & 0 \\ 0 & 0 \end{pmatrix}$, we see that $M_2(\mathbb{R})$ has zero divisors.

Exercise

1 Which of the following subsets of $M_2(\mathbb{R})$ are themselves rings with the $+$ and $\cdot$ from $(M_2(\mathbb{R}), +, \cdot)$?

$$H = \left\{ \begin{pmatrix} x & y \\ 0 & 0 \end{pmatrix} \middle| x, y \in \mathbb{R} \right\} \qquad K = \left\{ \begin{pmatrix} 0 & 0 \\ 0 & r \end{pmatrix} \middle| r \in \mathbb{R} \right\}$$

$$J = \left\{ \begin{pmatrix} 1 & a \\ 0 & b \end{pmatrix} \middle| a, b \in \mathbb{R} \right\} \qquad L = \left\{ \begin{pmatrix} a & b \\ c & d \end{pmatrix} \middle| a, b, c, d \in \mathbb{Z} \right\}$$

We can generalize this example in two directions. First we note that $+$ and $\cdot$ are defined if $a_1 + a_2$ and $a_1 a_2 + b_1 b_2$ (etc.) make sense, so in particular we don't need the reals; any ring will do as far as definitions go. If $(S, +, \cdot)$ is a ring, let $M_2(S)$ denote the ring of all 2×2 matrices with coefficients in S. We have to verify again that $(M_2(S), +)$ is an abelian group, that the distributive laws hold, and that $\cdot$ is associative for $M_2(S)$. This is better done as exercises and is left as such.

Exercises

2 If S is a ring, show that $(M_2(S), +)$ is an abelian group.

3 Verify one of the distributive laws for $(M_2(S), +, \cdot)$.

4 Verify that $\cdot$ is associative for $M_2(S)$.

5 Let $R_1 = M_2(\mathbb{Z}_2)$, $R_2 = M_2(R_1)$, ..., $R_{k+1} = M_2(R_k)$. What is $|R_k|$?

6 Find an example of an infinite noncommutative ring with no identity and with zero divisors.

7 Show that $\mathcal{U}(M_2(\mathbb{Z}_2)) = GL(2, \mathbb{Z}_2) \simeq \text{Sym}(3)$, where

$$GL(2, \mathbb{Z}_2) = \left\{ \begin{pmatrix} a & b \\ c & d \end{pmatrix} \,\middle|\, a, b, c, d \in \mathbb{Z}_2, \, ad - bc \neq 0 \right\}$$

8 What are $|M_2(\mathbb{Z}_p)|$ and $|GL(2, \mathbb{Z}_p)|$?

The remaining text in this section is not used in the rest of the book. Exercises 16 to 21 are used occasionally in Chaps. 8 and 10; they present the quaternions, which are an interesting example of historical importance. These exercises are independent of the rest of the section.

†As a second generalization, we note that any matrix in $M_2(\mathbb{R})$ can be regarded as a homomorphism from $(\mathbb{R}^2, +)$ to $(\mathbb{R}^2, +)$. This has been done previously as Exercise 10 in Sec. 1.10.

$$(x_1, y_1) \begin{pmatrix} a & b \\ c & d \end{pmatrix} + (x_2, y_2) \begin{pmatrix} a & b \\ c & d \end{pmatrix}$$

$$= (ax_1 + c_1 y_1, \, bx_1 + dy_1) + (ax_2 + cy_2, \, bx_2 + dy_2)$$

$$= (a(x_1 + x_2) + c(y_1 + y_2), \, b(x_1 + x_2) + d(y_1 + y_2))$$

$$= (x_1 + x_2, \, y_1 + y_2) \begin{pmatrix} a & b \\ c & d \end{pmatrix}$$

If we let $p_1 = (x_1, y_1)$, $p_2 = (x_2, y_2)$, and $M = \begin{pmatrix} a & b \\ c & d \end{pmatrix}$, this can be written as

$$p_1 M + p_2 M = (p_1 + p_2) M$$

A homomorphism from a group G to the same group G is called an **endomorphism**. Thus every element of $M_2(\mathbb{R})$ can be interpreted as an endomorphism of the group $(\mathbb{R}^2, +)$.

If α is any endomorphism of an abelian group $(A, +)$, then $\alpha(a + b) = \alpha(a) + \alpha(b)$, which closely resembles the left distributive law. In fact, the following exercise holds.

Exercise

9   If $(R,+,\cdot)$ is a ring, $x \in R$, and $f_x: R \to R$ is given by $f_x(r) = xr$, then f_x is an endomorphism of $(R,+)$. f_x is a monomorphism if and only if x is not a left zero divisor.

Let us now take a set E of endomorphisms of an abelian group $(A,+)$ such that if β and γ are in E, so is $\beta \circ \gamma$, where $\circ$ is composition of functions. We want to make E into a ring, but we have to define an appropriate $\oplus$ operation. We define by $\beta \oplus \gamma$ by $(\beta \oplus \gamma)(a) = \beta(a) + \gamma(a)$ for all a in A. We are now ready to prove $(E,\oplus,\circ)$ is a ring. First of all we see that we should add some more conditions on E. We shall need $\beta \oplus \gamma$ in E for all β and γ in E. Also, for $(\beta \oplus 0_E)(a) = \beta(a)$ for all a in A, we shall need $0_E(a) = 0_A$, the identity in A. Then if $\beta \in E$, we shall need $-\beta \in E$, where $-\beta$ is defined by $(-\beta)(a) = -[\beta(a)]$. Putting all these requirements together, we have the following proposition.

> **Proposition 1**   Let E be a set of endomorphisms of an abelian group $(A,+)$. Suppose also that E is closed under composition of functions and that $(E,\oplus)$ is a group. Then $(E,\oplus,\circ)$ is a ring.
>
> PROOF $(\beta \oplus \gamma)a = \beta(a) + \gamma(a) = \gamma(a) + \beta(a) = (\gamma \oplus \beta)(a)$ for all a in A, so $\beta \oplus \gamma = \gamma \oplus \beta$, and $(E,\oplus)$ is an abelian group. We prove one distributive law:
>
> $$[(\gamma \oplus \delta) \circ \beta](a) = (\gamma \oplus \delta)(\beta(a)) = \gamma(\beta(a)) + \delta(\beta(a))$$
> $$= (\gamma \circ \beta)(a) + (\delta \circ \beta)(a) = ((\gamma \circ \beta) \oplus (\delta \circ \beta))(a)$$
>
> for all a in A. Thus $(\gamma \oplus \delta) \circ \beta = (\gamma \circ \beta) \oplus (\delta \circ \beta)$. The other distributive law is similar and is left as an exercise.
>
> Composition of functions is closed by hypothesis and is associative as always (see Sec. 1.5); thus $(E,+,\circ)$ is a ring.   ////

The philosophical importance of this class of rings is pointed out in the exercises.

Exercises

10   Verify that $(E,\oplus,\circ)$ satisfies the right distributive law.

The following is a theorem analogous to Cayley's theorem, which we shall meet in Chap. 4.

11   If R is a ring, $r \in R$, then we have shown in Exercise 9 that f_r is an endomorphism of $(R,+)$.
    (a)   Show that if $\overline{R} = \{f_r \mid r \in R\}$, then $(\overline{R},\oplus,\circ)$ is a ring.

(b)   Define $\phi: R \to \bar{R}$ by $\phi(r) = f_r$. Show that ϕ is a homomorphism of $(R, +)$ onto $(\bar{R}, \oplus)$, and $\phi(r_1 r_2) = \phi(r_1)\phi(r_2)$.

[That is, R is ring homomorphic to $(\bar{R}, \oplus, \circ)$. See Sec. 2.4.] This shows that any ring $(R, +, \cdot)$ is homomorphic to a subring of $(E, \oplus, \circ)$, where E is the set of all endomorphisms of $(R, +)$.

12   Continuing the notation of Exercise 11, show that ϕ is one-to-one if either R satisfies the right cancellation law or R has 1. (Thus, in one sense, the study of rings with 1 is reduced to a special case in the study of endomorphism rings since every such ring is isomorphic to an endomorphism ring.)

13   Show that $M_2(\mathbb{R})$ is a ring by using Proposition 1.

14   Show that every endomorphism of $(\mathbb{Z}, +)$ can be written as f_n for some n in $\mathbb{Z}$.

15   (a)   Show that some endomorphism of $(\mathbb{R}^2, +)$ cannot be given as $f_{(x_1, y_1)}$.
(b)   Can each $f_{(x_1, y_1)}$ be given as an element of $M_2(\mathbb{R})$?

In the next few exercises we present the ring of quaternions. This ring was invented by Hamilton in 1843 and was a long-sought generalization of the complex numbers. As will be seen, the quaternions are almost a field, lacking only the commutative law for multiplication. This abandonment of commutativity was one of the key steps in the development of abstract algebra, as it forced mathematicians to an examination of the axioms they were implicitly using.

16   (Informal version)   Let $\mathscr{Q}$ be all elements of the form $a \cdot 1 + b \cdot i + c \cdot j + d \cdot k$, where a, b, c, d are in $\mathbb{R}$, and $i^2 = j^2 = k^2 = ijk = -1$. As examples of addition and multiplication we have

$$(1 + 2i + 3j - 5k) + (3 + 3i + 3j + 4k) = 4 + 5i + 6j - k$$

and

$$(1 + 2i + 3j - 5k) \cdot (3 + 3i + 3j + 4k) = 8 + 36i - 11j - 14k$$

(a)   Show that $ij = k$ $jk = i$ $ki = j$

and $ji = -k$ $kj = -i$ $ik = -j$

(You may use the associative law.)

(b)   Show that $\mathscr{Q}$ is a ring.

(c)   Show that $1 = 1 + 0 \cdot i + 0 \cdot j + 0 \cdot k$ is the multiplicative identity for $\mathscr{Q}$.

17   If $q = a + bi + cj + dk$ is a quaternion as above, then $\bar{q} = a - bi - cj - dk$ is the *conjugate* of q. Show that $q \cdot \bar{q} = a^2 + b^2 + c^2 + d^2 + 0 \cdot i + 0 \cdot j + 0 \cdot k$. Using this, show that $q^{-1} = \bar{q}/\Delta$, where $\Delta = a^2 + b^2 + c^2 + d^2$, and thus that every nonzero element of $\mathscr{Q}$ has a multiplicative inverse. This completes the proof that $\mathscr{Q}$ is a division ring. This is the prime example of a division ring that is not a field.

18 If we wish to formalize this presentation, we should imitate the procedure used for $\mathbb{C}$. We replace the informal $q = a + bi + cj + dk$ by (a,b,c,d). If we do this, what 4-tuples should we use for the additive identity, multiplicative identity, $\bar{q}$, and $q\bar{q}$? How would the addition and multiplication laws now be stated?

19 If q is a quaternion, we define $\delta(q) = q\bar{q}$. Show that $\delta(q_1 q_2) = \delta(q_1)\delta(q_2)$ so that δ can be interpreted as a homomorphism from $\mathcal{Q} - (0)$ to $\mathbb{R} - (0)$.

20 Let $q_1 = a_0 + a_1 i + a_2 j + a_3 k$ and $q_2 = b_0 + b_1 i + b_2 j + b_3 k$. Then writing out $\delta(q_1)\,\delta(q_2) = \delta(q_1 q_2)$ yields

$$
\begin{aligned}
(a_0{}^2 &+ a_1{}^2 + a_2{}^2 + a_3{}^2)(b_0{}^2 + b_1{}^2 + b_2{}^2 + b_3{}^2) \\
&= (a_0 b_0 - a_1 b_1 - a_2 b_2 - a_3 b_3)^2 + (a_0 b_1 + a_1 b_0 + a_2 b_3 - a_3 b_2)^2 \\
&\quad + (a_0 b_2 - a_1 b_3 + a_2 b_0 + a_3 b_1)^2 + (a_0 b_3 + a_1 b_2 - a_2 b_1 + a_3 b_0)^2.
\end{aligned}
$$

This is the four-square identity of Lagrange, which he used in his proof that every positive integer is the sum of at most four squares of integers. (See Chap. 7 for the two-square theorem.) First, write $103 = 10^2 + 1^2 + 1^2 + 1^2$. Second, find a similar expression for 23. Third, find a similar expression for 2,369.

21 (Historical) Hamilton invented his quaternions as an extension of the complex numbers. To do this, he had to abandon the commutative law and to go to four dimensions. For many years mathematicians sought a three-dimensional analog of $\mathbb{C}$, and this exercise shows why a good analog cannot be found.

Let T be the set of all elements of the form $a + bi + cj$, where a, b, c are in $\mathbb{R}$, and $i^2 = j^2 = -1$. Assume $ij = r + si + tj$, where r, s, and t are constant real numbers. Show then that $-i = (ij)j = rj + sij + tj^2$. Conclude that $s^2 = -1$, which is impossible for any real number. (We do use the associative and distributive laws and that $a + bi + cj = a' + b'i + c'j$ only if $a = a'$, $b = b'$, and $c = c'$. It is not necessary for the multiplication to be commutative.)

22 It is easy to see that $\mathscr{U}(\mathcal{Q}) = \mathcal{Q} - \{0\}$, since every nonzero element has a multiplicative inverse. Let $H = \{1, -1, i, -i, j, -j, k, -k\}$, and show that H forms a nonabelian group of order 8. Construct the subgroup lattice for H, and show every subgroup of H is normal in H.

2.4 SUBRINGS, IDEALS, AND RING HOMOMORPHISMS

In this section we develop concepts strongly analogous to those of subgroup, normal subgroup, quotient group, and homomorphism. We are able to proceed more quickly for four reasons. First, we need not repeat the material on sets and functions. Second, we can often use the results proved for groups. Third, we are a little swifter going through routine verifications. Fourth, there is no result analogous to Lagrange's theorem for rings.

Definition   A subset S of a ring $(R,+,\cdot)$ is a **subring** if $(S,+,\cdot)$ is itself a ring.

Proposition 1   A nonempty subset S of a ring $(R,+,\cdot)$ is a subring if for all x and y in S, $x \pm y$ and xy are in S.

PROOF   In showing that S is a subring, we do not need to discuss the commutative, distributive, and associative laws. They all hold in R, and the restriction to S will not affect them. If x and y are in S, then $x + y$ is in S by hypothesis, $x - x = 0$ is in S, and thus $0 - x = -x$ is in S. Therefore, $(S,+)$ is a group. $(S,\cdot)$ is associative automatically and closed by hypothesis. Therefore, $(S,+,\cdot)$ is a ring.   ////

EXAMPLE A   In the diagram, any ring is a subring of any ring above it that it is connected to by one or more upward lines.

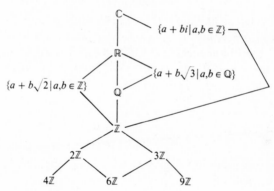

EXAMPLE B   If $S = \{a/p^n \,|\, a,n \in \mathbb{Z}, \ p$ is a prime$\}$, then S is a subring of $\mathbb{Q}$. We check this by noting that with, say, $n \geq m$,

$$\frac{a}{p^n} \pm \frac{b}{p^m} = \frac{a \pm bp^{n-m}}{p^n}$$

which is in S since $a \pm bp^{n-m} \in \mathbb{Z}$. Also

$$\frac{a}{p^n} \cdot \frac{b}{p^m} = \frac{ab}{p^{n+m}}$$

and thus, by Proposition 1, S is a subring of $\mathbb{Q}$.

When we discussed groups we first discussed normal subgroups, then quotient groups, and then homomorphisms. Here we follow the same pattern. First we introduce ideals, then quotient rings, and then ring homomorphisms.

Definition   A subring I of a ring R is an **ideal** if $i \cdot r$ and $r \cdot i$ are in I for all r in R and i in I.

These conditions, which are stronger than just multiplicative closure, are called the *right* and *left absorption laws*, as multiplication of an element in R by an element in I "absorbs" the product into I. An alternative way to write the absorption laws is $RI \subseteq I$ and $IR \subseteq I$. Here

$$RI = \left\{ \sum_{\substack{\text{finite} \\ \text{sums}}} r_j x_j \,\middle|\, r_j \in R,\, x_j \in I \right\}$$

EXAMPLE C   For a fixed integer n, $n\mathbb{Z}$ is an ideal in $\mathbb{Z}$. If na and nb are any two elements of $n\mathbb{Z}$, then $na \pm nb = n(a \pm b)$. If c is any element of $\mathbb{Z}$, then $c(na) = (na)c = n(ac)$, so the absorption laws hold and $n\mathbb{Z}$ is an ideal. In fact, every ideal of $\mathbb{Z}$ is one of the $n\mathbb{Z}$. This can be seen by noting that $(\mathbb{Z}, +)$ is cyclic, so every subgroup of $(\mathbb{Z}, +)$ is cyclic (Sec. 1.7) and thus has a generator n. Hence we see that any subgroup of $(\mathbb{Z}, +)$ is $(n\mathbb{Z}, +)$, and each $(n\mathbb{Z}, +, \cdot)$ is an ideal.

EXAMPLE D $\mathbb{Q}$ has no ideals except $\{0\}$ and $\mathbb{Q}$. Since $\{0\}$ and the ring are always ideals, we refer to them as *improper*. Thus, what we wish to show is that $\mathbb{Q}$ has no proper ideals. Let I be an ideal in $\mathbb{Q}$ with $I \neq \{0\}$. Let $a \in I - \{0\}$. Then, since $a \neq 0$ and $\mathbb{Q}$ is a field, $a^{-1} \in \mathbb{Q}$. Thus $a \cdot a^{-1} = 1$ is in I by the absorption property. Using this property again, we see that $q \cdot 1 = q$ is in I for all q in $\mathbb{Q}$, and thus $I = \mathbb{Q}$.

Exercises

1   Show that if D is a division ring, then D has no proper ideals. In particular, this is true for fields.

2   Find all four ideals contained in the ring $\mathbb{Q} \times \mathbb{Q}$. Find four more subrings that are not ideals.

3   If A and B are ideals in R with $A \cap B = \{0\}$, show that $ab = 0$ for all a in A, b in B.

EXAMPLE E   Let us examine the ring $\mathbb{Z} \times \mathbb{Z}$ of all ordered pairs of integers, where $(a,b) + (c,d) + (a + c,\ b + d)$ and $(a,b) \cdot (c,d) = (ac,\ bd)$. It is easy to see that the following are subrings:

1 $S_1 = \{(n,n) \,|\, n \in \mathbb{Z}\}$
2 $S_2 = \{(3n,0) \,|\, n \in \mathbb{Z}\} = 3\mathbb{Z} \times (0)$ or $3\mathbb{Z} \times 0\mathbb{Z}$
3 $S_3 = \{(n,m) \,|\, n,m \in \mathbb{Z}$ and $|n| + |m| \in 2\mathbb{Z}\}$
4 $S_4 = \{(3n,5m) \,|\, n,m \in \mathbb{Z}\} = 3\mathbb{Z} \times 5\mathbb{Z}$

Which of these are ideals? If I is an ideal of $\mathbb{Z} \times \mathbb{Z}$ and (a,b) is in I, then so are $(a,0) = (1,0)(a,b)$ and $(0,b) = (0,1)(a,b)$. The various $(a,0)$ form an ideal of $\mathbb{Z} \times \{0\}$

which must be $n\mathbb{Z} \times 0$. The $(0,b)$ form the ideal $\{0\} \times m\mathbb{Z}$ in $\mathbb{Z}$. Therefore any ideal of $\mathbb{Z} \times \mathbb{Z}$ has the form $n\mathbb{Z} \times m\mathbb{Z}$. Hence S_2 and S_4 are ideals, and S_1 and S_3 are not.

Proposition 2 and Definition If I is an ideal of R, then we can make the right cosets $I + r$ into a ring R/I, the **quotient ring** of R over I, if we define

$$(I + r) + (I + s) = I + r + s$$

$$(I + r)(I + s) = I + rs$$

PROOF Using Proposition 7 of Sec. 1.8, we know that $(R/I, +)$ is a group, since $(I, +)$ is normal in $(\mathbb{R}, +)$. Also, $(I + r) + (I + s) = I + r + s = I + s + r = (I + s) + (I + r)$, since $(R, +)$ is abelian, and thus $(R/I, +)$ is abelian.

Assuming that $\cdot$ is well-defined on R/I, we can quickly verify the remaining properties:

$$((I + r)(I + s))(I + t) = I + (rs)t = I + r(st)$$
$$= (I + r)((I + s)(I + t))$$

$$(I + r)((I + s) + (I + t)) = I + r(s + t) = I + rs + rt$$
$$= (I + rs) + (I + rt)$$
$$= (I + r)(I + s) + (I + r)(I + t)$$

The other distributive law follows in a similar manner.

To establish that $\cdot$ is well-defined, we note that if we multiply $I + r$ and $I + s$ as subsets of R we obtain

$$(I + r) \cdot (I + s) = II + rI + Is + rs \subseteq I + I + I + rs$$
$$= I + rs$$

Here we have used the absorption laws for I.

Thus $I + rs$ is the coset of I containing $\{(i + r) \cdot (j + s) \mid i, j \in I\} = \Delta$. If $I + r = I + r'$ and $I + s = I + s'$, we also have

$$\Delta = \{(i + r')(j + s') \mid i, j \in I\}$$

and $\Delta \subseteq I + r's'$ so $I + rs = I + r's'$, and $\cdot$ is a well-defined operation in R/I. Therefore $(R/I, +, \cdot)$ is again a ring. /////

We now proceed to connect the concept of ring with that of "transportation" between one ring and another.

Definition If $(R, +, \cdot)$ and $(S, \oplus, \odot)$ are rings, a function ϕ from R to S is a **ring homomorphism** if $\phi(r + r') = \phi(r) \oplus \phi(r')$ and $\phi(r) \odot \phi(r') = \phi(rr')$ for all r and r' in R. ϕ is a **monomorphism** if ϕ is one-to-one. If there is a monomorphism, ϕ, of R onto S, then R and S are **isomorphic,** and ϕ is an **isomorphism.**

EXAMPLE F   The matrices $\left\{ \begin{pmatrix} a & b \\ -b & a \end{pmatrix} \middle| a, b \in R \right\}$ are isomorphic to $\mathbb{C}$, the complex numbers. We look at the function $\phi(a + bi) = \begin{pmatrix} a & b \\ -b & a \end{pmatrix}$. To check addition, we have

$$\phi((a + bi) + (c + di)) = \phi((a + c) + (b + d)i)$$
$$= \begin{pmatrix} a + c & b + d \\ -b - d & a + c \end{pmatrix} = \begin{pmatrix} a & b \\ -b & a \end{pmatrix} + \begin{pmatrix} c & d \\ -d & c \end{pmatrix}$$
$$= \phi(a + bi) + \phi(c + di)$$

Similarly,

$$\phi((a + bi)(c + di)) = \phi(ac - bd + (ad + bc)i) = \begin{pmatrix} ac - bd & ad + bc \\ -ad - bc & ac - bd \end{pmatrix}$$
$$= \begin{pmatrix} a & b \\ -b & a \end{pmatrix} \begin{pmatrix} c & d \\ -d & c \end{pmatrix} = \phi(a + bi)\phi(c + di) \qquad \text{and } \phi$$

is a ring homomorphism. Since $\phi(a + bi) = \begin{pmatrix} a & b \\ -b & a \end{pmatrix}$, we have ϕ is onto. If $\phi(a + bi) = \phi(c + di)$, then $\begin{pmatrix} a & b \\ -b & a \end{pmatrix} = \begin{pmatrix} c & d \\ -d & c \end{pmatrix}$ so $a = c$, $b = d$, and thus $a + bi = c + di$. Whence, ϕ is an isomorphism.

EXAMPLE G   Let us look at all possible homomorphisms from $\mathbb{Z}$ to $\mathbb{Z}$. If $\phi(1) = k$, then $\phi(1 + 1) = \phi(1) + \phi(1) = 2k$, $\phi(3) = 3k$, and $\phi(n) = kn$. For $(\mathbb{Z}, +)$ this describes all group homomorphisms from $(\mathbb{Z}, +)$ to $(\mathbb{Z}, +)$. If ϕ is to be a ring homomorphism, it must also satisfy $2k = \phi(2) = \phi(2 \cdot 1) = \phi(2)\phi(1) = (2k)k$. Therefore, $2k = 2k^2$, so $k = 0$ or 1. Thus either $\phi(n) = 0$ for all n or $\phi(n) = n$ for all n, and these are the only two possible ring homomorphisms from $\mathbb{Z}$ to $\mathbb{Z}$.

Exercises

4   If p is a prime, show that p divides (evenly) the binomial coefficient $\binom{p}{k}$, where $1 \leq k \leq p - 1$.

5   Show that the binomial theorem holds for any commutative ring. That is, show that $(a + b)^n = \sum_{i=0}^{n} \binom{n}{i} a^{n-i} b^i$, where the interpretation of ka for k in $\mathbb{Z}^+$, a in the ring, is

$$\underbrace{a + a + \cdots + a}_{k \text{ times}}$$

Definition Suppose there is a least positive integer n such that $nr = 0$ for all r in a ring R. Then R is said to have **characteristic** n. Otherwise R is of characteristic **zero**.

$\mathbb{Z}_n$ is an example of a ring with characteristic n. So is $\mathbb{Z}_n[x]$, the ring of polynomials with coefficients in $\mathbb{Z}_n$. (See Sec. 2.6.)

Exercises

6 What is an example of a noncommutative ring of characteristic n, where $n \neq 0$?

7 If R is a commutative ring of characteristic p (p a prime), show that $\phi(x) = x^p$ is a ring homomorphism from R to R. This is called the *Frobenius homomorphism*. (*Hint:* Use Exercises 4 and 5.)

Fermat's little theorem Using Exercise 7, we can give a short proof that $a^p = a$ for all a in $\mathbb{Z}_p$. Define $\phi: \mathbb{Z}_p \to \mathbb{Z}_p$ to be the Frobenius homomorphism, $\phi(x) = x^p$. Obviously, $\phi(1) = 1$, and thus $\phi(2) = \phi(1 + 1) = \phi(1) + \phi(1) = 2$. Continuing, we have $a = \phi(a) = a^p$ for all a in $\mathbb{Z}_p$. Alternatively, this says $p \mid a^p - a$ for all integers a. For instance, $7 \mid 1{,}000^7 - 1{,}000$. See Chap. 7 for further discussion of this topic.

Exercises

8 Show that if $R = \{a + b\sqrt{3} \mid a, b \in \mathbb{Z}\}$ then the mapping $\phi(a + b\sqrt{3}) = a - b\sqrt{3}$ is a ring isomorphism.

9 If $R = \{a + b\sqrt{3} \mid a, b \in \mathbb{Q}\}$ and $S = \{a + b\sqrt{7} \mid a, b \in \mathbb{Q}\}$ is $\phi(a + b\sqrt{3}) = a + b\sqrt{7}$ a ring isomorphism?

10 If (a, b) is any point in the plane $\mathbb{R}^2$, find a 2×2 matrix M such that $(a, b)M = (0, 0)$ but with $M \neq \begin{pmatrix} 0 & 0 \\ 0 & 0 \end{pmatrix}$. Show that this M maps the whole line $\{(ra, rb) \mid r \in \mathbb{R}\}$ to $(0, 0)$.

11 If M is a 2×2 matrix with real coefficients considered as a function from $\mathbb{R}^2$ to $\mathbb{R}^2$, show that:

(a) The range of M is all of $\mathbb{R}^2$ if and only if M is nonsingular.

(b) The range of M is a line or $(0, 0)$ if and only if M is singular.

12 Show that every 2×2 matrix over $\mathbb{R}$ other than $\begin{pmatrix} 0 & 0 \\ 0 & 0 \end{pmatrix}$ is either nonsingular or a zero divisor.

13 Is every nonzero element of $\mathbb{Z}_n$ either a unit or a zero divisor? What is the situation for $\mathbb{Z}$?

14 Prove that $\mathbb{Z}_p$, for p prime, is a field.

We now want to collect some facts about the structure carried by a ring homomorphism. As in the case of a group homomorphism, we call this collection the *correspondence* theorem. As then, it is recommended that the proof be regarded as a series of exercises.

Proposition 3 (the correspondence theorem) If R and $\bar{R}$ are rings, and ϕ is a ring homomorphism from R onto $\bar{R}$, then

(a) $\phi(0_R) = 0_{\bar{R}}$.

(b) $\phi(-x) = -\phi(x)$.

(c) If S is a subring of R, then $\phi(S)$ is a subring of $\bar{R}$.

(d) If $\bar{S}$ is a subring of $\bar{R}$, then $\phi^{-1}(\bar{S})$ is a subring of R.

(e) If S is an ideal of R, then $\phi(S)$ is an ideal of $\bar{R}$.

(f) If $\bar{S}$ is an ideal of $\bar{R}$, then $\phi^{-1}(\bar{S})$ is an ideal of R.

(g) If R has a multiplicative identity 1_R, then $\phi(1_R)$ is the multiplicative identity of $\bar{R}$.

(ϕ onto is needed only for parts *e* and *g*.)

PROOF Since $(R,+)$ and $(\bar{R},+)$ are groups, and ϕ is certainly a group homomorphism of $(R,+)$ onto $(\bar{R},+)$, the correspondence theorem for groups proves parts *a* and *b*. Similarly, the correspondence theorem for groups says that $\phi(S)$ and $\phi^{-1}(\bar{S})$ are groups under addition. To finish part *c*, we need only show that $\phi(S)$ is closed under multiplication, since $\phi(S)$ as a subset of $\bar{R}$ automatically satisfies the associative and distributive laws. If $\bar{r}_1 = \phi(s_1)$ and $\bar{r}_2 = \phi(s_2)$ are any two elements of $\phi(S)$, then $\bar{r}_1 \cdot \bar{r}_2 = \phi(s_1)\phi(s_2) = \phi(s_1 s_2) \in \phi(S)$, proving part *c*.

If r_1 and r_2 are in $\phi^{-1}(\bar{S})$, then $\phi(r_1) = \bar{s}_1$ and $\phi(r_2) = \bar{s}_2$, with $\bar{s}_1$ and $\bar{s}_2$ in $\bar{S}$. Thus,

$$\phi(r_1 r_2) = \phi(r_1)\phi(r_2) = \bar{s}_1 \cdot \bar{s}_2 \in \bar{S}$$

so $r_1 r_2 \in \phi^{-1}(\bar{S})$ and $\phi^{-1}(\bar{S})$ is a subring, proving part *d*.

If S is an ideal of R, then we want to show that $\phi(S)$ is an ideal of $\bar{R}$. We examine $\phi(s)\bar{r}$ and $\bar{r}\phi(s)$. Since ϕ is onto, $\bar{r} = \phi(r)$ for some $r \in R$. Thus $\phi(s)\bar{r} = \phi(s)\phi(r) = \phi(sr)$, and $\bar{r}\phi(s) = \phi(r)\phi(s) = \phi(rs)$. But S is an ideal so rs and sr are in S, and thus $\phi(s)\bar{r}$ and $\bar{r}\phi(s)$ are in $\phi(S)$ proving part *e*.

If $\bar{S}$ is an ideal of $\bar{R}$ and $x \in \phi^{-1}(\bar{S})$, then we let $\bar{s} = \phi(x)$ and look at rx. But $\phi(rx) = \phi(r) \cdot (\bar{s}) \in \bar{S}$, since $\bar{S}$ is an ideal. Therefore $rx \in \phi^{-1}(\bar{S})$ and similarly, $xr \in \phi^{-1}(\bar{S})$, proving part *f*.

If $\bar{r} \in \bar{R}$ then $\bar{r} = \phi(r)$, since ϕ is onto. Thus

$$\phi(1_R)\bar{r} = \phi(1_R)\phi(r) = \phi(1_R \cdot r) = \phi(r) = \bar{r}$$

Similarly, $\bar{r} \cdot \phi(1_R) = \bar{r}$, so $\phi(1_R)$ is the identity for $\bar{R}$.   /////

Definition If $\phi: R \to S$ is a ring homomorphism from the ring R to the ring S, then **kernel** $\phi = \ker(\phi) = \{r \mid r \in R$ and $\phi(r) = O_S\}$. Thus, this kernel is just the kernel of ϕ considered as a group homomorphism from $(R, +)$ to $(S, +)$.

Corollary ker (ϕ) is an ideal of R, where $\phi: R \to S$ is a ring homomorphism from the ring R to the ring S.

PROOF We make three observations. $\{O_S\}$ is an ideal in S. $\ker(\phi) = \phi^{-1}\{O_S\}$ is thus an ideal of R by part f of the correspondence theorem. The proof of part f did not require ϕ to be onto.   /////

We are now able to translate information about homomorphisms to information about kernels and corresponding quotient rings.

Fundamental theorem of ring homomorphisms If ϕ is a ring homomorphism from the ring R onto the ring $\bar{R}$, then there exists a ring isomorphism from R/I onto $\bar{R}$, where $I = \ker(\phi)$. Further there is a one-to-one correspondence between the ideals of $\bar{R}$ and those of R that contain I.

PROOF Considering ϕ as a group homomorphism and using the fundamental theorem of group homomorphisms, we see that defining $\alpha: R/I \to \bar{R}$ by $\alpha(I + r) = \phi(r)$ gives us an α that is one-to-one onto, well-defined, and such that

$$\alpha(I + r) + \alpha(I + s) = \alpha(I + (r + s))$$

With this much given to us, we need only check that α is a multiplicative homomorphism.

$$\alpha((I + r)(I + s)) = \alpha(I + rs) \qquad \text{definition of multiplication in } R/I$$

$$= \phi(rs) \qquad \text{definition of } \alpha$$

$$= \phi(r)\phi(s) \qquad \phi \text{ is a ring homomorphism}$$

$$= \alpha(I + r)\alpha(I + s) \qquad \text{definition of } \alpha$$

As with groups, ϕ induces a one-to-one correspondence between the ideals in R containing I and the ideals in $\bar{R}$.   /////

EXAMPLE H If $R = \mathbb{R} \times \mathbb{R} \times \mathbb{R}$ and $S = \mathbb{R} \times \mathbb{R}$, and we define ϕ by $\phi(x,y,z) = (x,y)$, then ϕ is a ring homomorphism since

$$\phi((x_1,y_1,z_1) + (x_2,y_2,z_2)) = \phi(x_1 + x_2, y_1 + y_2, z_1 + z_2)$$
$$= (x_1 + x_2, y_1 + y_2)$$
$$= (x_1,y_1) + (x_2,y_2)$$
$$= \phi(x_1,y_1,z_1) + \phi(x_2,y_2,z_2)$$
$$\phi((x_1,y_1,z_1) \cdot (x_2,y_2,z_2)) = \phi(x_1x_2,y_1y_2,z_1z_2) = (x_1x_2,y_1y_2)$$
$$= (x_1,y_1)(x_2\ y_2)$$
$$= \phi(x_1\ y_1,z_1)\phi(x_2,y_2,z_2)$$
$$\ker(\phi) = \{(x,y,z) \mid \phi(x,y,z) = (0,0)\} = \{(0,0,z) \mid z \in \mathbb{R}\}$$

Since for any (x,y) in $\mathbb{R}^2$, $\phi(x,y,z) = (x,y)$, ϕ is onto, and we can apply the fundamental theorem which says $\{(x,y,z)\}/\{(0,0,z)\} \simeq \{(x,y)\}$, for x,y,z in $\mathbb{R}$. Alternatively, we could say that $\mathbb{R}^3/\mathbb{R} \simeq \mathbb{R}^2$.

Proposition 4 A ring homomorphism $\phi: R \to \bar{R}$ is one-to-one if and only if $\ker(\phi) = \{0_R\}$.

    PROOF ϕ is one-to-one $\Leftrightarrow \phi(r) = \phi(s)$ always implies $r = s \Leftrightarrow \phi(r) - \phi(s) = \phi(r - s) = 0_R$ always implies $r - s = 0_R \Leftrightarrow r - s \in \ker(\phi)$ always implies $r - s = 0_R \Leftrightarrow x \in \ker(\phi)$ always implies $x = 0_R \Leftrightarrow \ker(\phi) = \{0_R\}$.

An alternative proof is to observe that ϕ is a group homomorphism from $(R,+)$ to $(\bar{R},+)$. This result already has been obtained for groups. (See Exercise 22 of Sec. 1.13.)

EXAMPLE I If F is a field and ϕ is a ring homomorphism from F to R, then ϕ is either a monomorphism or ϕ maps all of F to 0_R. In fact, this is true if F is a division ring. Note that F has only two ideals, $\{0_F\}$ and F, as we saw in Exercise 1 of Sec. 2.4. If $\ker(\phi) = \{0_F\}$, then ϕ is one-to-one. If $\ker(\phi) = F$, all of F is mapped to 0_R.

Exercises

15   Find all four ideals in $R = \mathbb{R} \times \mathbb{R}$. If ϕ is given by $\phi(x,y) = x + 2y$, then is ϕ a group homomorphism of $(\mathbb{R} \times \mathbb{R},+)$ to $(\mathbb{R},+)$? Is ϕ a ring homomorphism?

16   If $[\ \ ]$ is the function given by letting $[r]$ equal the greatest integer less than r, then show that $\phi(r) = r - [r]$ is a group homomorphism from $(\mathbb{R},+)$ to $(\mathbb{R}/\mathbb{Z},+)$. Why can't this be a ring homomorphism?

It may be worth commenting on the origin of the term "ideal." In fact, rings and ideals sound a bit like the name for an optimistic modern-marriage course. In a ring R with 1, every element a in R corresponds to an ideal aR. An arithmetic of ideals can be set up, where $I_1 + I_2$ is the smallest ideal containing both I_1 and I_2, and $I_1 I_2 = \{ij \mid i \in I_1, j \in I_2\}$. However, not all ideals in R have to be of the form aR. These other ideals were at one time thought of as coming from some "ideal number" $\mathring{a}$, where $\mathring{a}$ was not in R. Actually, in many such cases all the ideals concerned are in $\mathbb{C}$, and a correct ideal number may well exist and even be easy to determine. Ideals in this context seem to have arisen before the concept of ring was formalized. The modern tendency is largely to ignore the numbers and just treat the ideals themselves as we have done in this section. The concept of an ideal is due to Dedekind and Kummer, although many special cases were known long before their work. Both Boyer [12] and Kline [47] discuss the history of ideal numbers, unique factorization, and Fermat's conjecture.

In Chap. 6 on domains, some of these topics, such as principal ideal domains, are discussed in more detail.

Definition If $\{R_i\}_{i=1}^n$ are rings, then we can form a new ring $\oplus \sum_{i=1}^n R_i$ which is the **direct sum** of the R_i. We define addition and multiplication as follows:

$$(r_1, r_2, \ldots, r_n) + (s_1, s_2, \ldots, s_n) = (r_1 + s_1, r_2 + s_2, \ldots, r_n + s_n)$$

and
$$(r_1, r_2, \ldots, r_n) \cdot (s_1, s_2, \ldots, s_n) = (r_1 s_1, r_2 s_2, \ldots, r_n s_n)$$

where all r_i and s_i are in R_i.

Exercise

17 If each $|R_i| = n_i < \infty$, what is $|\oplus \sum_{i=1}^n R_i|$?

Suppose we want to extend this construction to an infinite set of rings $\{R_\alpha\}_{\alpha \in \mathscr{A}}$. The informal approach is just to define

$$(\ldots r_{,\alpha}, \ldots) + (\ldots, s_\alpha, \ldots) = (\ldots, r_\alpha + s_\alpha, \ldots)$$

and
$$(\ldots, r_\alpha, \ldots) \cdot (\ldots, s_\alpha, \ldots) = (\ldots, r_\alpha s_\alpha, \ldots)$$

The more formal approach is to replace $(\ldots, r_\alpha, \ldots)$ with a function f on $\mathscr{A}$ so that $f(\alpha) = r_\alpha \in R_\alpha$ for all $\alpha \in \mathscr{A}$. This ring is called the direct *product* of the R_α and is denoted $\prod_{\alpha \in \mathscr{A}} R_\alpha$.

If we add an extra restriction that all but a finite number of the r_α are 0, then we obtain a subring which we call the *direct sum* of the R_α and again denote $\oplus \sum_{\alpha \in \mathscr{A}} R_\alpha$.

If $\mathscr{A}$ is a finite set, then we are back to the first definition as well as having $\oplus \sum_{\alpha \in \mathscr{A}} R_\alpha = \prod_{\alpha \in \mathscr{A}} R_\alpha$.

Exercises

18 Show that $\oplus\sum_{\alpha \in \mathscr{A}} R_\alpha$ is an ideal in $\prod_{\alpha \in \mathscr{A}} R_\alpha$.

19 If $\mathscr{A}$ is infinite and each $R_\alpha \neq (0)$, show that $\oplus\sum_{\alpha \in \mathscr{A}} R_\alpha$ cannot be a ring with identity.

20 Is $\mathbb{Z}_{15} \simeq \mathbb{Z}_3 \oplus \mathbb{Z}_5$? Is $\mathbb{Z}_9 \simeq \mathbb{Z}_3 \oplus \mathbb{Z}_3$? For what n is $\mathbb{Z}_{2n} \simeq \mathbb{Z}_2 \oplus \mathbb{Z}_n$?

21 If $\mathscr{A}$ contains more than one element and all $R_\alpha \neq (0)$, show that $\oplus\sum_{\alpha \in \mathscr{A}} R_\alpha$ must have zero divisors and thus cannot be either an integral domain, a division ring, or a field.

2.5 FACTORING OUT MAXIMAL IDEALS IN A COMMUTATIVE RING

Definition   An ideal M is **maximal** in a ring R if no proper ideal of R contains M properly and M is not the whole ring. Alternatively, if N is an ideal of R and $M \subseteq N \subseteq R$, then either $M = N$ or $N = R$ (where again $M \neq R$).

EXAMPLE A   If $R = (\mathbb{Z}, +, \cdot)$ is the ring of integers, then $M = p\mathbb{Z}$ is a maximal ideal if p is prime. This follows since if $p\mathbb{Z} \subseteq n\mathbb{Z} = N \subseteq \mathbb{Z}$, then $n|p$, so $n = 1$ or p, and $n\mathbb{Z} = \mathbb{Z}$ or $p\mathbb{Z}$.

Exercise

1 Find the maximal ideals in $R = 2\mathbb{Z}$, in $\mathbb{Z}_{30}$, and in $\mathbb{Q}_p = \{a/b \,|\, a,b \in \mathbb{Z}, p \nmid b, p \text{ prime}\}$.

We now use the homomorphism theorem to prove the next theorem, and spend the rest of the section exploiting the theorem in several different areas of mathematics.

Theorem 1   If R is a commutative ring with 1, then M is a maximal ideal if and only if R/M is a field.

PROOF R/M is automatically a commutative ring with 1 since R is. Thus we need only show that any nonzero element of R/M has a multiplicative inverse. Let $x + M$ be any nonzero element of R/M. Going back to R, we look at $N = \{xr + m \,|\, r \in R, \ m \in M\}$, and we can show (see Exercise 2) that this is an ideal containing M. $x = x \cdot 1 + 0 \in N$ but $x + M \neq M$, so $x \notin M$, and N contains M properly. Thus $N = R$ and $1 \in N$. So $1 = xr_0 + m_0$ for some $r_0 \in R$ and $m_0 \in M$. Therefore, $1 + M = xr_0 + M = (x + M)(r_0 + M)$ and $(x + M)^{-1} = r_0 + M$ in R/M.

If R/M is a field then, by Exercise 1 of Sec. 2.4, R/M contains no proper ideals. The fundamental theorem applied to the canonical homomorphism from R to R/M says any ideal in R must map onto an ideal in R/M; thus no ideal in R can properly contain M unless it is R itself. This completes the proof of the theorem. ////

Exercises

2 (a) Let R be a commutative ring with 1, I an ideal, and x an element in R. Show that $\langle x,I \rangle = \{xr + i \mid r \in R, i \in I\}$ is an ideal in R.

(b) If R is not necessarily commutative, how should part a be changed to obtain the same sort of result?

(c) If R is commutative but not necessarily with 1, show that $I' = \langle x,I \rangle = \{nx + xr + i \mid n \in \mathbb{Z}, r \in R, i \in I\}$ is an ideal, where $nx = x + x + \cdots + x$ (n copies) for n positive.

3 Let R be a ring, and let I be the smallest ideal containing all elements $rr' - r'r$, where r and r' are in R.

(a) Show that R/I is commutative.

(b) Show that if J is an ideal with R/J commutative, then $J \supseteq I$.

(c) What is a stronger version of Theorem 1 that uses this commuting-factor ideal?

APPLICATION 1 (*Construction of the complex numbers from the real numbers*) We outline this construction.

1 $\mathbb{R}$ is a field, so $\mathbb{R}[x]$, the ring of polynomials over $\mathbb{R}$, is a commutative ring with 1. (Polynomial rings are discussed more formally in the next section.)

2 We want to solve $x^2 + 1 = 0$, so, to make $x^2 + 1 = 0$ in some setting, we set up the quotient ring $R = \mathbb{R}[x]/\langle x^2 + 1 \rangle$. Here $\langle x^2 + 1 \rangle = $ the smallest ideal of $\mathbb{R}[x]$ containing $x^2 + 1 = \{p(x)(x^2 + 1) \mid p(x) \in \mathbb{R}[x]\}$.

3 $\langle x^2 + 1 \rangle = M$ is a maximal ideal in $\mathbb{R}[x]$.

4 $\bar{R} = R[x]/M$ is a field by the theorem.

5 Let $\overline{x^2 + 1} = \bar{x}^2 + \bar{1}$ be the image of $x^2 + 1$ in $\bar{R}$. Then $\bar{x}^2 + \bar{1} = \bar{0}$ in $\bar{R}$, since $x^2 + 1 \in M$. In effect we now have $i = \bar{x}$.

The details of step 3 will be discussed in Chap. 6, where euclidean and principal ideal domains are discussed. Otherwise, this construction is not hard to complete.

Some modification of this procedure will allow us to construct, over any field F and for any polynomial $p(x) \in F[x]$, a field E containing F such that $p(e) = 0$ for some e in E.

APPLICATION 2 (*Maximal ideals in C[0,1]*) We saw earlier that $(C[0,1], +, \cdot)$ is a commutative ring with 1. Although this ring more or less belongs to the analysts rather than the algebraists, it is still a natural qeustion to ask what the maximal ideals are in this ring. If $\alpha \in [0,1]$, then $M_\alpha = \{f(x) | f(x) \in C[0,1]$ and $f(\alpha) = 0\}$ is an ideal of $C[0,1]$. A very rough picture of this ideal would be as follows:

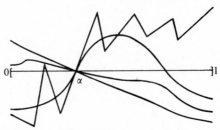

We next show that M_α is maximal. If $N \supseteq M_\alpha$ then there exists $n(x) \in N - M_\alpha$, so $n(\alpha) \neq 0$. If $g(x)$ is any element of $C[0,1]$, then $g(x) = [g(\alpha)/n(\alpha)]n(x) + r(x)$. Obviously, $r(x)$ is in M_α so $g(x)$ is in N, $N = C[0,1]$, and M_α is maximal.

Exercises

4 We can give a quicker proof that M_α is maximal as follows: Let $\phi: C[0,1] \to \mathbb{R}$ be given by $\phi(f(x)) = \phi(\alpha)$. Show that ϕ is a homomorphism and is onto. Note that $\mathbb{R}$ is a field, and use Theorem 1.

To show every maximal ideal has this form is much harder and is left to the ambitious reader.

5 Find a maximal ideal M in $R = \left\{ \begin{pmatrix} a & b \\ 0 & c \end{pmatrix} \middle| a,b,c \in \mathbb{R} \right\}$. Which field or ring is isomorphic to R/M?

6 Discuss the maximal ideals in $\mathbb{C}[x]$, the ring of polynomials over $\mathbb{C}$.

In courses called "foundations of the real numbers" or even "foundations of mathematics," the following path is often followed. After a brief discussion of sets and functions, the Peano axioms are introduced to develop the positive integers. Then there is a bit of historical discussion, and groups are defined. In the following lecture, the negative integers and 0 turn out to be necessary so that the positive integers can be in a group. Then fields are defined, and soon the rationals are defined so that the integers can be in a field. Then, to complete the rationals, either Dedekind cuts or Cauchy sequences are introduced. This is the subject we want to discuss.

APPLICATION 3 Assume that the field $\mathbb{Q}$ is known; using $\mathbb{Q}$, we want to construct $\mathbb{R}$. Let $\mathbb{Q}^*$ denote the set of all convergent sequences of rationals. That is,

$$(q_1, q_2, \ldots, q_1, \ldots) \in \mathbb{Q}^*$$

if it satisfies the Cauchy criterion (that is, $\forall \varepsilon > 0 \; \exists N \ni$ if $n,m > N$ then $|q_n - q_m| < \varepsilon$). [It is temping to say that $(q_2,q_2,...,q^i,...) \in \tilde{Q}$ if $\lim_{n \to \infty} q_n$ exists, but this amounts to using $\mathbb{R}$ to define $\mathbb{R}$.]

It is straightforward to verify that $\mathbb{Q}^*$ is a commutative ring with 1, if we define + and · by $(q_1,q_2,q_2,...) + (r_1,r_2,r_3,...) = (q_1 + r_1, q_2 + r_2, q_3 + r_3,...)$ and $(q_1,q_2,q_3,...) \cdot (r_1,r_2,r_3...) = (q_1 r_1, q_2 r_2, q_3 r_3,...)$. Here the additive identity is $(0,0,0,...)$, and the multiplicative identity is $(1,1,1,...)$. Some of the details here are left to the industrious reader.

Next we define a null sequence as one converging to 0; it can be shown that the null sequences form not only a subring but an ideal of $\mathbb{Q}^*$. We call this ideal N^*. Next N^* can be shown to be maximal in $\mathbb{Q}^*$. All this takes some verification but now automatically gives us the fact that $\mathbb{Q}^*/N^*$ is a field. In a typical course on foundations of the real numbers, they do not have Theorem 1, so all eleven of the field axioms need be verified (or passed over lightly with some sort of hand-waving).

To finish, each real number r is associated with the coset $(q_1,q_2,q_3...) + N^*$, where $\lim_{n \to \infty} q_n = r$.

Exercises

7   *Let $q_1 = 1$ and $q_n = \frac{1}{2}(q_{n-1} + 2/q_{n-1})$ for $n = 2, 3, \ldots$. Show that $\lim_{n \to \infty} q_n = \sqrt{2}$. [First compute q_2, q_3, and q_4 to see if this is reasonable. Then look at $f(x) = \frac{1}{2}(x + 2/x)$ on $S = [1,\infty)$. Show that q_n are a Cauchy sequence and, if $\lim_{n \to \infty} q_n = q$, then $q = \frac{1}{2}(q + 2/q)$.] In fact, any $q_1 > 0$ will work. This method, incidentally, is fast, easy to learn, and is built into many computers. What modification is needed to find $\sqrt{k}$?

Definition   If R is a commutative ring, then an ideal I is **prime** if $ab \in I$ implies $a \in I$ or $b \in I$.

8   Which of the following ideals are prime?
   (a) $R = \mathbb{Z}$ and $I_1 = 5\mathbb{Z}$. Also try $I_2 = 6\mathbb{Z}$.
   (b) $R = \mathbb{Z}$ and $I = (0)$.
   (c) $R = \mathbb{R}^2$, $I_1 = \mathbb{R} \times 0$ or $I_2 = ((0,0))$.
   (d) $\mathbb{R} = 3\mathbb{Z}, I = 12\mathbb{Z}$.
9   Using Theorem 1, give an alternative proof that $\mathbb{Z}_p$ is a field, where p is a prime.
10   Show that P is a prime ideal in R, where R is a commutative ring with 1 if and only if R/P is an integral domain.
11   Prove that, in a commutative ring with 1, every maximal ideal is prime.
12   Give an example of a prime ideal that is not maximal.
13   Show that any finite integral domain is a field.
14   If R is a finite commutative ring, show that every prime ideal is maximal.

2.6   POLYNOMIAL RINGS

In this section we discuss polynomial rings. Though interesting in themselves, these rings are a link between classical algebra and modern algebra. This link will be discussed in this section as well as in Chaps. 6 and 10.

Several questions arise naturally. When can a polynomial be factored? When are we guaranteed a solution as the quadratic formula guarantees us a solution with quadratic polynomials? What fields come into play in this solution process? We shall discuss the first question in this section, the second two in Chap. 10. The whole process of manufacturing a solution to a polynomial $f(x)$ over a field F comes down to studying the two homomorphisms

$$F \xrightarrow{\ \alpha\ } F[x] \xrightarrow{\ \beta\ } \frac{F[x]}{I} \quad \text{where} \quad I = \langle f(x) \rangle$$

This is interesting because we enlarge F a little by first enlarging it a lot up to $F[x]$, and then factoring out an ideal cleverly chosen so as to create desirable properties in the quotient ring.

Throughout this section F denotes a field. In this section we discuss $F[x]$ and its relatives, and it is time for a definition.

Definition (informal)   If R is a ring, then $R[x]$, the **ring of polynomials over R**, is the set of all elements of the form $\sum_{i=0}^{n} r_i x^i$ with r_i in R.

Addition and multiplication are defined by

$$\sum_{i=0}^{n} r_i x^i + \sum_{i=0}^{n} s_i x^i = \sum_{i=0}^{n} (r_i + s_i) x^i$$

and

$$\sum_{i=0}^{n} r_i x^i \cdot \sum_{i=0}^{m} s_i x^i = \sum_{i=0}^{n+m} t_i x^i \quad \text{where} \quad t_i = \sum_{j=0}^{i} r_j s_{i-j}$$

[This is just the addition and multiplication we have used since high school. When convenient, we add some $0 \cdot x^i$ terms. For instance,

$$(3 + 2x^2) + (2 + 2x + 5x^3)$$
$$= (3 + 0 \cdot x + 2 \cdot x^2 + 0 \cdot x^3) + (2 + 2x + 0 \cdot x^2 + 5x^3)$$
$$= 5 + 2x + 2x^2 + 5x^3.]$$

Exercise

1   Multiply the following polynomials:
(a)   In $\mathbb{Z}_3[x]$, multiply $\bar{1} + \bar{1}x + \bar{2}x^2$ by $\bar{2} + \bar{2}x^4$.
(b)   In $\mathbb{Z}_{12}[x]$, multiply $\bar{6}x + \bar{3}x^3$ by $\bar{4}x^2 + \bar{8}x^3$.
(c)   In $M_2(\mathbb{R})[x]$, multiply

$$\begin{pmatrix} 1 & 0 \\ 0 & 1 \end{pmatrix} + \begin{pmatrix} 1 & 1 \\ 0 & 1 \end{pmatrix} x + \begin{pmatrix} 1 & 3 \\ 0 & 4 \end{pmatrix} x^2 \quad \text{by} \quad \begin{pmatrix} 1 & 4 \\ 0 & 2 \end{pmatrix} x + \begin{pmatrix} 1 & 0 \\ 1 & 1 \end{pmatrix} x^3$$

(d) $\mathbb{Q}[x]$ is itself a ring, so $(\mathbb{Q}[x])[y]$ is a polynomial ring. In $(\mathbb{Q}[x])[y]$, multiply $(3 + 2x^2) + (1 + x + x^5)y$ by $1 + xy + (x^2 + x)y^2$.

Proposition 1 The ring of polynomials over a ring R is itself a ring.

REMARK To save space, we shall use summation notation to write polynomials as $\sum_{i=0}^{\infty} r_i x^i$, with the understanding that only a finite number of coefficients will be nonzero. If we remove this finiteness restriction, we obtain $R[[x]]$, the ring of formal power series over R.

PROOF Both addition and multiplication are closed operations by their definitions. The additive identity is $\hat{0} = 0 + 0 \cdot x + 0 \cdot x^2 + \cdots = \sum_{i=0}^{\infty} 0 \cdot x^i$, since it satisfies

$$\hat{0} + \sum_{i=0}^{\infty} r_i x^i = \sum_{i=0}^{\infty} (0 + r_i)x^i = \sum_{i=1}^{\infty} r_i x^i = \sum_{i=1}^{\infty} r_i x^i + \hat{0}$$

Since addition in R is both associative and commutative, addition in $R[x]$ is also both associative and commutative. For instance,

$$\sum_{i=0}^{\infty} r_i x^i + \sum_{i=0}^{\infty} s_i x^i = \sum_{i=0}^{\infty} (r_i + s_i)x^i = \sum_{i=0}^{\infty} (s_i + r_i)x^i$$

$$= \sum_{i=0}^{\infty} s_i x^i + \sum_{i=0}^{\infty} r_i x^i$$

The additive inverses for $\sum_{i=0}^{\infty} r_i x^i$ is $\sum_{i=0}^{\infty} (-r_i)x^i$, and thus $(R[x], +)$ is an abelian group.

To verify the associative law for multiplication, we proceed as follows:

$$\left(\sum_{i=0}^{\infty} r_i x^i \sum_{j=0}^{\infty} s_j x^j\right) \sum_{k=0}^{\infty} t_k x^k = \sum_{m=0}^{\infty} \left(\sum_{n=0}^{m} r_n s_{m-n}\right) x^m \cdot \sum_{k=0}^{\infty} t_k x^k$$

$$= \sum_{p=0}^{\infty} \left[\sum_{q=0}^{p} \left(\sum_{n=0}^{q} r_n s_{q-n}\right) t_{p-q}\right] x^p$$

$$= \sum_{p=0}^{\infty} \sum_{i+j+k=0}^{p} r_i s_j t_k x^p$$

where i, j, k are nonnegative integers and this is a symmetric arrangement in the $r_i, s_j,$ and t_k. Thus,

$$\left(\sum_{i=0}^{\infty} r_i x^i \sum_{j=0}^{\infty} s_j x^j\right) \sum_{k=0}^{\infty} t_k x^k = \sum_{p=0}^{\infty} \left(\sum_{i+j+k=0}^{p} r_i s_j t_k\right) x^p$$

$$= \sum_{i=0}^{\infty} r_i x^i \left(\sum_{j=0}^{\infty} s_j x^j \sum_{k=0}^{\infty} t_k x^k\right)$$

The distributive laws are similar, easier, and left as an exercise.

////

Exercises

2 Verify the associative law for addition and the distributive law in $R[x]$.

Among the next four exercises, prove the true statements, and provide a counterexample for the false.

3 R has 1 if and only if $R[x]$ has 1.
4 R is commutative if and only if $R[x]$ is commutative.
5 R is an integral domain if and only if $R[x]$ is also.
6 F is a field if and only if $F[x]$ is a field.

Our next project is to find a more rigorous definition for $R[x]$. Two items are bothersome in the definition we have been using. We are looking at elements "of the form $\sum_{i=0}^{\infty} r_i x^i$," and neither "of the form" nor "x" has been defined. We eliminate both difficulties by replacing $\sum_{i=0}^{\infty} r_i x^i$ with the infinite sequence $(r_0, r_1, r_2, r_3, \dots)$, again adding the provision that only a finite number of terms are nonzero.

If we make this translation into sequences, we find that

$$(r_0, r_1, r_2, \dots, r_i, \dots) + (s_0, s_1, s_2, \dots, s_i, \dots) = (r_0 + s_0, r_1 + s_1, r_2 + s_2, \dots, r_i + s_i, \dots)$$

$$(r_0, r_1, r_2, \dots, r_i, \dots) \cdot (s_0, s_1, s_2, \dots, s_i, \dots) =$$
$$(r_0 s_0, r_0 s_1 + r_1 s_0, r_0 s_2 + r_1 s_1 + r_2 s_0, \dots, t_i, \dots)$$

with $t_i = r_0 s_i + r_1 s_{i-1} + \cdots + r_i s_0 = \sum_{j=0}^{i} r_j s_{i-j}$. The additive identity is now $(0,0,0,\dots)$, and x is $(0,1,0,0,\dots)$ if 1 is in R.

Exercises

7 (a) Multiply $1 + x + 2x^3$ by $5 + 4x^3$ in $\mathbb{Q}[x]$.
   (b) Multiply $(1,1,0,2,0,0,0,\dots)$ by $(5,0,0,4,0,0,\dots)$, and show that the expected answer is the one obtained.
8 Using this definition of $R[x]$ by infinite sequences, prove that both associative laws hold.

At this point we revert to the familiar notation for polynomials. There is one point, however, that the definition by infinite sequences does clear up.

Proposition 2 In the ring $R[x]$ we have $\sum_{i=0}^{\infty} r_i x^i = \sum_{i=0}^{\infty} s_i x^i$ only if $r_0 = s_0, r_1 = s_1, r_2 = s_2, r_3 = s_3, \dots$.

PROOF We really want $(r_0, r_1, r_2, \dots) = (s_0, s_1, s_2 \dots)$, and this is true only when $r_0 = s_0, r_1 = s_1, r_2 = s_2, \dots$. ////

Another reasonable way to define equality of polynomials in $R[x]$ is to view polynomials as functions. Over an infinite field this is equivalent to having all the coefficients equal as above. The next exercise, however, shows that the two possible definitions do not always coincide.

Exercise

9 Show that in $\mathbb{Z}_3[x]$ the polynomials $\bar{1}x^3 + \bar{2}x + \bar{2}$ and $\bar{2}$ are equal as functions.

We now want to reexamine factoring, one of the major topics of high-school algebra. After some definitions, we shall examine factoring over a field and then over $\mathbb{Q}$. If $p(x) = a(x)b(x)$ in $R[x]$, we can denote this by $a(x)|p(x)$.

Definition A nonzero polynomial $p(x) = \sum_{i=0}^{\infty} r_i x^i$ in $R[x]$ has **degree** n if $p(x) = r_0 + r_1 x + \cdots + r_n x^n$ and $r_n \neq 0$. A polynomial $p(x)$ in $R[x]$ is **reducible** over R if there are polynomials $a(x)$ and $b(x)$ in $R[x]$, both of lower degree than $p(x)$, such that $p(x) = a(x)b(x)$. If no such polynomials exist, $p(x)$ is **irreducible.**

Exercises

10 If F is a field, show that the units of the polynomial ring $F[x]$ are just the nonzero constant polynomials.

11 Show that $p(x)$ is irreducible in $F[x]$, for F a field, if $p(x) = a(x)b(x)$ in $F[x]$ implies $a(x)$ or $b(x)$ is a unit.

The concept of irreducibility depends heavily on the ring involved. For instance, $x^2 - 3$ is irreducible over $\mathbb{Q}$ but is reducible over $\mathbb{R}$, since $x^2 - 3 = (x - \sqrt{3})(x + \sqrt{3})$ in $\mathbb{R}[x]$.

12 Is $x^4 + 5x^2 + 6$ reducible over $\mathbb{Q}$? Over $\mathbb{R}$? Is $x^4 + 1$ reducible over $\mathbb{Q}$? Over $\mathbb{R}$? Over $\mathbb{C}$?

Let F be any field, and observe that if $b(x)$ is a nonzero polynomial in $F[x]$, and $a(x)$ is any polynomial in $F[x]$, then

$$\frac{a(x)}{b(x)} = c(x) + \frac{r(x)}{b(x)}$$

where the degree of $r(x)$ is less than that of $b(x)$ or $r(x) = 0$.

To prove this we first observe that, if $a(x)$ is of smaller degree than $b(x)$, then we can take $c(x) = 0$ and $r(x) = a(x)$. We may assume

$$a(x) = \sum_{i=1}^{n} a_i(x) \qquad a_n \neq 0$$

$$b(x) = \sum_{i=1}^{m} b_i x^i \qquad b_m \neq 0 \text{ and } n \geq m$$

Now $c_{n-m} = a_n/b_m$, so $a_1(x) = a(x) - c_{n-m}x^{n-m}b(x)$ has smaller degree than $a(x)$.

We now work on $a_1(x)$ in the same manner, and we continue until we have $a(x) - (c_{n-m}x^{n-m} + c_{n-m-1}x^{n-m-1} + \cdots + c_0)b(x) = r(x)$, where $r(x)$ has degree less than $b(x)$ or $r(x) = 0$. This is just our high-school division of polynomials and is called the *division algorithm for polynomials*.

Exercises

13 (a) Find $c(x)$ and $r(x)$ in $\mathbb{Q}[x]$ so that $x^4 + x^3 + 3x^2 + x + 4 = c(x)(x^2 + 1) + r(x)$, where the degree of $r(x)$ is less than the degree of $x^2 + 1$.

   (b) Find $c(x)$ and $r(x)$ in $\mathbb{Z}_5[x]$ so that $\bar{1}x^4 + \bar{1}x^3 + \bar{3}x^2 + \bar{1}x + \bar{4} = c(x)(\bar{1}x^2 + \bar{1}) + r(x)$, where the degree of $r(x)$ is less than the degree of $\bar{1}x^2 + \bar{1}$.

14 Consider $x^2 + x + 1$ and $2x + 3$ in $\mathbb{Z}[x]$, and show no division algorithm exists for $\mathbb{Z}[x]$. Of course, $\mathbb{Z}$ is not a field.

The division algorithm gives us a method for detecting factors of degree 1. We say r_0 in R is a *root* of $p(x)$ in $R[x]$ if $p(r_0) = 0$.

Proposition 3 Let r_0 be a root of $f(x)$ in $F[x]$, where F is a field. Then $x - r_0 | f(x)$. The converse is obvious.

   PROOF   We let $a(x) = f(x)$ and $b(x) = x - r_0$ in the division algorithm. Then $f(x) = c(x)(x - r_0) + r(x)$. Then either $r(x)$ has degree 0 or $r(x)$ actually equals 0. Either way,

$$f(x) = c(x)(x - r_0) + k$$

for some k in F. Substituting r_0 for x yields $k = 0$, and

$$f(x) = c(x)(x - r_0) \qquad ////$$

Corollary A quadratic or cubic polynomial in $F[x]$ is irreducible if and only if it has no roots in F.

   PROOF   If $p(x)$ is the polynomial and $p(x)$ factors properly to $p(x) = a(x)b(x)$, then we may assume $a(x)$ has degree 1 and $b(x)$ has degree 1 or 2. If $a(x) = f_0 + f_1x$, we can multiply by f_1^{-1} and relabel so that $a(x) = x - r_0$. This r_0 will be a root of $p(x)$.

   Conversely, if $p(x)$ is irreducible, it can have no root r_1 since, if it did, $x - r_1$ would be a factor of $p(x)$. $\qquad ////$

WARNING   This does not apply to polynomials higher than cubics. The polynomial $x^4 + x^2 + 1 = p(x)$ in $\mathbb{Q}[x]$ has no roots, since $p(x) \geq 1$ for all x in $\mathbb{Q}$. However,

$$x^4 + x^2 + 1 = (x^2 + x + 1)(x^2 - x + 1)$$

EXAMPLE   (a)   We can show that $\bar{1}x^3 + \bar{1}x + \bar{1} = p(x)$ is irreducible over $\mathbb{Z}_5$, since $p(\bar{0})$, $p(\bar{1})$, $p(\bar{2})$, $p(\bar{3})$, and $p(\bar{4})$ are all nonzero.

(b)   The polynomial $\bar{1}x^3 + \bar{3}$ is reducible over $\mathbb{Z}_5$, since $\bar{3}$ is a root and $\bar{1}x - \bar{3} = \bar{1}x + \bar{2}$ is a factor. In fact,

$$\bar{1} \cdot x^3 + \bar{3} = (\bar{1}x + \bar{2})(\bar{1}x^2 + \bar{3}x + \bar{4})$$

The general question of irreducibility is very difficult, as the warning indicates. Coding theory, a modern branch of applied mathematics, makes extensive use of irreducible polynomials over finite fields. Two references to this material are Levinson [56] and Berlekamp [8].

At this point we shall concentrate on factorization in $\mathbb{Q}$ and $\mathbb{Z}$.

Definition   If $p(x) = \sum_{i=0}^{n} a_i x^i$ is a polynomial in $\mathbb{Z}[x]$, then the **content** of $p(x)$ is the greatest common divisor of the coefficients a_i. We denote the content of $p(x)$ as $C(p(x))$. If $p(x)$ has content 1, then $p(x)$ is **primitive**.

For example, the content of $2x^3 - 6x + 8$ is 2, while $15x^2 + 10x + 6$ is primitive.

Proposition 4 (Gauss)   The product of primitive polynomials is primitive.

PROOF   Let $a(x) = \sum_{i=0}^{n} a_i x^i$ and $b(x) = \sum_{i=0}^{m} b_i x^i$ be primitive polynomials, and assume $a(x)b(x)$ is not primitive. Then there is some prime p such that $p \mid C(a(x)b(x))$. However, p cannot divide all the a_i nor all the b_i, since $a(x)$ and $b(x)$ are primitive. Pick k to be the smallest integer such that $p \nmid a_k$, and let l be the smallest integer such that $p \nmid b_l$. Now we examine the x^{k+l} term in $a(x)b(x)$ and see that it equals

$$\underbrace{[(a_0 b_{k+l} + a_1 b_{k+l-1} + \cdots + a_{k-1}b_{l+1})}_{\text{divisible by } p \text{ since } p \mid a_i} + \underbrace{a_k b_l}_{\substack{\text{not} \\ \text{divisible} \\ \text{by } p}} + \underbrace{(a_{k+1}b_{l-1} + \cdots + a_{k+l}b_0)}_{\text{divisible by } p \text{ since } p \mid b_j}$$

But this whole coefficient must be divisible by p since $p \mid C(a(x)b(x))$, and we have a contradiction.   ////

Corollary 1   If $f(x)$ and $g(x)$ are polynomials in $\mathbb{Z}[x]$, then

$$C(f(x)C(g(x)) = C(f(x)g(x))$$

REMARK   Note that if $f(x)$ has content c it can be written as $f(x) = cf^\circ(x)$, where $f^\circ(x)$ is also in $\mathbb{Z}[x]$.

Exercise

15   Prove Corollary 1.

Proposition If $f[x]$ is in $\mathbb{Z}[x]$ and is irreducible over $\mathbb{Z}$, then it is also irreducible over $\mathbb{Q}$.

PROOF If $f(x)$ were not primitive, we could factor out its content; this would not affect its factorization in $\mathbb{Q}[x]$. Thus we assume $f(x)$ is primitive and $f(x) = a(x)b(x)$ is a proper factorization in $\mathbb{Q}[x]$.

Let c be the least common multiple of the denominators of the coefficients a_i in $a(x) = \sum_{i=0}^{n} a_i x^i$. Let d be the least common multiple of the denominators of the coefficients of $b(x)$. Thus $cdf(x) = ca(x) \cdot db(x)$ is a factorization in $\mathbb{Z}[x]$. If C_a is the content of $ca(x)$ and C_b is the content of $db(x)$, then we have

$$cdf(x) = C_a \tilde{a}(x) C_b \tilde{b}(x)$$

where $\tilde{a}(x)$ and $\tilde{b}(x)$ are primitive and in $\mathbb{Z}[x]$. Since $f(x)$ is primitive, the content of the left side of this equation is cd. Since $\tilde{a}(x)$ and $\tilde{b}(x)$ are primitive, so is $\tilde{a}(x)\tilde{b}(x)$, and the content of the right-hand side is $C_a C_b$. Thus, $cd = C_a C_b$ and

$$f(x) = \tilde{a}(x)\tilde{b}(x)$$

is the desired factorization in $\mathbb{Z}[x]$.   ////

This theorem is of such a nature that an example is necessarily contrived. For instance, $x^2 + 5x + 6 = (\frac{1}{3}x + 1)(3x + 6)$ is quickly adjusted back to a factorization in $\mathbb{Z}[x]$.

Our next proposition gives us a very handy test for irreducibility. Unfortunately, it is a sufficient but far from necessary condition.

Proposition (Eisenstein's criterion) If $a(x) = a_0 + a_1 x + a_2 x + \cdots + a_n x^n$ is a polynomial in $\mathbb{Z}[x]$, and p is a prime such that $p \nmid a_n, p \mid a_0, a_1, \ldots, a_{n-1}$, and $p^2 \nmid a_0$, then $a(x)$ is irreducible in $\mathbb{Q}[x]$.

PROOF By the last proposition it suffices to show $a(x)$ is irreducible in $\mathbb{Z}[x]$. Assume $a(x) = b(x)c(x)$ is a proper factorization of $a(x)$ in $\mathbb{Z}[x]$.

Let $b(x) = \sum_{i=0}^{l} b_i x^i$, and $c(x) = \sum_{i=0}^{m} c_i x^i$. The prime p cannot divide either b_l or c_m since $p \nmid a_n = b_l c_m$. Also, p divides exactly one of b_0 and c_0, since $p^2 \nmid a_0 = b_0 c_0$.

Assume $p \mid b_0$ and that k is the integer such that $p \mid b_0, p \mid b_1, \ldots, p \mid b_{k-1}, p \nmid b_k$. But

$$p \mid a_k = \underbrace{b_0 c_k + b_1 c_{k-1} + b_2 c_{k-2} + \cdots}_{\text{divisible by } p \text{ since } p \mid b_i} + \underbrace{b_k c_0}_{\substack{\text{not divisible} \\ \text{by } p}}$$

This is the contradiction, and we cannot factor $a(x)$. Observe that we could reverse the conditions on a_0 and a_n; that is, we could specify that

$p \nmid a_0$, $p \mid a_n$, $p^2 \nmid a_n$, and $p \mid a_i$ for $i = 1,2,...,n-1$. The proof would be essentially the same. ////

Exercises

16 Which of the following polynomials in $\mathbb{Z}[x]$ are irreducible?
   (a) $x^n + p$, p a prime   (b) $x^3 + 3x^2 + 81x + 6$
   (c) $2x^4 + 4x^3 + 12x^2 + 16x + 3$ (d) $x^2 + 5x + 7$

17 (a) If $f(x)$ is a polynomial and $g(x) = f(x+1)$, show that $f(x)$ is irreducible if and only if $g(x)$ is irreducible.

   (b) Show that if p is a prime then $p \mid \binom{p}{k}$ for $k = 1,2,...,p-1$.

   (c) Show that $1 + x + x^2 + x^3$ is reducible over $\mathbb{Q}$.

Using Exercise 17, we can prove the following result: Let $\Phi_p(x) = 1 + x + x^2 + \cdots + x^{p-1}$, where p is prime. We want to show that $\Phi_p(x)$ is irreducible over $\mathbb{Q}$. $\Phi_p(x)$ is irreducible if

$$\Phi_p(x+1) = 1 + (x+1) + (x+1)^2 + \cdots + (x+1)^{p-1}$$
$$= [(x+1)^p - 1]/[(x+1) - 1]$$

is. But

$$\Phi_p(x+1) = \frac{1}{x}\left[x^p + \binom{p}{1}x^{p-1} + \binom{p}{2}x^{p-2} + \cdots + \binom{p}{p-1}x \right]$$
$$= x^{p-1} + \binom{p}{1}x^{p-2} + \cdots + \binom{p}{p-1}$$

Now we apply Eisenstein's criterion, which is legitimate since $\binom{p}{p-1} = p$.
Thus $\Phi_p(x+1)$ and, in turn, $\Phi_p(x)$ are irreducible.

Later on we can generalize these last propositions to any unique factorization domain. The proofs for the more general case are much the same as here. We can also give shorter proofs utilizing a homomorphism from $\mathbb{Z}[x]$ to $\mathbb{Z}_p[x]$. Here we have tried to give an uncluttered presentation, and the refinements wait in the exercises at the end of Chap. 6.

Exercise

18 Let $\mathbb{Q}_{2Z}[x]$ be all polynomials of the form $a_0 + a_2 x^2 + a_4 x^4 + \cdots + a_{2k} x^{2k}$, where the a_i are all in $\mathbb{Q}$. Is $\mathbb{Q}_{2Z}[x]$ a subring of $\mathbb{Q}[x]$? Is it an ideal? What about $\mathbb{Q}_{0,3,4,...}[x] = \{a_0 + a_3 x^3 + a_4 x^4 + \cdots \mid a_i \in \mathbb{Q}\}$? How can these examples be generalized?

3

VECTOR SPACES

3.1 BASIC DEFINITIONS

In order to go further in modern algebra it is necessary to learn a few basic concepts of linear algebra. The pattern here is similar to that followed with groups and rings. We define vector spaces, subspaces, and vector space homomorphisms, and prove a homomorphism theorem. The new concepts are those of basis and dimension, and the result that each homomorphism is expressible as a matrix.

Since this material should be covered at the beginning of any linear algebra course, we shall quote some theorems instead of proving them. We shall concentrate on seeing how the theorems apply and, in particular, how they apply to modern algebra. We shall also be much freer in skipping routine verifications, as the general pattern for these should indeed be routine by now.

Definition A **vector space** $(V, +, \cdot, F)$ over a field F is a set V together with two operations $+$ and $\cdot$ and a field F such that

(a) $(V, +)$ is an abelian group.

(b) $\alpha \cdot v \in V$ for all $\alpha \in F$, $v \in V$ (scalar multiplication).

(c) $(\alpha_1 + \alpha_2) \cdot v = \alpha_1 \cdot v + \alpha_2 \cdot v$ for all $\alpha_1, \alpha_2 \in F$, and $v \in V$.

(d) $\alpha \cdot (v_1 + v_2) = \alpha \cdot v_1 + \alpha \cdot v_2$ for all $\alpha \in F$, and $v_1, v_2 \in V$.

(e) If 1 is the multiplicative identity for F, then $1 \cdot v = v$ for all $v \in V$.

(f) $\alpha\beta(v) = \alpha(\beta v)$ for all $\alpha, \beta \in F$, and $v \in V$.

Several things should be noted. One, we do not require that elements of V can be multiplied together. Two, as with groups and rings, we shall usually shorten $\alpha \cdot v$ to αv. Three, this $\cdot$ is a function from $F \times V$ to V. Axioms c and d are just distributive laws connecting $+$ and $\cdot$. This multiplication is often called **scalar multiplication** or **multiplication by a scalar**, especially when some other multiplication is around. Elements of F are often called **scalars**.

EXAMPLE A Let $\mathbb{R}^2$ be all ordered pairs (r_1, r_2) with r_1 and r_2 in $\mathbb{R}$. Then $(\mathbb{R}^2, +, \cdot, \mathbb{R})$ is a vector space if we define $+$ and $\cdot$ as follows:

$$(r_1, r_2) + (r_3, r_4) = (r_1 + r_3, r_2 + r_4)$$

$$r \cdot (r_1, r_2) = (rr_1, rr_2)$$

We saw in Chap. 1 that $(\mathbb{R}^2, +)$ is an abelian group. The rest of the vector-space axioms are verified as follows:

$$\begin{aligned}
(r_1 + r_2)(r,s) &= ((r_1 + r_2)r, (r_1 + r_2)s) = (r_1 r + r_2 r, r_1 s + r_2 s) \\
&= (r_1 r, r_1 s) + (r_2 r, r_2 s) \\
&= r_1(r,s) + r_2(r,s) \qquad \forall r_1, r_2 \text{ in } \mathbb{R} \text{ and } (r,s) \text{ in } \mathbb{R}^2
\end{aligned}$$

$$\begin{aligned}
r_1[(r,s) + (t,u)] &= r_1(r + t, s + u) = (r_1(r + t), r_1(s + u)) \\
&= (r_1 r + r_1 t, r_1 s + r_1 u) = (r_1 r, r_1 s) + (r_1 t, r_1 u) \\
&= r_1(r,s) + r_1(t,u) \qquad \forall r_1 \in \mathbb{R} \text{ and } (r,s), (t,u) \in \mathbb{R}^2
\end{aligned}$$

$$1(r,s) = (1 \cdot r, 1 \cdot s) = (r,s)$$

$$r_1 r_2(r,s) = (r_1 r_2 r, r_1 r_2 s) = r_1(r_2 r, r_2 s) = r_1[r_2(r,s)]$$

We have at various times given various algebraic structures to the set $\mathbb{R}^2$. It is a group under $+$. Depending on how we define multiplication, it can be several rings, $(\mathbb{R} \times \mathbb{R}, +, \cdot)$ and $\mathbb{C}$ being two of the possibilities. Now, with this new scalar multiplication, it can also be a vector space.

EXAMPLE B We can generalize Example 1 in two directions without really making the verifications harder. We can use any field F instead of $\mathbb{R}$, and use ordered n-tuples instead of ordered pairs.

Thus F^n is the set of ordered n-tuples of elements of F. Let $+$ and $\cdot$ be defined by

$$(f_1, f_2, \ldots, f_n) + (g_1, g_2, \ldots, g_n) = (f_1 + g_1, f_2 + g_2, \ldots, f_n + g_n)$$

and

$$f \cdot (f_1, f_2, \ldots, f_n) = (ff_1, ff_2, \ldots, ff_n)$$

With these definitions, $(F^n, +, \cdot, F)$ is a vector space over F. Since the verifications are so close to that of Example A, we shall only verify Axiom c here

$$(f + g)(f_1, f_2, \ldots, f_n) = ((f + g)f_1, (f + g)f_2, \ldots, (f + g)f_n)$$
$$= (ff_1 + gf_1, ff_2 + gf_2, \ldots, ff_n + gf_n)$$
$$= (ff_1, ff_2, \ldots, ff_n) + (gf_1, gf_2, \ldots, gf_n)$$
$$= f(f_1, f_2, \ldots, f_n) + g(f_1, f_2, \ldots, f_n)$$

Once we define dimension and vector-space isomorphism, this example will become very important, as all other finite-dimensional vector spaces turn out to be isomorphic to one of this form.

EXAMPLE C $C[0,1]$, the continuous real-valued functions on the interval $0 \le x \le 1$, has already been shown to be an abelian group under addition and a commutative ring with 1. We shall use the fact that $(C[0,1], +)$ is an abelian group, but for the moment do not need to know anything about rings. Define $\cdot$ by making $r \cdot f$ the function taking x to $rf(x)$ [that is, $r \cdot f(x) = rf(x)$, which is so simple as to be confusing]. If f is continuous so is rf, and Axiom b is satisfied. Since $(r + s)f(x) = rf(x) + sf(x)$ $\forall x \in [0,1]$, we have $(r + s)f = rf + sf$ $\forall r, s \in \mathbb{R}, f \in C[0,1]$. Similarly,

$$r[f_1(x) + f_2(x)] = rf_1(x) + rf_2(x) \qquad \forall x \in [0,1]$$

and thus $\qquad r(f_1 + f_2) = rf_1 + rf_2 \qquad \forall r \in \mathbb{R}$ and $f_1, f_2 \in C[0,1]$

Also, $1 \cdot f(x) = f(x)$ $\forall x \in [0,1]$, so $1f = f$, and $(C[0,1], +, \cdot, \mathbb{R})$ is a vector space over $\mathbb{R}$.

EXAMPLE D   Let $P_2(x)$ be all polynomials with degree less than or equal to 2 and coefficients from a field F. If we define addition as usual and scalar multiplication by $d(ax^2 + bx + c) = dax^2 + dbx + dc$, where $d, a, b, c \in F$, then, indeed, we have a vector space over F. Here we do not have a ring, because multiplication of these polynomials may lead to a cubic or quartic polynomial.

EXAMPLE E   Let F be a subfield of K (a well-known case being $\mathbb{R}$ as a subfield of $\mathbb{C}$). Then K forms a vector space over F. To check this we note that $(K, +)$ is a group; $fk = k' \in K$ $\forall f \in F$, $k \in K$. $(f_1 + f_2)k = f_1 k + f_2 k$ and $f(k_1 + k_2) = fk_1 + fk_2$ $\forall f, f_1, f_2 \in F$ since $k, k_1, k_2 \in K$ and the distributive laws hold in K. $1k = k$ is obvious. We have needed less than the total amount of information given by K being a field. We know that $k_1 \cdot k_2 \in K$, but we only need that $f \cdot k_2 \in K$. However, we can already see that vector spaces are potentially very useful in studying fields.

EXAMPLE F $M_2(F)$ is a vector space over F. Here we define $+$ and $\cdot$ as follows:

$$\begin{pmatrix} a_1 & a_2 \\ a_3 & a_4 \end{pmatrix} + \begin{pmatrix} b_1 & b_2 \\ b_3 & b_4 \end{pmatrix} = \begin{pmatrix} a_1 + b_1 & a_2 + b_2 \\ a_3 + b_3 & a_4 + b_4 \end{pmatrix}$$

and

$$f \cdot \begin{pmatrix} a_1 & a_2 \\ a_3 & a_4 \end{pmatrix} = \begin{pmatrix} fa_1 & fa_2 \\ fa_3 & fa_4 \end{pmatrix}$$

We have seen that $(M_2,(F),+)$ is an abelian group with $\begin{pmatrix} 0 & 0 \\ 0 & 0 \end{pmatrix}$ as its identity. We again shall verify one of the distributive laws, just to show that nothing extraordinary occurs. Axiom c goes as follows:

$$(f_1 + f_2)\begin{pmatrix} a_1 & a_2 \\ a_3 & a_4 \end{pmatrix} = \begin{pmatrix} (f_1 + f_2)a_1 & (f_1 + f_2)a_2 \\ (f_1 + f_2)a_3 & (f_1 + f_2)a_4 \end{pmatrix}$$

$$= \begin{pmatrix} f_1 a_1 + f_2 a_1 & f_1 a_2 + f_2 a_2 \\ f_1 a_3 + f_2 a_3 & f_1 a_4 + f_2 a_4 \end{pmatrix}$$

$$= \begin{pmatrix} f_1 a_1 & f_1 a_2 \\ f_1 a_3 & f_1 a_4 \end{pmatrix} + \begin{pmatrix} f_2 a_1 & f_2 a_2 \\ f_2 a_3 & f_2 a_4 \end{pmatrix}$$

$$= f_1 \begin{pmatrix} a_1 & a_2 \\ a_3 & a_4 \end{pmatrix} + f_2 \begin{pmatrix} a_1 & a_2 \\ a_3 & a_4 \end{pmatrix}$$

Note that

$$f\begin{pmatrix} a & b \\ c & d \end{pmatrix} = \begin{pmatrix} fa & fb \\ fc & fd \end{pmatrix} = \begin{pmatrix} f & 0 \\ 0 & f \end{pmatrix}\begin{pmatrix} a & b \\ c & d \end{pmatrix}$$

so that multiplying by $\begin{pmatrix} f & 0 \\ 0 & f \end{pmatrix}$ is the same as scalar multiplication. As a result, matrices of the form $\begin{pmatrix} f & 0 \\ 0 & f \end{pmatrix}$ are called *scalar matrices*.

Definition Any element v in a vector space is called a **vector.**

The reader who has taken calculus may remember vectors as "directed line segments" or as arrows of uncertain origin. These other definitions are derived from definitions of a particular vector space and are fine as long as one doesn't mistake them for the general case. The general pattern is to use $\mathbb{R}^2$ in calculus and $\mathbb{R}^n$ in advanced calculus. The reason most calculus books settle for an example is that the general definition of vector spaces has 22 parts (11 for a field, 5 for an abelian group, 6 for vector spaces themselves) and would represent a considerable side excursion. It is very difficult to define a vector without defining a vector space, and the reader may find it instructive (or entertaining) to see how various calculus texts struggle with this task.

Following the pattern set for groups, rings, and fields, we now define a subspace.

Definition   If $(V,+,\cdot,F)$ is a vector space over F, and S is a subset of V, then S is a **subspace** if $(S,+,\cdot,F)$ is itself a vector space over F (again using the $+$ and $\cdot$ used for V).

As before, all distributive, associative, and commutative laws will still hold when we restrict our interest to a subset of V. This implies that we need only check additive closure and inverses and closure under scalar multiplication. One economical way to do this is as follows:

Proposition 1   Let S be a nonempty subset of V, where $(V,+,\cdot,F)$ is a vector space. S is a subspace if and only if

$$s_1 \pm s_2 \in S \qquad \forall s_1, s_2 \in S$$

$$f \cdot s \in S \qquad \forall f \in F \text{ and } s \in S$$

PROOF   First, what does $s_1 - s_2$ mean? Since $(V,+)$ is a group under addition, $-s$ is the inverse of s, and $s_1 - s_2$ is a short way of writing $s_1 + (-s_2)$.

Taking the $+$ in $\pm$, we see that S is closed. Since S is nonempty, some s is in S. Thus $s - s = 0$ is in S, and then we have $0 - s = -s$ in S. Thus $(S,+)$ has 0 and inverses and is closed, and so is an abelian group. $f \cdot s$ is in S by hypothesis, and $1s = s$ since this holds in V. Hence, $(S,+,\cdot,F)$ is a subspace.   ////

In fact, we could even weaken $s_1 \pm s_2 \in S$ to $s_1 + s_2 \in S$ by letting $f = -1$ when we want inverses.

Using Proposition 1 as a criterion, we now look at several examples of subspaces and near-subspaces. Let V be the vector space $(\mathbb{R}^2,+,\cdot,\mathbb{R})$.

EXAMPLE G'   Let $S = \{(x,y) \mid x + y = 1\}$. Is S a subspace of $\mathbb{R}^2$? We can picture S as the line $y + x = 1$ in the cartesian plane. However, both $(3,-2)$ and $(5,-4)$ are in S. But $(3,-2) + (5,-4) = (8,-6) \notin S$. Also, $(3,-2) - (5,-4) \notin S$ and $-(3,-2) \notin S$, so this S is certainly not a subspace. An even quicker observation is that $(0,0)$ is not in S, so S cannot be a subspace.

EXAMPLE G"   Let $S = \{(x,y) \mid x + y = 0\}$. Is this S a subspace? Trying out some definite numbers as in Example G' does not lead to a contradiction, so we use Proposition 1. Clearly S is nonempty. If (x_1,y_1) and (x_2,y_2) are in S, then $x_1 + y_1 = 0 = x_2 + y_2$. We look at $(x_1,y_1) \pm (x_2,y_2) = (x_1 \pm x_2, \ y_1 \pm y_2)$. Since $x_1 \pm x_2 + y_1 \pm y_2 = x_1 + y_1 \pm (x_2 + y_2) = 0 \pm 0 = 0$, we have $(x_1,y_1) \pm (x_2,y_2) \in S$.

Similarly, $r(x_1,y_1) = (rx_1,ry_1)$ and $rx_1 + ry_1 = r(x_1 + y_1) = r \cdot 0 = 0$ (we shall prove $r \cdot 0 = 0$ early in the next section), so $r(x_1,y_1)$ is in S. Thus, the proposition applies, and S is a subspace.

†EXAMPLE H   Let $V = (C[0,1], +, \cdot, \mathbb{R})$ from Example C. Let S be all twice-differentiable functions in V that satisfy $d^2 y/dx^2 + y = 0$. (For instance, it is easy to see that $\cos x$ and $\sin x$ are in S.) We want to see if S is a subspace of V.

If $f(x)$ and $g(x)$ are in S, then we first want to look at

$$\frac{d^2}{dx^2}[f(x) \pm g(x)] + [f(x) \pm g(x)] = \frac{d^2}{dx^2}f(x) + f(x) \pm \frac{d^2}{dx^2}g(x) + g(x)$$

$$= 0 \pm 0 \qquad \text{since } f(x), g(x) \text{ are in } S$$

$$= 0$$

Also, $$\frac{d^2}{dx^2}[rf(x)] + rf(x) = r\left[\frac{d^2}{dx^2}f(x) + f(x)\right]$$

$$= r \cdot 0 = 0 \qquad \text{for any } r \text{ in } \mathbb{R}$$

Thus the proposition holds, and the solutions of $d^2 y/dx^2 + y = 0$ form a subspace of $C[0,1]$. (We have again used that $r \cdot 0 = 0$, which is yet to be proved.)

Exercises

1   Complete the verification that $P_2(x)$ in Example D is a vector space.
2   †Show that $D[0,1]$ is a vector space over $\mathbb{R}$, where $D[0,1]$ is the set of all functions differentiable everywhere on the interval $0 \leq x \leq 1$. [*Hint:* Define $+$ first; then show $(D[0,1], +)$ is an abelian group. You can use any theorems from calculus.]
3   †Show that functions satisfying $y'' + 5y' + 6y = 0$ form a vector space over $\mathbb{R}$. Show that this is a subspace of $D[0,1]$.
4   Which of the following are subspaces of $\mathbb{R}^3$?
   (a) $S = \{(a,b,c) \mid a + b + c = 0\}$ (b) $S = \{(a,b,c) \mid a + b + c = 2\}$
   (c) $S = \{(a,0,c)\}$ (d) $S = \{(a,b,c) \mid a^2 + b^2 = c^2\}$
5   Let $V = \{(a_1, a_2, a_3, \ldots) \mid a_i \in \mathbb{R}\}$ be all infinite sequences of real numbers. Show that V is a vector space if $+$ and $\cdot$ are chosen appropriately. (V could be called $\mathbb{R}^\infty$.)
6   Which of the following are subspaces of V in Exercise 5?
   (a) $S = \{(0,b,0,d,0,f,0,\ldots)\}$
   (b) $S = \{(a_1, a_2, a_3, a_4, \ldots) \mid a_i \leq 1 \; \forall i\}$
7   Let $F = \{(a_1, a_2, a_3, \ldots) \mid a_i = a_{i-1} + a_{i-2}, a_1, a_2 \text{ arbitrary}, a_i \in \mathbb{R}\}$. Show that F is a subspace of V in Exercise 5. This is called the *Fibonacci subspace* since, if we pick $a_1 = a_2 = 1$, then we get the sequence

$$(1,1,2,3,5,8,13,21,34,\ldots),$$

which is the famous Fibonacci sequence.

8   Let $(V,+,\cdot,F)$ satisfy all the axioms except that $(V,+)$ is not necessarily abelian. Show that the other axioms imply that $(V,+)$ must indeed be abelian.

9   Let $(V,+,\cdot,R)$ satisfy all the axioms for a vector space except that we only require that R be a ring with 1 instead of being a field. Then V is called a **left R-module**. If $(V,+,\cdot,\mathbb{Z})$ is a left $\mathbb{Z}$-module, $(V,+)$ is an abelian group, by definition. Show conversely that any abelian group $(A,+)$ can be turned into a left $\mathbb{Z}$-module by defining $\cdot$ correctly.

10   A nonempty set S is a subspace of $(V,+,\cdot,F)$ if $v_1,v_2 \in S \Rightarrow \alpha_1 v_1 + \alpha_2 v_2 \in S$ $\forall v_1,v_2 \in V$ and $\alpha_1,\alpha_2 \in F$. Prove that this definition is equivalent to the first definition by showing that the axioms of either definition can be used to prove the axioms of the other as theorems.

11   If S is any set and V is any vector space over F, look at Hom (S,V), the set of all functions from S to V. If g and h are in Hom (S,V), that is, are functions from S to V, then what is a good definition for $g + h$? What is a good definition for $\alpha \cdot f$? If you have made good enough choices, you can now show that (Hom $(S,V),+,\cdot,F)$ is also a vector space over F.

3.2   BASES AND DIMENSION

In this section we establish some elementary but useful properties of vector spaces. To avoid confusion, we let 0_F and 1_F be the additive and multiplicative identities of the underlying field F. Letting 0_V be the additive identity of $(V,+)$, we have the following proposition:

> **Proposition 1**   If $(V,+,\cdot,F)$ is a vector space over the field F, then the following hold:
> (a) $0_F \cdot v = 0_V$ for all v in V.
> (b) $\alpha \cdot 0_V = 0_V$ for all α in F.
> (c) $(-\alpha) \cdot v = -(\alpha \cdot v)$.
>
> PROOF (a)
> $$(1_F + 0_F) \cdot v = 1_F \cdot v = v = v + 0_V$$
> $$1_F \cdot v + 0_F \cdot v = v + 0_F \cdot v$$

Canceling v from the left yields $0_F \cdot v = 0_V$.   ////

Exercises

1   Prove parts b and c of Proposition 1.
2   Using the above notation, prove that $-1_F \cdot v = -v$.

3   We know that $(V, +)$ is a commutative group. Exploit this fact to show that the following statements are true by quoting the appropriate propositions from Chap. 1:

(a) $v + w = v$ implies that $w = 0_V$.

(b) $(V, +)$ has a unique additive identity 0_V.

(c) v has a unique inverse $-v$.

(d) If $v + v = v$ then $v = 0_V$.

Definition   If S is a nonempty subset of a vector space V, then $\langle S \rangle = \{\alpha_1 v_1 + \alpha_2 v_2 + \cdots + \alpha_k v_k \,|\, \alpha_i$ in F, v_i in $S\}$ is called the **span** of S, or the **subspace** of V **generated** by S.

As an example in the vector space $\mathbb{R}^3$, the span of $S = \{(0,1,0), (1,1,0)\}$ is the xy plane. We see this by noting that

$$\begin{aligned}
\langle S \rangle &= \{\alpha(0,1,0) + \beta(1,1,0) \,|\, \alpha, \beta \text{ in } \mathbb{R}\} \\
&= \{(0,\alpha,0) + (\beta,\beta,0) \,|\, \alpha, \beta \text{ in } \mathbb{R}\} \\
&= \{(\beta, \alpha + \beta, 0) \,|\, \alpha, \beta \text{ in } \mathbb{R}\} = \{(c,d,0) \,|\, c,d \text{ in } \mathbb{R}\}
\end{aligned}$$

which is the xy plane in $\mathbb{R}^3$. Note that this is a plane containing the origin.

Exercises

4   Show that in $\mathbb{R}^3$ the span of $\{(1,1,1)\}$ is a line containing the origin. Show that the span of $\{(2,2,2),(\pi,\pi,\pi),(7,7,7)\}$ is the same line.

5   Find the span of $S = \{1, x^2, x^4, x^6, \ldots\}$ in $P(F)$, where $P(F)$ is the vector space of all polynomials over a field F.

Proposition 2   If S is a nonempty subset of a vector space V, then $\langle S \rangle$ is a subspace of V.

PROOF   If $s' = \alpha_1 v_1 + \alpha_2 v_2 + \cdots + \alpha_k v_k$ is a typical element of $\langle S \rangle$, and each v_i is in S, then

$$\begin{aligned}
\beta s' &= \beta(\alpha_1 v_1 + \alpha_2 v_2 + \cdots + \alpha_k v_k) \\
&= (\beta\alpha_1)v_1 + (\beta\alpha_2)v_2 + \cdots + (\beta\alpha_k)v_k
\end{aligned}$$

which is also in S.

If $s' = \alpha_1 v_1 + \alpha_2 v_2 + \cdots + \alpha_k v_k$ and $s'' = \beta_1 w_1 + \beta_2 w_2 + \cdots + \beta_l w_l$ with all the v_i and w_i in S, then

$$s' + s'' = \alpha_1 v_1 + \alpha_2 v_2 + \cdots + \alpha_k v_k + \beta_1 w_1 + \beta_2 w_2 + \cdots + \beta_l w_l$$

Some of these v_i and w_j may coincide, but this is not prohibited. We need only write $s' + s''$ as a finite linear sum of elements of S, which we have done. Using Proposition 1 of the last section, we have proved this proposition.   ////

Another way of looking at $\langle S \rangle$ is provided by the next proposition.

Proposition 3 If S is a nonempty subset of V, then $\langle S \rangle$ is the intersection of all subspaces containing S.

PROOF   We want to show $\langle S \rangle = \bigcap_\alpha W_\alpha$, where $\{W_\alpha\}$ is the collection of all subspaces containing S. First note that $\langle S \rangle$ certainly contains S, since $s = 1 \cdot s \in \langle S \rangle$ for all s in S. Thus $\langle S \rangle \supseteq \bigcap_\alpha W_\alpha$, since $\langle S \rangle$ must be one of the W_α. Conversely, if W_β is some subspace containing S it must, as a subspace, contain all αs, where $\alpha \in F$, $s \in S$. It must also contain all finite sums $\alpha_1 s_1 + \alpha_2 s_2 + \cdots + \alpha_k s_k$, but this means $\langle S \rangle \subseteq W_\beta$, and, as W_β is an arbitrary subspace in $\{W_\alpha\}$, $\langle S \rangle \subseteq \bigcap_\alpha W_\alpha$.   ////

A concept complementary to that of span is that of linear independence. Linear independence concerns uniqueness of representation, while spanning concerns the possibility of representation.

Definition   Let S be a subset of a vector space V. S is **linearly independent** if no element of S can be represented as a linear sum of other elements in S. Thus, if we can find α_i such that $s = \alpha_1 s_1 + \alpha_2 s_2 + \cdots + \alpha_k s_k$, where $s, s_1, s_2, \ldots, s$ are all in S and $s \neq s_i$, then S is *not* linearly independent. In this case S is called **linearly dependent.**

Again taking $\mathbb{R}^3$ as an example, we see that $S = \{(1,0,1), (0,1,0), (2,0,0)\}$ is linearly independent. For instance, if

$$(1,0,1) = \alpha(0,1,0) + \beta(2,0,0) = (2\beta, \alpha, 0)$$

we would find that $1 = 0$.

Similarly, $(0,1,0) = \alpha(1,0,1) + \beta(2,0,0)$ leads to a quick contradiction, as does $(2,0,0) = \alpha(1,0,1) + \beta(0,1,0)$. However, $T = \{(1,1,1), (0,1,2), (2,1,0)\}$ is linearly dependent, since $(2,1,0) = 2(1,1,1) + (-1)(0,1,2)$.

Exercise

6   If $S = \{1 + x^2, x + x^3, 3 + x\}$, is S linearly independent in $P(\mathbb{R})$?

Our next proposition puts the linear-independence condition into a more symmetric form which is often used as the definition.

Proposition 4   A set S is linearly independent if, whenever $s_1, s_2, \ldots, s_k$ are distinct elements of S and

$$0_V = \alpha_1 s_1 + \alpha_2 s_2 + \cdots + \alpha_k s_k \tag{*}$$

then $0 = \alpha_1 = \alpha_2 = \cdots = \alpha_k$. (The point here is *not* that 0_V can be represented, but conversely that it can be represented only one way, the obvious way.)

PROOF   Let $0_V = \alpha_1 s_1 + \alpha_2 s_2 + \cdots + \alpha_k s_k$, where not all the α_i are 0. For convenience assume $\alpha_1 \neq 0$. Then $-\alpha_1 s_1 = \alpha_2 s_2 + \cdots + \alpha_k s_k$, and

$$s_1 = \frac{-\alpha_2}{\alpha_1} s_2 + \cdots + \frac{-\alpha_k}{\alpha_1} s_k$$

which contradicts linear independence. Thus the condition (*) is necessary. Since $s = \alpha_1 s_1 + \alpha_2 s_2 + \cdots + \alpha_k s_k$ can be written as $0 = \alpha_1 s_1 + \alpha_2 s_2 + \cdots + \alpha_k s_k + (-1)s$, we see that condition (*) is also sufficient for linear independence.   ////

We now use this proposition to show that no three vectors in $\mathbb{R}^2$ are linearly independent. Look at $\{(x_1,y_1), (x_2,y_2), (x_3,y_3)\}$.

If $\alpha(x_1,y_1) + \beta(x_2,y_2) + \gamma(x_3,y_3) = (0,0)$ we have $\alpha x_1 + \beta x_2 + \gamma x_3 = 0$ and $\alpha y_1 + \beta y_2 + \gamma y_3 = 0$. We have three unknowns α, β, γ and two linear equations, which can always be solved. [If γ is fixed, we just have two lines which either are parallel or cross. Fix γ (or α or β) to guarantee crossing.]

Exercises

7   In this exercise we want to rephrase linear independence in terms of homomorphisms. Let V be a vector space over a field F, and let $S = \{s_1,s_2,\dots,s_k\}$ be the set we are interested in. We know that $F^{(k)}$ is also a vector space over F. We can define $\theta: F^{(k)} \to V$ by $\theta(\alpha_1,\alpha_2,\dots,\alpha_k) = \alpha_1 s_1 + \alpha_2 s_2 + \cdots + \alpha_k s_k$. Show that θ is one-to-one if and only if S is linearly independent. [Check Exercise 23 of Sec. 1.13, and remember that $(F^{(k)},+)$ and $(V,+)$ are groups.]

8   In $P(F)$, show that $S = \{1,x^2,x^4,\dots\}$ is a linearly independent set.

9   If S is a linear independent set and T is a subset of S, is T itself linearly independent?

The next definition allows us to reintroduce finiteness to the world of vector spaces.

Definition   If V is a vector space and S is a nonempty linearly independent subset of V spanning V, then S is a **basis** of V. The number of elements of a basis S is the **dimension** of V.

FACT A   Every vector space has a basis.

FACT B   All the bases of a vector space V have the same cardinality, so that dimension is a well-defined concept.

Proofs can be found in Stoll and Wong [83], Goldhaber and Ehrlich [30], and Jacobson, vol. II [42]. Proofs for the finite-dimensional case can be found in Herstein [38], Halmos [33], Brinkmann and Klotz [13], and many linear-algebra texts.

EXAMPLE $\mathbb{C}$ is a vector space over $\mathbb{R}$, and $S = \{1,i\}$ is a basis. Thus $\mathbb{C}$ has dimension 2 over $\mathbb{R}$. To see that S is a basis, we check first that S spans $\mathbb{C}$. Since any element of $\mathbb{C}$ can be written as $a \cdot 1 + b \cdot i$, where a and b are in $\mathbb{R}$, S spans $\mathbb{C}$. If $a \cdot 1 + b \cdot i = 0$, then $a = b = 0$, so S is linearly independent. $S' = \{i + 1, i - 1\}$ and $S'' = \{3, 7i + 12\}$ are also bases, although not as convenient as S.

Exercises

10   Show that $S = \{1,\sqrt{2}\}$ is a basis for the field $F = \{a + b\sqrt{2} \,|\, a,b \in \mathbb{Q}\}$ as a vector space over $\mathbb{Q}$. What is the dimension of F over $\mathbb{Q}$?

11   Show that $S = \left\{ \begin{pmatrix} 1 & 0 \\ 0 & 0 \end{pmatrix}, \begin{pmatrix} 0 & 1 \\ 0 & 0 \end{pmatrix}, \begin{pmatrix} 0 & 0 \\ 1 & 0 \end{pmatrix}, \begin{pmatrix} 0 & 0 \\ 0 & 1 \end{pmatrix} \right\}$ is a basis for $M_2(\mathbb{R})$ over $\mathbb{R}$. What is another basis? What is the dimension of $M_2(\mathbb{R})$ over $\mathbb{R}$? (Of $M_n(\mathbb{R})$ over $\mathbb{R}$?)

Bases are not always finite. For instance, $S = \{1,x,x^2,x^3,...\}$ is a basis for $P(\mathbb{R})$. Since any polynomial over F can be written as $\alpha_0 \cdot 1 + \alpha_1 \cdot x + \cdots + \alpha_n \cdot x^n$, S spans $P(\mathbb{R})$. But $0 = \alpha_0 \cdot 1 + \alpha_1 \cdot x + \cdots + \alpha_n \cdot x^n$ implies $0 = \alpha_0 = \alpha_1 = \cdots = \alpha_n$, so the set S is linearly independent and thus a basis. This verification is typical in that S is infinite, but we only need work with a finite number of elements of S at a time.

12   Show that $\{(1,0,0), (0,1,0), (0,0,1)\}$ is a basis for $\mathbb{R}^3$. This gives us the comforting fact that $\mathbb{R}^3$ has dimension 3. How can this be generalized?

At this point we have developed a sufficient amount of vector-space theory to study field extensions and Galois theory. The remaining sections are included for the convenience of anyone wishing a review of the first part of a linear-albegra course. Also, this remaining material is necessary for an understanding of Chap. 9 on group representations and for certain of the exercises throughout the book.

13   If S is a linearly independent set, and if $v \in \langle S \rangle$, show that v can be represented in only one way as a linear combination of elements of S with non-zero coefficients. (Assume $v = \alpha_1 s + \alpha_2 s_2 + \cdots + \alpha_k s_k = \beta_1 s'_1 + \beta_2 s'_2 + \cdots + \beta_m s'_m$ with all s_i and s'_j in S and all $\alpha_i, \beta_j \in F - \{0\}$.)

14 Find the dimension of the Fibonacci subspace of Exercise 7 of the last section. These are the sequences $\{(a_1, a_2, a_3, \ldots) \mid a_{i+2} = a_i + a_{i+1}, i \geq 1\}$.

3.3 LINEAR TRANSFORMATIONS

In this section we discuss vector-space homomorphisms, which are usually called *linear transformations*. After that we discuss isomorphism, classify finite-dimensional vector spaces up to isomorphism, prove a correspondence theorem, and prove a fundamental theorem for vector-space homomorphisms. There are two differences from the analogous situation for rings. First, every subspace is the kernel of a homomorphism. Thus we do not need a special class like ideals or normal subgroups. Second, we obtain a complete classification theorem, which certainly we do not have for groups or rings.

Definition Let V and W both be vector spaces over a field F. A function T from V to W is a **linear transformation** (or **vector-space homomorphism** or **transformation**) if

LT1 $T(v_1 + v_2) = T(v_1) + T(v_2)$ for all v_1, v_2 in V.
LT2 $T(\alpha v) = \alpha T(v)$ for all v in V and all α in F.

As before, T is a **monomorphism** if it is also one-to-one. If T is both one-to-one and onto, it is a **vector-space isomorphism** or, more simply, an **isomorphism**. If there is an isomorphism between two vector spaces, they are said to be **isomorphic.**

By virtue of condition LT1, T is a group homomorphism from $(V, +)$ to $(W, +)$. This will simplify many of the proofs in this section. At this point it might be worthwhile for the student to close this book and write out the preceding definitions. As examples of linear transformations we have the following:

EXAMPLE A Let $V = \mathbb{C}$ and $W = \mathbb{R}$, where both are considered as vector spaces over $\mathbb{R}$. If we let $T(a + bi) = a$ for all a, b in $\mathbb{R}$, then T is a function from $\mathbb{C}$ to $\mathbb{R}$.

$$T((a + bi) + (c + di)) = T(a + c + (b + d)i)$$
$$= a + c = T(a + bi) + T(c + di)$$

Also, $$T(r(a + bi)) = T(ra + (rb)i) = ra$$
$$= rT(a + bi) \qquad \text{for all } r \text{ in } \mathbb{R}$$

Thus T is a linear transformation.

†EXAMPLE B Let $V = \mathbb{R}[x]$, the vector space of all polynomials with real coefficients. Also let $W = \mathbb{R}[x]$, and let $T(f(x)) = f'(x) = df(x)/dx$. Since

$$\frac{d}{dx}[f(x) + g(x)] = \frac{df(x)}{dx} + \frac{dg(x)}{dx}$$

and

$$\frac{d}{dx}[cf(x)] = c\frac{df(x)}{dx}$$

differentiation is a linear transformation.

EXAMPLE C Let $V = \mathbb{R}^3$, $W = \mathbb{R}^2$, and $T(a,b,c) = (a + b + c, 0)$. Then T is a function from V to W. We check that

$$T(a_1,b_1,c_1) + T(a_2,b_2,c_2) = (a_1 + b_1 + c_1, 0) + (a_2 + b_2 + c_2, 0)$$
$$= (a_1 + a_2 + b_1 + b_2 + c_1 + c_2, 0)$$
$$= T((a_1,b_1,c_1) + (a_2,b_2,c_2))$$

Also,
$$T(r(a,b,c)) = T(ra,rb,rc) = (ra + rb + rc, 0)$$
$$= r(a + b + c, 0) = rT(a,b,c)$$

Thus T is a linear transformation.

EXAMPLE D If $V = W = \mathbb{R}^2$ and $T(a,b) = (a^2, b^2)$, then $T((1,0) + (2,3)) = T(3,3) = (9,9)$. But $T(1,0) + T(2,3) = (1,0) + (4,9) = (5,9) \neq (9,9)$, and T is not a linear transformation.

Exercises

1 Let $V = W = \mathbb{R}^2$, and let $T(a,b) = (b, a + b)$. Is T a linear transformation from V to W?

2 Find a function T from a vector space V to a vector space W such that LT1 holds but LT2 does not. [You might try $V = W = \mathbb{C}$ or $\mathbb{Q}(\sqrt{2})$.]

3 Let $T(x) = x$ if x is rational, and $T(x) = 0$ if x is irrational. Make $\mathbb{R}$ into a vector space in such a way that T satisfies LT2 but not LT1.

4 Let $V = W = \mathbb{R}^3$. Which of the following functions are linear transformations from V to W?
   (a) $T(a,b,c) = (a,b,0)$ (b) $T(a,b,c) = (a^2 + b^2 + c^2, 0, 0)$
   (c) $T(a,b,c,) = (|a|,|b|,|c|)$ (d) $T(a,b,c) = (a + b + c, b + c, c)$

5 Determine which of the following are linear transformations:
   (a) $V = \mathbb{R}$, $W = \mathbb{R}^3$, $T(a) = (a,a,3a)$.
   (b) $V = \mathbb{R}^3$, $W = \mathbb{R}$, $T(a,b,c) = 3a + b$.
   (c) $V = \mathbb{R}^3$, $W = \mathbb{R}$, $T(a,b,c) = \sqrt{a^2 + b^2 + c^2}$.

(d) $V = W = P_2(F)$, the polynomials of degree 2 or less; $T(ax^2 + bx + c)$ $= 3ax + b$. (If the 3 were replaced by a 2, which example would we be back to?)

6 This exercise is for students who have had an intermediate or advanced calculus course. Other students may either change $C[0,1]$ to $P(\mathbb{R})$ or omit the exercise.

(a) Define $T(f(x)) = \int_0^1 f(x)\,dx$, where $V = C[0,1]$ and $W = \mathbb{R}$. Is T a function? Is T a linear transformation?

(b) If $V = P(\mathbb{R})$ and we define $T: V \to V$ by $T(f(x)) = \int f(x)\,dx$, then what choice of constant in the integration must we make in order that T be a linear transformation?

The big classification theorem for vector spaces is as follows:

Theorem 1 If V and W are both vector spaces over F of the same dimension, then V and W are isomorphic as vector spaces.

PROOF Let $B = \{b_i\}$ and $C = \{c_i\}$ be bases for V and W. Fact A from the last section guarantees the existence of B and C. For V and W to have the same dimension, B and C have the same cardinality, and there exists a one-to-one onto function T from B to C. Relabeling if we need to, we can assume $T(b_i) = c_i$. We can now define T on all of V by

$$T(\textstyle\sum \alpha_i b_i) = \sum \alpha_i c_i$$

(where we are only considering finite sums).

T is a linear transformation, since

$$T\left(\sum_i \alpha_i b_i + \sum_i \beta_i b_i\right) = T\left(\sum_i (\alpha_i + \beta_i)b_i\right)$$
$$= \sum_i (\alpha_i + \beta_i)c_i$$
$$= \sum_i \alpha_i c_i + \sum_i \beta_i c_i$$
$$= T\left(\sum_i \alpha_i b_i\right) + T\left(\sum_i \beta_i b_i\right)$$

and
$$T(\alpha \sum \alpha_i b_i) = T(\sum (\alpha\alpha_i)b_i) = \sum (\alpha\alpha_i)c_i$$
$$= \alpha \sum \alpha_i c_i = \alpha T(\sum \alpha_i b_i)$$

T is onto, since a typical element of W is $\sum \alpha_i c_i$. We show that T is one-to-one, since otherwise $\ker(T) \neq 0$. If $\sum_i \alpha_i b_i \neq 0$ is in $\ker(T)$, then $\sum_i \alpha_i c_i = 0$ with some of the $\alpha_i \neq 0$, which contradicts the fact that $\{c_i\}$ is a basis.

The use of subscripts here is deliberately vague. The basic idea is that the union of two finite sets is finite and there seems little use in devoting three pages to index set notation just to clean up the notation.   ////

Exercises

7   Go through the last proof, and replace $\sum_i$ by $\sum_{i=1}^n$. This will give a proof for the finite-dimensional case that is not vague in the use of subscripts.

8   (a)   Complete the verification that T is onto in the last proof.

(b)   Give a different verification that T is one-to-one by assuming $T(\sum_{i=1}^n \alpha_i b_i) = T(\sum_{i=1}^n \beta_i b_i)$ with some $\alpha_k \neq \beta_k$. (This verification takes an extra line or two but does not require any notion of kernel.)

9   $V = \mathbb{R}^3$ and $W = P_2(\mathbb{R})$ both have dimension 3 over $\mathbb{R}$. What is an isomorphism T from V onto W?

10   $\mathbb{C}$, as a vector space over $\mathbb{R}$, is isomorphic to some $\mathbb{R}^{(k)}$. Which positive integer k is it? Write down an isomorphism.

11   Show that $\mathbb{Q}(\sqrt{2})$ and $\mathbb{Q}(\sqrt{3})$ are not isomorphic as rings but are isomorphic as vector spaces over $\mathbb{Q}$.

At this point we pursue the analogy with groups and rings further and write down a correspondence theorem for vector spaces.

Theorem 2 (the correspondence theorem)   If V and $\overline{V}$ are vector spaces over F, and T is a linear transformation from V to $\overline{V}$, then the following statements hold:

(a) $T(0_V) = 0_W$.

(b) $T(-v) = -T(v)$.

(c)   If W is a subspace of V, then $T(W)$ is a subspace of W.

(d)   If $\overline{W}$ is a subspace of $\overline{V}$, then $T^{-1}(\overline{W})$ is a subspace of V.

PROOF   Parts a and b are true since $(V,+)$ and $(W,+)$ are groups and T is a homomorphism. Part c is left as an exercise, and Part d goes as follows:

Let v_1 and v_2 belong to $T^{-1}(\overline{W})$. Then $T(v_1)$ and $T(v_2)$ are in $\overline{W}$, and thus $T(v_1) + T(v_2) = T(v_1 + v_2)$ is in $\overline{W}$. Going backward, this means that $v_1 + v_2$ is in $T^{-1}(\overline{W})$. Also, if $T(v_1)$ is in $\overline{W}$, then so is $\alpha T(v_1) = T(\alpha v_1)$; so if v_1 is in $T^{-1}(\overline{W})$, then so is αv_1. Thus $T^{-1}(\overline{W})$ is a subspace.   ////

Several items should be noted. First, Part b really is not vital, since $\alpha T(v) = T(\alpha v)$ for all α in F, including $\alpha = -1$. Second, there is no analog here of normal subgroup or ideal. This is not needed, since the next proposition holds.

Proposition   If W is any subspace of vector space V over F, then V/W is a vector space over F, where the elements of V/W are the cosets $v + W$ and

$$\alpha(v + W) = \alpha v + W \quad \text{for all } \alpha \text{ in } F$$

PROOF   Since $(W, +)$ is a normal subgroup of the abelian group $(V, +)$, $(V/W, +)$ is an abelian group. The remaining verifications are as follows:

$$(\alpha + \beta)(v + W) = (\alpha + \beta)v + W = \alpha v + \beta v + W$$
$$= (\alpha v + W) + (\beta v + W)$$
$$\alpha((v_1 + W) + (v_2 + W)) = \alpha(v_1 + v_2 + W) = \alpha v_1 + \alpha v_2 + W$$
$$= \alpha v_1 + W + \alpha v_2 + W$$
$$\alpha\beta(v + W) = \alpha\beta v + W = \alpha(\beta v + W) = \alpha(\beta(v + W))$$
$$1(v + W) = 1 \cdot v + W = v + W$$

These all hold for all α and β in F and all v, v_1, and v_2 in V.   ////

Definition   As before, if W is a subspace of V, then the linear transformation $\eta : V \rightarrow V/W$ is called the **canonical homomorphism**, where $\eta(v) = v + W$.

We also have the fundamental theorem for vector spaces.

Theorem   If V and $\bar{V}$ are vector spaces over F, and T is a linear transformation of V onto $\bar{V}$, then there is a vector-space isomorphism T° from V/W onto $\bar{V}$, where $W = \ker(T) = \{v \mid T(v) = 0_{\bar{V}}\}$.

PROOF   Since $(V, +)$ and $(W, +)$ are groups and T is a homomorphism of these groups, we define $T^\circ(v + W) = T(v)$. From the fundamental theorem of group homomorphisms we learn that T° is well-defined, one-to-one, onto and a group homomorphism. Since all groups here are additive, this means

$$T^\circ(v_1 + W + v_2 + W) = T^\circ(v_1 + W) + T^\circ(v_2 + W)$$

Thus the only fact not handed to us by the theorem on groups is "commutativity with scalars," which is shown by

$$T^\circ(\alpha v + W) = T(\alpha v) = \alpha T(v) \qquad \text{since } T \text{ is linear}$$
$$= \alpha T^\circ(v + W)$$

This completes the proof.   ////

Definition   In linear algebra the kernel of a linear transformation T is often called the **null space** of T. The dimension of this null space or kernel is called the **nullity** of T. The dimension of $T(V)$, the range of T, is called the **rank** of T.

Exercises

12   If $V = W = P_2(\mathbb{R}) = $ all polynomials over $\mathbb{R}$ of degree 2 or less, and $T(f(x)) = df(x)/dx$, what are the null space, range, nullity, and rank of T?

13 Show that a linear transformation T is one-to-one if and only if the nullity of T is 0.

At this point we introduce another fact which we use but do not prove. It is a stronger version of Fact A.

FACT C   If W is a subspace of a vector space V, then for any basis $\{w_i\}$ of W there exist vectors $\{v_i\}$ in V such that $\{w_i\} \cup \{v_i\}$ is a basis of V. This is called *completing a basis*. [If $W = \{0\}$, we are back to Fact A.]

14 (a) If W is the xy plane, $V = \mathbb{R}^3$, $\{w_1, w_2\} = \{(0,1,0), (1,1,0)\}$ what is a vector v_1 so that $\{w_i\} \cup \{v_1\}$ is a basis for V.
    (b) If $W = P_2(\mathbb{R})$, $V = P_4(\mathbb{R})$ and $\{w_i\} = \{1, x, x^2\}$, which $\{v_i\}$ will complete the $\{w_i\}$ to a basis for V?

15 If W is a subspace of V, $\{w_i\}$ is a basis for W, and $\{v_i\} \cup \{w_i\}$ is a basis for V, show that $\{v_i + W\}$ is a basis for V/W. Thus the dimension of $V = $ the dimension of $W + $ the dimension of V/W.

16 If T is a linear transformation from V to $\overline{V}$, show that the dimension of V $= $ nullity of $T + $ rank of T (use Exercise 15 and the last theorem).

17 Let $V = \overline{V} = P_4(\mathbb{R})$, and let $T(f(x)) = df(x)/dx$. Verify that Exercise 16 holds.

3.4 MATRICES

Our aim in this section is to work with the fact that every linear transformation can be represented as a matrix, and vice versa. This is the most important fact in linear algebra and means that information about vector-space homomorphisms can be put in terms of matrices and thus can be computed.

We start by expressing elements in terms of a basis. For example, $B = \{(1,0,0), (0,1,0), (0,0,1)\}$ is an extremely convenient basis for $\mathbb{R}^3$. $(5, -3, 1)$ is an element of $\mathbb{R}^3$, and, in terms of B, we have

$$(5, -3, 1) = 5(1,0,0) + (-3,)(0,1,0) + 1(0,1,1)$$

$B' = \{(1,0,0), (1,1,0), (1,1,1)\}$ is another basis for $\mathbb{R}^3$, and by fiddling around we see that

$$(5, -3, 1) = 8(1,0,0) + (-4)(1,1,0) + 1(1,1,1)$$

Let us call the triple $(8, -4, 1)$ the *B'-representation vector* for $(5, -3, 1)$. Thus, $(5, -3, 1)$ is the B-representation vector for $(5, -3, 1)$.

Exercise

1 (*a*) $B'' = \{(1,1,0), (0,1,1), (1,0,1)\}$. Represent $(5, -3, 1)$ in terms of B''.
    (*b*) Represent $(1,1,1)$ in terms of B, B', and B''.

As a second example let us take $B = \{1, x, x^2, x^3, ...\}$ as a basis for $P(\mathbb{R})$. Then $x^2 + 5x + 6$ can be represented as $6 \cdot 1 + 5 \cdot x + 1 \cdot x^2 + 0 \cdot x^3 + 0 \cdot x^4 + \cdots = x^2 + 5x + 6$. Thus, the B-representation vector for $x^2 + 5x + 6$ is $(6,5,1,0,0,0,...)$. Note that you have to pick a definite order for the elements in the basis B before talking about B-representation vectors. It does not matter what ordering is picked, but it must be adhered to.

Exercises

2 Define a B-representation vector [or, to be fussier, a $(B\text{-}O)$-representation vector, where O is an ordering of B].

3 (*a*) $B' = (1, x + 1, (x + 1)^2, (x + 1)^3, ...)$ is also a basis for $P(\mathbb{R})$. Express $x^2 + 5x + 6$ in terms of B'.
    (*b*) †If you have studied calculus, review Taylor-series expansions, and redo part *a*.

4 Pick two bases for $\mathbb{R}^4$, and express $(1,2,3,4)$ in terms of both of them.

At this point we describe the procedure for writing a linear transformation as a matrix. Let V and W be vector spaces, and let $T: V \to W$ be a linear transformation. Also let B_V and B_W be ordered bases of V and W. Find the B_W-representation vector for $T(b_1)$, where b_1 is the first element of B_V. Then find the B_W-representation vectors of $T(b_2)$, $T(b_3)$, ... The B_W-representation vector for $T(b_i)$ is the ith row of the desired matrix.

For an example, let $V = \mathbb{R}^3$, $W = \mathbb{R}^2$, and $T(x,y,z) = (x,y)$. Picking $B_V = \{(1,0,0), (0,1,0), (0,0,1)\}$ and $B_W = \{(1,0), (1,1)\}$, we have that $T(b_1) = T(1,0,0) = (1,0) = 1(1,0) + 0(1,1)$. Thus the B_W-representation vector for $T(b_1)$ is $(1,0)$, and so far the matrix is $\begin{pmatrix} 1 & 0 \\ ? & ? \\ ? & ? \end{pmatrix}$.

$$T(b_2) = T(0,1,0) = (0,1) = -1(1,0) + 1(1,1). \text{ This gives us } \begin{pmatrix} 1 & 0 \\ -1 & 1 \\ ? & ? \end{pmatrix}.$$

$$T(b_3) = T(0,0,1) = (0,0) = 0(1,0) + 0(1,1), \text{ so the matrix is } \begin{pmatrix} 1 & 0 \\ -1 & 1 \\ 0 & 0 \end{pmatrix}.$$

Exercises

5   Redo the last example, changing T so that $T(x,y,z) = (y,2z)$.

6   Let $V = \mathbb{R}^3$ with $B_V = ((1,0,0),\ (0,1,0),\ (0,0,1))$, and $W = \mathbb{R}^3$ with $B_W = ((1,1,1),\ (1,1,0),\ (1,0,0))$. If $T: V \to W$ is given by $T(x,y,z) = (y,x,-z)$, then what matrix represents T?

As a second example, let $V = P_4(\mathbb{R})$ and $W = P_3(\mathbb{R})$. Picking the most natural bases, we let $B_V = (1,x,x^2,x^3,x^4)$ and $B_W = (1,x,x^2,x^3)$. $T: V \to W$ is just differentiation; that is, $T(f(x)) = f'(x)$. Computing, we have

$$
\left.
\begin{array}{l}
T(1) = 0 = 0 \cdot 1 + 0 \cdot x + 0 \cdot x^2 + 0 \cdot x^3 \\
T(x) = 1 = 1 \cdot 1 + 0 \cdot x + 0 \cdot x^2 + 0 \cdot x^3 \\
T(x^2) = 2x = 0 \cdot 1 + 2 \cdot x + 0 \cdot x^2 + 0 \cdot x^3 \\
T(x^3) = 3x^2 = 0 \cdot 1 + 0 \cdot x + 3 \cdot x^2 + 0 \cdot x^3 \\
T(x^4) = 4x^3 = 0 \cdot 1 + 0 \cdot x + 0 \cdot x^2 + 4 \cdot x^3
\end{array}
\right\}
\Rightarrow
\begin{pmatrix}
0 & 0 & 0 & 0 \\
1 & 0 & 0 & 0 \\
0 & 2 & 0 & 0 \\
0 & 0 & 3 & 0 \\
0 & 0 & 0 & 4
\end{pmatrix}
$$

Exercises

7   †Let $V = W$ be all functions of the form $a \cos 3x + b \sin 3x + ce^{2x}$, where a, b, c are real constants. Let $B_V = B_W = (\cos 3x,\ \sin 3x,\ e^{2x})$, and let $T(f(x)) = f'(x)$. What is the appropriate matrix for T? If $T_0(f(x)) = \int f(x)\,dx$, what is M_{T_0}, the matrix for T_0? Show that

$$
M_T \circ M_{T_0} =
\begin{pmatrix}
1 & 0 & 0 \\
0 & 1 & 0 \\
0 & 0 & 1
\end{pmatrix}
$$

(The reader unfamiliar with matrix multiplication should postpone this problem for several pages.)

8   (a)   If $K \supseteq F$ are fields, then K is a vector space over F. Show that for any a in K we have a linear transformation $T_a: K \to K$, where $T_a(k) = ka$ for all k in K. Informally, this might be expressed as "right multiplication is a linear transformation over F."

   (b) $\mathbb{C}$ is an extension of $\mathbb{R}$. Find matrices for T_i, T_{2i+3}, T_1.

9   Let $V = \mathbb{R}^2$ and $W = \mathbb{R}^3$. If $B_V = ((1,0),(0,1))$, $B_W = ((0,0,1),(0,1,0),(1,0,0))$, and $T(x,y) = (x,\ x+y,\ x+2y)$, what is the matrix representing T?

This ability to transform linear transformations into matrices, although not involved, is extremely important. That is why we have concentrated on examples and exercises and postponed any deeper consideration of what is happening. Assuming this skill has been assimilated, we make a few remarks. One is that almost always both V and W must be finite for this procedure to be meaningful and useful. The second is that simplifications often take place. Often, V will be the same as W, sometimes with B_V and B_W also coinciding and sometimes not. If V and W coincide, T can be pictured as moving the elements in V around.

10   Let $V = W$ and $T = I$ be the identity map. For any two bases B_V and B_W, the matrix representing T is called the *change-of-basis matrix*.
(a)   If $V = W = \mathbb{R}^3$, $B_V = ((1,2,3), (0,1,0), (0,0,3))$, and

$$B_W = ((1,0,0), (0,1,0), (0,0,1))$$

find the change-of-basis matrix.
(b)   If $V = W = P_2(\mathbb{R})$, $B_V = (1, 1 + x, 1 + x^2)$, and
$$B_W = (1, 1 + x, (1 + x)^2)$$
find the change-of-basis matrix.

Exercises 11 to 15 illustrate how to convert some concepts in geometry into 2×2 matrices. For Exercises 11 to 15 we assume throughout that $V = W = \mathbb{R}^2$ and that $B_V = B_W = ((1,0), (0,1))$ (the *standard basis*).

11   If T_a is a $90°$ rotation counterclockwise of V, then what is the matrix representing T_a. (It is true that T_a is a linear transformation.) Call this matrix M_a. What matrix corresponds to a $180°$ rotation?
12   If T_b is reflection through the x axis, what is the matrix M_b representing b?
13   Show that $M_a{}^4 = M_b{}^2 = \begin{pmatrix} 1 & 0 \\ 0 & 1 \end{pmatrix}$. Also show that $M_b{}^{-1}M_a M_b = M_a{}^3$. Thus we have the dihedral group D_8 here.
14   If T_c is the linear transformation collapsing every point to the point on the x axis directly above or below it, find the matrix representing T_c.
15   If a reflection about the x axis is followed by a reflection about the y axis, what motion results? Show that multiplication of matrices yields the same result.
16   If V has dimension n and W has dimension m over F, what size matrices will represent linear transformations from V onto W.

What happens if we have $T_1: V \to W$ and $T_2: W \to X$, where T_1 and T_2 are linear transformations? Then $T_2 \circ T_1$ is a linear transformation from V to X and is also representable as a matrix. If we know M_{T_1} and M_{T_2}, then we can easily find $M_{T_2 \circ T_1}$ by multiplying matrices. Let (a_{ij}) denote an $n \times m$ matrix (that is, n rows and m columns) whose (i,j)th entry is a_{ij}. Suppose (b_{ij}) similarly denotes an $m \times k$ matrix over the same field. Then we define $(a_{ij})(b_{ij}) = (c_{ij})$, where $c_{ij} = \sum_{k=1}^{m} a_{ik} b_{kj}$. This is the same "rows by columns" definition we saw in the earlier 2×2 definition. This (c_{ij}) is an $n \times k$ matrix. We cannot multiply an $n \times m$ matrix on the left with a $j \times l$ matrix on the right unless $m = j$. In less condensed notation, this is

$$
n\left\{\overbrace{\begin{pmatrix} a_{11} & a_{12} & \cdots & a_{1m} \\ a_{21} & a_{22} & \cdots & a_{2m} \\ \multicolumn{4}{c}{\cdots\cdots\cdots\cdots} \\ a_{n1} & a_{n2} & \cdots & a_{nm} \end{pmatrix}}^{m} \overbrace{\begin{pmatrix} b_{11} & b_{12} & \cdots & b_{1l} \\ b_{21} & b_{22} & \cdots & b_{2l} \\ \multicolumn{4}{c}{\cdots\cdots\cdots\cdots} \\ b_{m1} & b_{m2} & \cdots & b_{ml} \end{pmatrix}}^{l}\right\}m
$$

$$
= \overbrace{\begin{pmatrix} a_{11}b_{11} + a_{12}b_{21} + \cdots + a_{1m}b_{m1} & \sum_{k=1}^{m} a_{1k}b_{k2} & \cdots & \sum_{k=1}^{m} a_{1k}b_{kl} \\ \sum_{k=1}^{m} a_{2k}b_{k1} & \sum_{k=1}^{m} a_{2k}b_{k2} & \cdots & \sum_{k=1}^{m} a_{2k}b_{kl} \\ \multicolumn{4}{c}{\cdots\cdots\cdots\cdots\cdots\cdots\cdots\cdots} \\ \sum_{k=1}^{m} a_{nk}b_{k1} & \sum_{k=1}^{m} a_{nk}b_{k2} & \cdots & \sum_{k=1}^{m} a_{nk}b_{kl} \end{pmatrix}}^{l} \Big\} n
$$

For example,

$$
\begin{pmatrix} 2 & 3 & 0 \\ -1 & 5 & -2 \end{pmatrix}\begin{pmatrix} 1 & -2 \\ 4 & 7 \\ -5 & 6 \end{pmatrix}
$$

$$
= \begin{pmatrix} 2 \cdot 1 + 3 \cdot 4 + 0(-5) & 2(-2) + 3 \cdot 7 + 0 \cdot 6 \\ -1 \cdot 1 + 5 \cdot 4 - 2(-5) & (-1)(-2) + 5 \cdot 7 + (-2)6 \end{pmatrix}
$$

$$
= \begin{pmatrix} 14 & 17 \\ 29 & 25 \end{pmatrix}
$$

For two more examples, we have

$$
(1 \quad 2)\begin{pmatrix} 3 \\ 4 \end{pmatrix} = (1 \cdot 3 + 2 \cdot 4) = (11)
$$

but

$$
\begin{pmatrix} 3 \\ 4 \end{pmatrix}(1 \quad 2) = \begin{pmatrix} 3 & 6 \\ 4 & 8 \end{pmatrix}
$$

We also see that

$$
(1 \quad 3)\begin{pmatrix} 0 & 5 \\ -1 & 2 \end{pmatrix} = (1 \cdot 0 + 3(-1) \quad 1 \cdot 5 + 3 \cdot 2) = (-3 \quad 11)
$$

However, $\begin{pmatrix} 0 & 5 \\ -1 & 2 \end{pmatrix}(1 \quad 3)$ isn't defined, since the number of columns on the left doesn't equal the number of rows on the right.

Exercise

17 Suppose we have the following matrices:

$$A = \begin{pmatrix} 0 & 1 & 2 \\ 0 & 0 & 3 \end{pmatrix} \qquad B = \begin{pmatrix} 0 & 1 \\ 1 & 0 \end{pmatrix} \qquad C = \begin{pmatrix} 1 & -1 \\ 2 & 3 \\ 4 & 5 \end{pmatrix} \qquad D = \begin{pmatrix} 5 & 6 & 7 \end{pmatrix}$$

Compute (when possible) the following: *AB, BC, CD, AD, BD, AC, ABC, DCBA.*

To go backward from a matrix to a linear transformation is even simpler. If

$$M = \begin{pmatrix} a_{11} & a_{12} & \cdots & a_{1m} \\ a_{21} & a_{22} & \cdots & a_{2m} \\ \cdots\cdots\cdots\cdots\cdots\cdots \\ a_{n1} & a_{n2} & \cdots & a_{nm} \end{pmatrix}$$

is an $n \times m$ matrix with the a_{ij} in F, then $T_M \colon F^{(n)} \to F^{(m)}$ can be defined by $T_M(x_1, x_2, \ldots, x_n) = (\sum_{i=1}^{n} a_{i1}x, \sum_{i=1}^{n} a_{i2}x_i, \ldots, \sum_{i=1}^{n} a_{im}x_i)$. For example, if $M = \begin{pmatrix} 1 & 0 & 2 \\ -1 & 3 & 5 \end{pmatrix}$, then

$$T_M(x_1, x_2) = (x_1, x_2) \begin{pmatrix} 1 & 0 & 2 \\ -1 & 3 & 5 \end{pmatrix} = (x_1 - x_2, 0 + 3x_2, 2x_1 + 5x_2),$$

which is a function from $\mathbb{R}^2$ to $\mathbb{R}^3$. This function can be remembered by considering $(x_1, x_2, \ldots, x_n)$ as a $1 \times n$ matrix, writing it on the left, and using the rule for matrix multiplication.

Exercises

18 If $M = \begin{pmatrix} 1 & 0 & 1 \\ 2 & 1 & 2 \\ -1 & 0 & 3 \\ 4 & 1 & 5 \end{pmatrix}$, then what are V and W, where $T_M \colon V \to W$? What

is the rule for T_M?

19 ("Composition of functions corresponds to matrix multiplication.") Let V, W, and X be finite-dimensional vector spaces over F with bases B_V, B_W, and B_X, respectively. Let $T \colon V \to W$ and $S \colon W \to X$ be linear transformations. If M_T represents T and M_S represents S, then show that $M_T M_S$ represents $S \circ T \colon V \to X$. [This exercise tells us that $(M_T M_S)M_R = M_T(M_S M_R)$, since $(R \circ S) \circ T = R \circ (S \circ T)$ by the associativity of the composition of functions.]

Our next major fact, borrowed from a linear-algebra course, completes what we have been doing.

FACT D   Let V and W be finite-dimensional vector spaces over F. In addition, let B_F and B_W be bases for V and W, with $|B_V| = n$, $|B_W| = m$. Then any linear transformation from V to W is representable uniquely as an $n \times m$ matrix with entries from F. Conversely, any $n \times m$ matrix represents a unique linear transformation between these vector spaces with the given bases.

Having Exercise 19 and Fact D, we can do many things. For instance, if we have the situation $\mathbb{R}^2 = V = W$, and using the standard basis $B_V = B_W = ((1,0), (0,1))$, a rotation of the plane with angle α is given by $\begin{pmatrix} \cos \alpha & \sin \alpha \\ -\sin \alpha & \cos \alpha \end{pmatrix}$
Following this by a rotation through an angle β gives a total rotation of angle $\alpha + \beta$. Thus,

$$\begin{pmatrix} \cos (\alpha + \beta) & \sin (\alpha + \beta) \\ -\sin (\alpha + \beta) & \cos (\alpha + \beta) \end{pmatrix} = \begin{pmatrix} \cos \alpha & \sin \alpha \\ -\sin \alpha & \cos \alpha \end{pmatrix} \begin{pmatrix} \cos \beta & \sin \beta \\ -\sin \beta & \cos \beta \end{pmatrix} \quad (*)$$

Multiplying this out yields $\cos (\alpha + \beta) = \cos \alpha \cos \beta - \sin \alpha \sin \beta$, and

$$\sin (\alpha + \beta) = \cos \alpha \sin \beta + \sin \alpha \cos \beta$$

You may find Eq. (*) easier to memorize than the trig formulas. Letting either $\beta = \alpha$ or letting $\alpha = \beta = \frac{1}{2}\theta$ will yield other standard trig formulas.

If Hom (V,W) denotes all linear transformations from V to W, then we can define an addition in Hom (V,W) by $(T_1 + T_2)(v) = T_1(v) + T_2(v)$. In Hom (V,V) we can even define $T \cdot S$ by letting $T \cdot S(v) = T(S(v))$.

Exercises

20   Let V and W be vector spaces over F. Using the just-defined additions, show that Hom (V,W) is also a vector space over F.

21   Let V be a vector space over F. Show that Hom (V,V) is a ring with $+$ and $\cdot$ as defined above.

22   Show that Hom (V,W) is vector space isomorphic to the $n \times m$ matrices over F. Show that Hom (V,V) is ring isomorphic to the ring of $n \times n$ matrices over F. (Fact D may be used here, as well as Exercise 19.)

Let W be a subspace of V. Now Hom (V,W) is the set of all linear transformations T of V such that $T(V) \subseteq W$. If $S(V) \subseteq W$ and $T(V) \subseteq W$, then certainly $(S \cdot T)(V) = T(S(V)) \subseteq T(W) \subseteq W$, so Hom (V,W) is a subring of Hom (V,V). (Additive closure is easily verified.)

Pick a basis for V by first picking a basis for W, say, $B_W = (w_1, w_2, \ldots, w_k)$ and $B_V = (w_1, w_2, \ldots, w_k, v_{k+1}, \ldots, v_n)$. Then the matrices for elements of Hom (V, W) look like

$$\begin{pmatrix} a_{11} & a_{12} & \cdots & a_{1k} & 0 & 0 & \cdots & 0 \\ a_{21} & a_{22} & \cdots & a_{2k} & 0 & 0 & \cdots & 0 \\ \cdots\cdots\cdots\cdots\cdots\cdots\cdots\cdots\cdots\cdots\cdots \\ a_{n1} & a_{n2} & \cdots & a_{nk} & 0 & 0 & \cdots & 0 \end{pmatrix}$$

since $T(v_i) = a_{i1} w_1 + a_{i2} w_2 + \cdots + a_{ik} w_k + 0$.

We could, of course, directly verify that matrices of the form

$$\begin{pmatrix} a_{11} & a_{12} & \cdots & a_{1k} & 0 & 0 & \cdots & 0 \\ a_{21} & a_{22} & \cdots & a_{2k} & 0 & 0 & \cdots & 0 \\ \cdots\cdots\cdots\cdots\cdots\cdots\cdots\cdots\cdots\cdots\cdots \\ a_{n1} & a_{n2} & \cdots & a_{nk} & 0 & 0 & \cdots & 0 \end{pmatrix}$$

do form a ring, but this method of looking at subspaces, suitably generalized to modules instead of vector spaces, has provided an important motivation in ring-theory research in the last half century.

Exercise

23 Let l_1 and l_2 be distinct lines in $\mathbb{R}^2$. Find the matrices representing those linear transformations which interchange l_1 and l_2. Here, let b_1 be a nonzero vector on l_1, b_2 a nonzero vector on l_2 and $B = (b_1, b_2)$.

Definition If T is a linear transformation on V, then a nonzero vector v is an **eigenvector** if $T(v) = \alpha v$. The scalar α is called an **eigenvalue**. (If M is an $n \times n$ matrix and $vM = \alpha v$, we use the same terminology.)

Eigenvalues may equal zero even though eigenvectors are nonzero.

EXAMPLE A   Differentiation is a linear transformation on the vector space of differentiable functions. Since $D(e^{mx}) = me^{mx}$, we have that e^{mx} is an eigenvector and m is an eigenvalue.

EXAMPLE B   If $M = \begin{pmatrix} 1 & 3 \\ 3 & 1 \end{pmatrix}$, then both $(1,1)$ and $(1,-1)$ are eigenvectors. Their eigenvalues are 4 and -2.

FACT E   Every eigenvalue of M is a root of the polynomial given by the determinant of $M - xI$. Also, every root of det $(M - xI)$ is an eigenvalue. The proof can be found in any linear-algebra text as can material on determinants.

The determinant of $\begin{pmatrix} a & b \\ c & d \end{pmatrix}$ is $ad - bc$.

EXAMPLE C   Let $M = \begin{pmatrix} 3 & 0 & 0 \\ 0 & 1 & 4 \\ 0 & 4 & 1 \end{pmatrix}$ so that

$$M - xI = \begin{pmatrix} 3 & 0 & 0 \\ 0 & 1 & 4 \\ 0 & 4 & 1 \end{pmatrix} - \begin{pmatrix} x & 0 & 0 \\ 0 & x & 0 \\ 0 & 0 & x \end{pmatrix} = \begin{pmatrix} 3-x & 0 & 0 \\ 0 & 1-x & 4 \\ 0 & 4 & 1-x \end{pmatrix}$$

which has determinant $(3 - x)(x - 5)(x + 3)$, and 3, -3, 5 are the eigenvalues for M.

Exercises

†24 If M is an $n \times n$ matrix over $\mathbb{C}$, show that M has at least one eigenvalue.

25 Show that $M = \begin{pmatrix} 0 & 1 \\ -1 & 0 \end{pmatrix}$ has no eigenvalues (and therefore no eigenvectors) over $\mathbb{R}$.

†26 If M has 0 as an eigenvalue, show that M is singular. (It suffices to show that M is not one-to-one when regarded as a homomorphism on V.)

If our underlying field is $\mathbb{Z}_p$, then our finite-dimensional vector spaces are finite sets. This enables us to count such things as the number of elements in V, the number of subspaces, and the number of bases.

27 What is an example of an infinite-dimensional vector space over $\mathbb{Z}_p$?

28 (a) If V has dimension n over $\mathbb{Z}_p$, show that $|V| = p^n$.

  (b) If K is a finite field, then the smallest subfield of K is isomorphic to $\mathbb{Z}_p$ for some p. Thus K must have some dimension m over $\mathbb{Z}_p$. What is $|K|$?

  (c) If V has dimension n over K, where $|K| = q = p^m$, what is $|V|$?

  (d) What is the order of the following finite vector spaces (which are also rings):

$$\left\{ \begin{pmatrix} a & b \\ 0 & c \end{pmatrix} \;\middle|\; a, b, c \in \mathbb{Z}_p \right\} \quad \text{and} \quad \left\{ \begin{pmatrix} a & b \\ 0 & b \end{pmatrix} \;\middle|\; a, b \in \mathbb{Z}_p \right\}$$

29 Find the number of subspaces of each dimension of the following vector spaces: $\mathbb{Z}_2 \times \mathbb{Z}_2$, $\mathbb{Z}_3 \times \mathbb{Z}_3$, $\mathbb{Z}_p \times \mathbb{Z}_p$, $\mathbb{Z}_2 \times \mathbb{Z}_2 \times \mathbb{Z}_2$, $\mathbb{Z}_p^3$, $\mathbb{Z}_p^n$.

30 First recall the dimension of Hom $(F^{(n)}, F^{(m)})$ as a vector space over F (Exercise 16). Then find $|\text{Hom}\,(\mathbb{Z}_p^{(n)}, \mathbb{Z}_p^{(m)})|$.

31 What is the number of basis vectors in $\mathbb{Z}_p^{(n)}$? (*Hint:* v_1 can be chosen as any of the $p^n - 1$ nonzero vectors, v_2 as any of the $p^n - p$ vectors in $\mathbb{Z}_p^{(n)} - \langle v_1 \rangle$, v_3 as any of $p^n - p^2$ vectors in $\mathbb{Z}_p^{(n)} - \langle v_1, v_2 \rangle$.)

32 †What is $|GL(n,\mathbb{Z}_p)|$, where $GL(n,\mathbb{Z}_p)$ is the number of vector-space isomorphisms from $\mathbb{Z}_p^{(n)}$ to itself. Alternatively, $GL(n,\mathbb{Z}_p)$ is the group of nonsingular $n \times n$ matrices over $\mathbb{Z}_p$. [*Hint:* Let M be any nonsingular matrix in $GL(n,\mathbb{Z}_p)$, and let $(1,0,0,...,0)M = v_1$, $(0,1,0,...,0)M = v_2$, and so on. Use Exercise 31.]

4

GROUPS ACTING ON SETS

4.1 BASIC DEFINITIONS

The concept of a group acting on a set is a small generalization of the idea of a permutation group. It provides a viewpoint that is useful in attacking a wide variety of problems. The basic concept and some of the applications are presented here.

Definition A group $(G, \circ)$ **acts** on a set S if each g in G is a function from S to S and

(a) $(g \circ h)(s) = g(h(s))$ for all g, h in G, s in S.
(b) $I(s) = s$ for all s in S, where I is the identity of G.

We have already seen several examples of this. If $G = \mathrm{GL}(2, \mathbb{R})$ and $S = \mathbb{R}^2$, then $\begin{pmatrix} a & b \\ c & d \end{pmatrix} \in G$ takes $(x_0, y_0) \in S$ to $(ax_0 + cy_0, bx_0 + dy_0) \in S$.

We have

$$\left((x_0,y_0)\begin{pmatrix} a_1 & b_1 \\ c_1 & d_1 \end{pmatrix}\right)\begin{pmatrix} a_2 & b_2 \\ c_2 & d_2 \end{pmatrix}$$

$$= (a_1x_0 + c_1y_0, b_1x_0 + d_1y_0)\begin{pmatrix} a_2 & b_2 \\ c_2 & d_2 \end{pmatrix}$$

$$= ((a_1x_0 + c_1y_0)a_2 + (b_1x_0 + d_1y_0)c_2, (a_1x_0 + c_1y_0)b_2 + (b_1x_0 + d_1y_0)d_2)$$

$$= (x_0,y_0)\begin{pmatrix} a_1a_2 + b_1c_2 & a_1b_2 + b_1d_2 \\ c_1a_2 + d_1c_2 & c_1b_2 + d_1d_2 \end{pmatrix}$$

$$= (x_0,y_0)\left(\begin{pmatrix} a_1 & b_1 \\ c_1 & d_1 \end{pmatrix}\begin{pmatrix} a_2 & b_2 \\ c_2 & d_2 \end{pmatrix}\right)$$

So part a is satisfied. (We have written our matrix functions on the right to avoid using columns for points in $\mathbb{R}^2$.) Since

$$(x_0,y_0)I = (x_0,y_0)\begin{pmatrix} 1 & 0 \\ 0 & 1 \end{pmatrix} = (x_0,y_0)$$

part b is satisfied.

Exercise

1 Let $G = \mathrm{GL}\,(2,\mathbb{R})$, and let S now be the set of all lines through the origin of $\mathbb{R}^2$. Show that G acts on S.

Proposition 1 If G acts on S, each g in G is a permutation of S.

PROOF If g is in G, then g^{-1} is also in G. Thus g^{-1} is a function, and g^{-1} being a function is equivalent to g being one-to-one and onto (see Sec. 1.5). ////

Proposition 2 and definition If G acts on a set S, then there is a homomorphism from G into Sym (S). This homomorphism will always be denoted θ and will be called the **homomorphism** of G **acting on** S or the **action homomorphism**.

PROOF If g is an element of G, then call P_g the permutation of S given by g. Condition a then says

$$P_{g \circ h}(s) = (g \circ h)(s) = g(h(s)) = g(P_h(s)) = P_g(P_h(s))$$

so $P_{g \circ h} = P_g P_h$. Define $\theta(g) = P_g$. We now have $\theta(g \circ h) = P_{g \circ h} = P_g P_h = \theta(g)\theta(h)$, and we have the desired homomorphism. ////

Proposition 3 and definition If t is in S, then $G_t = \{g \mid g \in G, g(t) = t\}$ is called the **stability subgroup** of t.

PROOF   We need to prove that G_t is actually a subgroup. $I(t) = t$, so certainly $I \in G_t$. If $g \in G_t$ then we have $g(t) = t$; applying the function g^{-1} to both sides gives $g^{-1}(t) = g^{-1}(g(t)) = (g^{-1} \circ g)(t) = I(t) = t$. To finish, we show closure by assuming g and h are both in G_t. Then $(g \circ h)(t) = g(h(t)) = g(t) = t$, and $g \circ h \in G_t$.   ////

Definition   If, $t \in S$. where G acts on S, then the **orbit** of t under G is the set of all $g(t)$ where g ranges through the elements of G. This orbit is denoted $\mathcal{O}_G(t)$ or $\mathcal{O}(t)$. Alternatively, $\mathcal{O}(t) = \{g(t) \mid g \in G\}$.

Exercises

2 If $S = \{1,2,3,4\}$ and $G = \text{Sym}(4)$, find G_4 and $\mathcal{O}(4)$.

3 (a) If l is the line $y = 2x$ or, equivalently, $l = \{(x,2x) \mid x \in \mathbb{R}\}$, find $\mathcal{O}_l$ and G_l, where $G = \text{GL}(2,\mathbb{R})$ acts on the set of all lines in $\mathbb{R}^2$ through $(0,0)$.

   (b) If $G = \begin{pmatrix} \pm 1 & 0 \\ 0 & \pm 1 \end{pmatrix} \cup \begin{pmatrix} 0 & \pm 1 \\ \pm 1 & 0 \end{pmatrix}$, what are G_l and $\mathcal{O}_l$, where all eight possible $+$ and $-$ combinations are to be included?

4 Let $S = \{1,2,3,4,5,6,7\}$, and let G be the following group of permutations;

$$I, \begin{pmatrix} 1 & 2 & 3 & 4 & 5 & 6 & 7 \\ 2 & 3 & 4 & 1 & 6 & 5 & 7 \end{pmatrix}, \begin{pmatrix} 1 & 2 & 3 & 4 & 5 & 6 & 7 \\ 3 & 4 & 1 & 2 & 5 & 6 & 7 \end{pmatrix}, \text{ and}$$

$$\begin{pmatrix} 1 & 2 & 3 & 4 & 5 & 6 & 7 \\ 4 & 1 & 2 & 3 & 6 & 5 & 7 \end{pmatrix}.$$

Find G_1, G_5, G_7, $\mathcal{O}_1$, $\mathcal{O}_5$, and $\mathcal{O}_7$.

5 Let $H = \left\{ \begin{pmatrix} r & 0 \\ 0 & r' \end{pmatrix} \middle| r,r' > 0 \right\}$, and let H act on $S = \mathbb{R}^2$ by

$$(x,y)\begin{pmatrix} r & 0 \\ 0 & r' \end{pmatrix} = (rx, r'y).$$

Since H is a subgroup of $\text{GL}(2,\mathbb{R})$, this is an action. Find the nine orbits of H acting on $\mathbb{R}^2$. Also find $G_{(1,1)}$ and $G_{(1,0)}$.

Exercise 1 gives an example of a group acting on a set which is not a permutation group. Another such example is the following.

6 Let $G = \left\{ \begin{pmatrix} a & b \\ 0 & c \end{pmatrix} \middle| ac \neq 0 \right\}$ be the group of nonsingular upper triangular matrices with real coefficients. If $f = \begin{pmatrix} a & b \\ 0 & c \end{pmatrix}$, then let $\bar{f}(x) = (ax + b)/c$. Show that this defines an action of G on the real line $\mathbb{R}$. What is the kernel of the action homomorphism? What are G_0 and $\mathcal{O}(0)$? Here $\bar{f} = \theta(f)$, but $\bar{f}$ is a bit more convenient to use.

Proposition 3   If G acts on S, we say $s \sim t$ if $g(s) = t$ for some $g \in G$, where s and t are in S. Then $\sim$ is an equivalence relation, and the equivalence classes of $\sim$ are the orbits of G.

PROOF   Since $I \in G$ and $I(s) = s$ for all s in S, we have $s \sim s$ for all s in S. Assume now that $s \sim t$ so that there is some g in G with $g(s) = t$. Then g^{-1} is in G; applying the function g^{-1} to both sides yields $s = g^{-1}(g(s)) = g^{-1}(t)$, so $t \sim s$. If $s \sim t$ and $t \sim u$, then, using the definition of $\sim$ twice, we have $g(s) = t$ and $h(t) = u$, where g and h are both in G. Putting these together yields $(h \circ g)(s) = h(g(s)) = h(t) = u$. Since $h \circ g$ is in G, we have $s \sim u$.

We now have that $\bar{s} = \{t \mid s \sim t\} = \{t \mid t = g(s), g \in G\} = \mathcal{O}(s)$.   ////

Since equivalence classes partition the set, we know that $\mathcal{O}(s) \cap \mathcal{O}(t) = \varnothing$ or $\mathcal{O}(s) = \mathcal{O}(t)$. In particular, if $t \in \mathcal{O}(s)$ then $\mathcal{O}(t) = \mathcal{O}(s)$. Perhaps the next exercise shows the origin of the term orbit.

Exercises

7   Let G be the group of all 2×2 matrices of the form $\begin{pmatrix} \cos\theta & \sin\theta \\ -\sin\theta & \cos\theta \end{pmatrix}$.
If $f = \begin{pmatrix} \cos\theta & \sin\theta \\ -\sin\theta & \cos\theta \end{pmatrix}$ then let $f(x,y) = (x\cos\theta - y\sin\theta, x\sin\theta + y\cos\theta)$.
Thus f rotates the point (x,y) through an angle θ around the origin. Show that G is a group and that G acts on $\mathbb{R}^2$. What are $G_{(1,0)}$, $G_{(0,0)}$, $\mathcal{O}(1,0)$, and $\mathcal{O}(0,0)$? Describe the orbits of G. These orbits partition $\mathbb{R}^2$, which we knew a priori by Proposition 3.

8   If G acts on S and $g(s) = t$, then show that $G_s = g^{-1}G_t g$. In particular, G_s and G_t are isomorphic.

We now assume G is finite, and we can use the partition to do counting in S.

Theorem 1 (the basic theorem)   If a finite group G acts on a set S, then $|G| = |G_s| \, |\mathcal{O}(s)|$ for any s in S.

PROOF   Let $G(s \mapsto t)$ be the subset of G taking s to t. That is, $G(s \mapsto t) = \{g \mid g \in G \text{ and } g(s) = t\}$. Then $G = \bigcup_{t \in \mathcal{O}(s)} G(s \mapsto t)$, where the union is disjoint since each g in G takes s to exactly one other element.

Let t_1 and t_2 both belong to $\mathcal{O}(s)$, so that $s \sim t_1$ and $s \sim t_2$. Then $t_1 \sim t_2$ since $\sim$ is an equivalence relation (see Proposition 3). Let $h(t_1) = t_2$, and note that left multiplication by h takes an element of $G(s \mapsto t_1)$ to a unique element of $G(s \mapsto t_2)$. The element h^{-1} reverses this, so $|G(s \mapsto t_1)| = |G(s \mapsto t_2)|$ and $|G| = |\mathcal{O}(s)| \, |G(s \mapsto t)|$ for any t in $\mathcal{O}(s)$. Picking $t = s$, we have $G(s \mapsto s) = G_s$, so $|G| = |\mathcal{O}(s)| \, |G_s|$.   ////

As an application we reprove Lagrange's theorem. Let G act on the right cosets of some subgroup H. Thus g takes Hx to Hxg. Obviously all the cosets are in $\mathcal{O}(H)$, and $G_H = H$.

$$|G| = |G_H| \, |\mathcal{O}(H)| = |H| \text{ (number of right cosets of } G).$$

In this sense Theorem 1 is just an extension of Lagrange's theorem.

As a second application we use Theorem 1 to determine the order of a group. Let G be the group of all symmetries of a cube, that is, all rigid motions in three-space of a cube to itself. If the vertices of the cube are labeled as

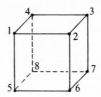

then it is easy to see that

$$\alpha = \begin{pmatrix} 1 & 2 & 3 & 4 & 5 & 6 & 7 & 8 \\ 2 & 3 & 4 & 1 & 6 & 7 & 8 & 5 \end{pmatrix} \quad \text{and} \quad \beta = \begin{pmatrix} 1 & 2 & 3 & 4 & 5 & 6 & 7 & 8 \\ 2 & 6 & 7 & 3 & 1 & 5 & 8 & 4 \end{pmatrix}$$

are in G. Now we try to figure out $\mathcal{O}(1)$. Obviously, by using α a number of times we can take corner 1 to positions 2, 3, and 4. If corner 1 is at position 2, we can use β to get to position 6. From position 6 we can use α several more times to proceed to positions 7, 8, and 5. Thus $\mathcal{O}(1) = \{1,2,\dots,8\}$.

What can we say about the subgroup G_1? Since 2 is next to 1, if 1 goes to 1 then 2 has to go to 2, 4, or 5, for any permutation preserving the cube. If, say, 2 goes to 4, then 4 must go to 5, and we have

$$\gamma = \begin{pmatrix} 1 & 2 & 3 & 4 & 5 & 6 & 7 & 8 \\ 1 & 4 & 8 & 5 & 2 & 3 & 7 & 6 \end{pmatrix}$$

I, γ, and γ^2 all belong to G_1, and since the location of 1 and 2 together determines the whole cube, we have that $G_1 = \{I, \gamma, \gamma^2\}$.

Now, using the basic theorem, we have $|G| = |\mathcal{O}(1)| \, |G_1| = 8 \cdot 3 = 24$.

Exercises

9 Find the order of the group of the regular tetrahedron:

10 How many long diagonals does a cube have? Do the positions of these long diagonals determine the position of the cube? Show that the group of rotations of the cube is isomorphic to Sym (4).

11 Which four motions of the cube will not interchange any of the three axes of a cube connecting midpoints of opposite sides? Show that these four motions form a subgroup V. The motions of the cube act on this set of three axes. Show that ker $(\theta) = V$ if θ is the action homomorphism. This yields a geometric proof that Sym $(4)/V \simeq$ Sym (3).

4.2 FIXED POINTS OF p-GROUPS

If $G_t = G$ or, equivalently, $g(t) = t$ for all g in G or, equivalently, $\mathcal{O}(t) = \{t\}$, then t is a *fixed point* of the action of G on S. We now examine a recurring situation which guarantees fixed points.

If P is a finite p-group acting on set S, then every orbit has length $1, p, p^2, \ldots,$ or $|P|$. The *length* of the orbit $\mathcal{O}(t)$ is just $|\mathcal{O}(t)|$. This follows, since $|\mathcal{O}(t)| \,|\, |P|$ by the basic theorem.

Theorem 1 (basic p-group theorem) If P is a finite p-group acting on a finite set S with $p \nmid |S|$, then P has at least one fixed point.

PROOF The action of P partitions S into orbits. The possible orbit lengths, by the previous theorem, are $1, p, p^2, \ldots, |P|$. If no orbit has length 1, then $p \,|\, |S| = |\mathcal{O}_1| + |\mathcal{O}_2| + \cdots + |\mathcal{O}_k|$, which contradicts our hypothesis. Thus some $|\mathcal{O}_i| = 1$, and $\mathcal{O}_i = \{t\}$, and t is a fixed point. ////

Exercise

1 If G is a group of order 55 acting on a set S of order 18, show that G must have a fixed point (in fact, at least two). (Imitate the proof of Theorem 1.)

Returning to the rotations of a cube discussed above, we have a group R of order 24. Any element of order 3 generates a cyclic group of order 3. Conversely, any subgroup of order 3 is cyclic and thus generated by an element. By the p-group theorem, any element of order 3 in R has two fixed points among the eight vertices. Geometrically, these points must be diametrically opposite. How many elements of order 3 are there in R? Since there are but four sets of opposite poles, there are four such subgroups, each with two nonidentity elements. So R has exactly eight elements of order 3.

We now set up another action. Let G be a group, and let $S = G$. We let G act on G by conjugation. That is, if g is in G, then $f_g(x) = gxg^{-1}$ for all x in

G. Is this an action? Since $f_g f_h(x) = f_g(hxh^{-1}) = g(hxh^{-1})g^{-1} = ghx(gh)^{-1} = f_{gh}(x)$, and $f_I(x) = IxI^{-1} = x$, this is indeed an action. The function f_g is called an *inner automorphism*.

Exercises

2   Show that the kernel of the action of G acting on G by conjugation is $Z(G)$, the center of G.

3   If H is a subgroup of G, let H act on the set G by conjugation. Show that g is a fixed point of H if and only if g is in $C_G(H)$, the centralizer of H in G.

This case of G acting on G by conjugation is of sufficient interest that a special terminology has developed. An orbit is called a *conjugate class*, and the stabilizer of a is just the *centralizer* $C_G(a)$ of a in G. Since we have verified that we have an action, we know that the conjugate classes partition G. This action of G on G by conjugation in general does not make G into a permutation group on itself.

Theorem 2   (the class equation) Let the conjugate classes of the finite group G be $\text{Cl}(a_1)$, $\text{Cl}(a_2)$, ..., $\text{Cl}(a_m)$, $\text{Cl}(a_{m+1})$, ..., $\text{Cl}(a_k)$, with $|\text{Cl}(a_1)| = |\text{Cl}(a_2)| = \cdots = |\text{Cl}(a_m)| = 1$, and $|\text{Cl } a_i)| > 1 \; \forall i > m$. Then

$$|G| = \sum_{i=1}^{k} |\text{Cl}(a_i)| = |Z(G)| + \sum_{i=m+1}^{k} |\text{Cl}(a_i)|$$

$$= |Z(G)| + \sum_{i=m+1}^{k} \frac{|G|}{|C_G(a_i)|} = \sum_{i=1}^{k} \frac{|G|}{|C_G(a_i)|}$$

PROOF   All we are doing here is counting out a partition. $\text{Cl}(a_i) = \mathcal{O}(a_i)$, so $|\text{Cl}(a_i)| = 1 \Leftrightarrow a_i$ is a fixed point $\Leftrightarrow g^{-1}a_i g = a_i$ for all g in $G \Leftrightarrow a_i \in Z(G)$. Also,

$$|G| = |\mathcal{O}(a_i)||G_{a_i}| = |\mathcal{O}(a_i)||C_G(a_i)| = |\text{Cl}(a_i)||C_G(a_i)| \qquad /\!/\!/\!/$$

Putting these remarks together, we get some interesting propositions.

Proposition 1   If P is a finite p-group, then $Z(P)$ is larger than just the identity.

PROOF   Let the finite p-group P act on the set $P^{\#}$ of all nonidentity elements of P by conjugation. Since $p \nmid |P^{\#}|$, P must have some fixed point x in $P^{\#}$. Thus $x \in Z(P)$, and $Z(P)$ is nontrivial. $\qquad /\!/\!/\!/$

Proposition 2 Every group of order p^2 (p a prime) is abelian.

PROOF By Lagrange's theorem and proposition 1, $|Z(P)| = p$ or p^2. If $|Z(P)| = p^2$, then $P = Z(P)$ and we are done. So we can assume $Z(P)$ has order p and pick a in $P - Z(P)$. Then $C_G(a)$ includes both a and $Z(P)$ so, by Lagrange's theorem again, $C_G(a) = G$. However, this can happen only if $a \in Z(P)$, and we have the concluding contradiction. In fact looking at this second sentence a bit, we can improve it to the following:

Proposition 3 If G is a group, then $G/Z(G)$ cannot be cyclic except in the obvious case where G is abelian so that $G = Z(G)$.

PROOF If $G/Z(G)$ is cyclic with $aZ(G)$ as a generator, then $C_G(a)$ contains $Z(G), aZ(G), \ldots, a^iZ(a), \ldots$ and thus all of G. Thus a is in $Z(G)$.
/////

Exercises

4 (a) Show that $Z(\mathrm{Sym}\,(3)) = \{e\}$.
   (b) What is the order of the center of a nonabelian group of order p^3? Does this check with D_8 and the quarternions?

5 Let N be a normal subgroup of a finite p-group P, and let $N^\#$ be the non-identity elements of N. Then show that P acts on $N^\#$ by conjugation. Also show that $p \nmid |N^\#|$, that P has fixed points in $N^\#$, and that $|N \cap Z(P)| > 1$. (Taking $N = P$, we obtain Proposition 1 as a special case.)

6 Find all finite groups G with exactly one, two, or three conjugate classes.

7 If G acts on S and the only orbit of this action is S itself, then we say G is *transitive* on S. Show that if G is transitive on S and G is finite, then $|S|\,|\,|G|$.

Exercises 8 to 10 constitute a diversion and are not essential in what follows, but they do present some interesting groups. Let P be a set (called the set of points) and L be a set of subsets of P (called the set of lines). Coll (P,L), the *group of collineations of P and L*, is the subgroup of Sym (P) that takes lines to lines. That is, if $\alpha \in$ Coll (P,L) and $l = \{p_1,p_2,\ldots,p_k\}$ is any line in L, then $\alpha(l) = \{\alpha(p_1), \alpha(p_2), \ldots, \alpha(p_k)\}$ is also in L.

8 Show that Coll (P,L) is actually a subgroup of Sym (P).

9 Let $P = (a,b,c,d,e,f,g,h,k)$, and let $L = \{\{a,b,c\}, \{d,e,f\}, \{g,h,k\}, \{a,d,g\}, \{b,e,h\}, \{c,f,k\}, \{a,e,k\}, \{b,f,g\}, \{d,h,c\}, \{g,e,c\}, \{h,f,a\}, \{d,b,k\}\}$. If we arrange the nine points in a square, and indicate lines by connecting points contained in the same line, we have a useful though unofficial picture:

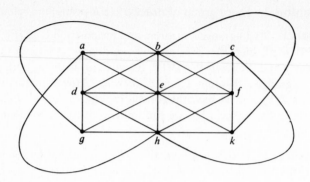

Looking at this, we guess that $\alpha = \begin{pmatrix} a & b & c & d & e & f & g & h & k \\ b & c & a & e & f & d & h & k & g \end{pmatrix}$ and

$\beta = \begin{pmatrix} a & b & c & d & e & f & g & h & k \\ d & e & f & g & h & k & a & b & c \end{pmatrix}$ are in Coll (P,L), as is

$$\gamma = \begin{pmatrix} a & b & c & d & e & f & g & h & k \\ a & k & e & b & g & f & c & h & d \end{pmatrix}.$$

Now, as in the case of the cube and tetrahedron (Exercise 8 of Sec. 4.1), we can ask for the order of Coll (P,L), which is the exercise.

10   Let $P = \{a,b,c,d,e,f,g\}$ and $L = \{a,c,e\}$, $\{b,d,e\}$, $\{c,d,g\}$, $\{a,b,g\}$, $\{c,f,b\}$, $\{a,f,d\}$, and $\{e,f,g\}$. The picture here is

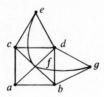

What is order of Coll (P,L)?

As one further application, we include a recent proof by McKay of an old theorem of Cauchy. This result is a special case of the first Sylow theorem, so its inclusion here is mainly for aesthetic value.

Theorem 3   If $p \mid |G|$, where p is a prime and G is a finite group, then G contains an element of order P.

PROOF   First consider $G^p = G \times G \times \cdots \times G$ (p copies). Let S equal those p-tuples $(g_1, g_2, \ldots, g_p)$ in G^p such that $g_1 g_2 \cdots g_p = e$. Since $g_1 g_2 \cdots g_{p-1} x = e$ has a unique solution for any choice of $g_1, g_2, \ldots, g_{p-1}$, we see that $|S| = |G|^{p-1}$; and, if $S^* = S - (e,e,\ldots,e)$, then $|S^*| = |G|^{p-1} - 1$. Thus $p \nmid |S^*|$.

Let $C = \langle c \rangle$ be a cyclic group of order p, and let C act on S^* by $c(a_1,a_2,a_3,\dots,a_p) = (a_2,a_3,\dots,a_p,a_1)$. We leave the verification that this is an action as an exercise, and proceed. We now have a p-group C acting on S^* with $p \nmid |S^*|$, so once again there must be a fixed point. A fixed point for $C = \langle c \rangle$ is one of the form $(a,a,a,\dots,a)$, so $a^p = e$, and the theorem is proved.   ////

Exercise

11 Show that $c(a_1,a_2,a_3,\dots,a_p) = (a_2,a_3,\dots,a_p,a_1)$ does define an action of $C = \langle c \rangle$ on S^* above.

4.3   THE BURNSIDE COUNTING THEOREM

In this section we discuss the Burnside counting theorem and a few of its applications. This theorem appears in Burnside's book of 1911 [14] on page 189. Polya [71] realized the applications of this theorem to enumeration problems and extended its use by introducing cycle indices and generating functions.

Theorem 1   (Burnside) If a finite group G acts on a finite set S and $\chi(g)$ is the number of elements in S fixed by g, then $(1/|G|) \sum_{g \in G} \chi(g) =$ the number of orbits of G acting on S.

PROOF   The procedure here is to count the number of pairs (g,s), where $g(s) = s$. We do this in two ways and equate the results. One way yields $\sum_{s \in S} |G_s|$; the other, $\sum_{g \in G} \chi(g)$. Thus,

$$\sum_{G \in g} \chi(g) = \sum_{S \in s} |G_s|$$

If $S = \mathcal{O}_1 \cup \mathcal{O}_2 \cup \cdots \cup \mathcal{O}_k$ is the partition of S into orbits, we have

$$\sum_{G \in g} \chi(g) = \sum_{s \in \mathcal{O}_1} |G_s| + \sum_{s \in \mathcal{O}_2} |G_s| + \cdots + \sum_{s \in \mathcal{O}_k} |G_s|$$

But if s and t are in the same orbit, $|G_s| = |G_t|$ by Exercise 7 in Sec. 4.1. Thus,

$$\sum_{g \in G} \chi(g) = |\mathcal{O}_1| |G_{s_1}| + |\mathcal{O}_2| |G_{s_2}| + \cdots + |\mathcal{O}_k| |G_{s_k}|$$

where $s_i \in \mathcal{O}_i$. By the basic theorem, each $|\mathcal{O}_i| |G_{s_i}| = |G|$, so

$$\sum_{s \in G} \chi(g) = k|G|,$$

where k is the number of orbits. This completes the proof.   ////

Using this theorem, we now can consider a counting problem. Let us assume we have wires set up as a regular tetrahedron, and at each of the six edges we can attach our choice of a 100-ohm resistor, a 75-watt light bulb, or a capacitor. In our supplies are at least six of each of these components. We want to know how many essentially different contraptions we can make if we allow rotations of the tetrahedron. First we take care of two preliminaries.

PRELIMINARY 1   It is straightforward to count the different (ignoring rotations) contraptions available. We have three choices available at each of six locations, so we have a set S of $3^6 = 729$ possible contraptions.

PRELIMINARY 2   What does the group of rotations of a regular tetrahedron look like? If we label the vertices of a tetrahedron as

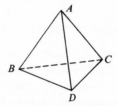

we find that the elements are I, (ABC), (ACB), (ABD), (ADB), (ACD), (ADC), (BCD), (BDC), $(AB)(CD)$, $(AC)(BD)$, and $(AD)(BC)$. Thus there are one element of order 1; eight elements of order 3, each fixing one vertex; and three elements of order 2, each with no fixed vertices. Let us call this group G.

Returning now to our problem, we find that we have a count from preliminary 1, but it is too high. For instance. there are six different contraptions with five resistors and one light bulb, but these are not essentially different if we allow rotations. In fact, these six elements of S form just one orbit under G. Further, each orbit under G gives us just one "essentially different" contraption, so the Polya-Burnside theorem is exactly what is needed here.

We now need only to compute the $\chi(g)$ for each g in G. $\chi(I) = 3^6$, since the identity fixes every contraption in S. If g is an element of order 3 (a 120° rotation about an axis through one vertex V and the center of the opposite triangle), then g fixes elements of the form

where r and s are any of the three choices. Thus

$$\chi(g) = 3^2$$

If h has order 2 and thus interchanges two pairs of vertices, we can assume $g = (EF)(GH)$.

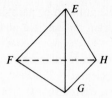

Then we have an arbitrary choice of three for the edges EF and HG. Side EG is taken to FH (and conversely), so we have a free choice for EG. But then FH must be the same choice. Similarly, we have a free choice for EH, but then no choice for FG. Thus $\chi(h) = 3^4$ here. Putting this all together, we obtain:

$$\text{Answer} = \text{number of orbits} = \frac{1}{|G|} \sum_{g \in G} \chi(g)$$
$$= \tfrac{1}{12}[\chi(I) + 8\chi(g) + 3\chi(h)] = \tfrac{1}{12}(3^6 + 8 \cdot 3^2 + 3 \cdot 3^4)$$
$$= 87$$

Exercises

1 Redo this example with n choices available at each side instead of three. This should incidentally give you a somewhat elaborate proof that $\tfrac{1}{12}(n^6 + 8n^2 + 3n^4)$ is an integer for all positive integers n.

2 Analyze the dihedral group of order 12 (that is, the group of symmetries of a regular hexagon) as a permutation group on its six vertices. "Analyze" here means first find the number of elements of each order, and then subdivide these geometrically. In particular, there are three kinds of elements of order 2.

3 At each carbon atom in a benzene molecule, either a —NH^3, a COOH, or a —OH radical can be attached. How many different compounds are possible? A benzene molecule can be viewed as six carbon atoms arranged in a hexagon:

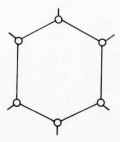

A plain molecule will have H's attached at each of the six carbon atoms. (For this problem, assume the benzene molecule lies in a plane.) The answer is 92.

4 If each side of a regular hexagon can be painted red, yellow, black, or green, how many essentially different designs are possible, allowing all symmetries of a regular hexagon?

5 If each side and both ends of a regular triangular prism can be painted one of six colors, how many essentially different combinations are possible?

6 If a tape contains 10 digits, each either 0 or 1, but the tape can be read indiscriminately from either end, then how many essentially different messages can be recorded on this tape?

For further development of this material, see Polya [71], Liu [58], or Harary and Palmer [34]. This theorem can be extended to give a count of the number of graphs with n vertices. The book by Harary and Palmer extensively discusses this and many other counting problems in graph theory and combinatorics. This book also contains an extensive list of unsolved counting problems.

There are two famous theorems, one in algebra and the other in geometry, that we can discuss at this point. These theorems will not be used again in this text but are interesting in themselves. The first says that the only finite groups of rotations in 3-space are the cyclic groups, dihedral groups, and the groups of the rotations of a regular tetrahedron, cube, and icosahedron. The second says that there are only five regular solids in 3-space, the tetrahedron, the cube, the octahedron, the dodecahedron, and the icosahedron.

To sketch a proof of the first, we quote without proof the following facts. For a detailed proof see Benson and Grove [7] or Yale [91].

1 Rotations in 3-space (about a fixed center point) form a group with composition of functions as the group operation.

2 (Euler) Each nonidentity rotation has exactly two fixed points, called poles. (This is true since every 3×3 orthogonal matrix over $\mathbb{R}$ has a real eigenvalue which must equal 1.)

3 The groups mentioned can be produced. In particular, a cyclic group of order n can be produced by fixing two poles (which we can call the north pole and the south pole) and letting the generator fix these poles and rotate the equator $(360/n)°$, say, clockwise. This group has two poles and n elements, and no element moves the north pole to the south pole. The dihedral group of order $2n$ contains this cyclic group. There are also n rotations, each having two poles on the equator [if n is odd, one for each $(360/n)k°$, for $k = 0, 1, \ldots, n - 1$; if n is even, one for each $(360/2n)k°$, for $k = 0, 1, \ldots, n - 1$]. Each such rotation interchanges the north and south poles.

To start the proof we let G be a finite group of rotations of 3-space, and let S be the set of poles of the nonidentity elements of G. Half of the count is very easy, since $\chi(I) = |S|$ and $\chi(g) = 2$ (using B) for all other g in G. Thus

$$|G|k = \sum_{g \in G} \chi(g) = |S| + (|G| - 1)2 \qquad (*)$$

where G acts on S, and k is the number of orbits of this action.

Now if G is not trivial, $2 \leq |S|$ and, since each nonidentity element of G has at most two poles, $|S| \leq 2(|G| - 1)$. Therefore,

$$2 + (|G| - 1)2 \leq k|G| \leq 2(|G| - 1) + (|G| - 1)2$$

and $k = 2$ or 3. We now put these two values for k back into Eq. (*). First note that $|G_{s_i}| \geq 2$, since both the identity and the rotation having s_i as a pole are in G_{s_i}.

CASE I: $k = 2$

$$2|G| = |S| + (|G| - 1)2$$
$$|S| = 2$$

and, calling the two elements of S the north and south poles, we arrive back at the cyclic groups.

CASE II: $k = 3$

$$3|G| = |S| + (|G| - 1)2$$

so

$$|G| + 2 = |S|$$

and $S = \mathcal{O}_1 \cup \mathcal{O}_2 \cup {}_3$ is a partition of S into orbits

$$|G| + 2 = |\mathcal{O}_1| + |\mathcal{O}_2| + |\mathcal{O}_3|$$

$$= \frac{|G|}{|G_{s_1}|} + \frac{|G|}{|G_{s_2}|} + \frac{|G|}{|G_{s_3}|} \qquad s_i \text{ in } \mathcal{O}_i$$

$$1 + \frac{2}{|G|} = \frac{1}{|G_{s_1}|} + \frac{1}{|G_{s_2}|} + \frac{1}{|G_{s_3}|} \qquad (**)$$

where we may as well assume

$$\frac{1}{|G_{s_1}|} \geq \frac{1}{|G_{s_2}|} \geq \frac{1}{|G_{s_3}|}$$

This tells us that $1/|G_{s_1}| = \frac{1}{2}$ since, were $1/|G_{s_1}|$ smaller, Eq. (**) could not hold; were it larger, we would not have $|G_{s_1}| \geq 2$. Hence,

$$\frac{1}{2} + \frac{2}{|G|} = \frac{1}{|G_{s_2}|} + \frac{1}{|G_{s_3}|}$$

which implies $1/|G_{s_2}| = \frac{1}{2}$ or $\frac{1}{3}$.

Subcase a: $1/|G_{s_2}| = \frac{1}{2}$ implies $|G_{s_3}| = |G|/2$. Let s_3 be the north pole; then this turns out to be a dihedral group.

Subcase b: $1/|G_{s_2}| = \frac{1}{3}$, which says $\frac{1}{6} + 2/|G| = 1/|G_{s_3}|$. So $1/|G_{s_3}| = \frac{1}{3}, \frac{1}{4}$, or $\frac{1}{5}$, yielding $|G| = 12$ (the tetrahedron), $|G| = 24$ (the cube), or $|G| = 60$ (the icosahedron). To verify that these work, I recommend getting models of each and checking that their groups have the right order, that they have the prescribed number of poles, and that these poles fall in the right number of orbits. One item to note is our use of Burnside's theorem to limit the number of possibilities.

To prove the second theorem using the first, we note that any regular polyhedron has a finite group of symmetries, each of which can be considered as a rotation in 3-space. The cube and the octahedron have the same group of symmetries since if you mark the center of each face of a cube, these marks form the vertices of an octahedron. The same process, starting with the octahedron, yields a cube. Similarly, the icosahedron and the dodecahedron have the same group of symmetries. If there were some other regular polyhedron, it would have to have the same group order as well as three orbits and the correct orbit sizes, and some case-by-case investigation yields the desired results. A second proof using no groups, but assuming geometry in a carefree manner, is outlined in the following exercises.

Exercises

7 Let V distinct points be placed on the plane, and each connected to at least one other by a line segment. Show that $V - E + F = 2$, where E the number of line segments, and F the number of enclosed regions plus 1 (for the region at ∞). One way to do this is by induction on E.

8 Show that, if a polyhedron in 3-space is stretched out and then squashed onto the plane in such a way as to leave all the vertices, lines, and all but one face intact, we get $V - E + F = 2$ for the polyhedron. (Indeed, this is true for any simply connected graph, not only those obtained from squashed polyhedra.)

9 Show, for a regular polyhedron, that there are but a finite number of possibilities for V, E, and F. (*Hint*: $2E = kV$, and $2E = lF$, where k is the number of edges meeting at a single vertex, and l is the number of edges about any face.)

Here are some other formulas similar to the Burnside counting theorem.

Definition If a group G acts on a set S and the only orbit of this action is S itself, then G is said to be **transitive**. If, for every two ordered pairs of distinct points (s_1, s_2) and (t_1, t_2) in S, there is a g in G such that $g(s_1) = t_1$ and $g(s_2) = t_2$, then G is **doubly transitive**.

Exercises

10   If G is transitive on S, show that G is doubly transitive if and only if G_t is transitive on $S - \{t\}$ for all t in S.

11   If G is transitive on a finite set S, then

$$\sum_{g \in G} [\chi(g)]^2 = t^* |G|$$

where $\chi(g)$ is the number of fixed points of g, and t^* is the number of orbits of G_t. In particular, if G is doubly transitive, then

$$\sum_{g \in G} [\chi(g)]^2 = 2|G|$$

Check this for Sym (4) and Alt (4).

We conclude with a curiosity:

12   (The bracelet-theoretic proof of the little Fermat theorem) Assume there are p beads placed at regular intervals about a circular bracelet.
  (*a*)   If each bead can be any of n different colors, show that the number of possible bracelets, ignoring rotations, is just n^p.
  (*b*)   If p is a prime and we allow rotations, use Burnside's theorem to find $(1/p)[n^p + (p-1)n]$ essentially different bracelets.
  (*c*)   Part b implies $p \mid n^p + (p-1)n$. Show that this implies $p \mid n^p - n$ or $n^p \equiv n \pmod{p}$.

4.4   SOME APPLICATIONS IN GROUP THEORY

We come now to some deeper properties of finite groups. We use the methods of groups acting on sets, where, if G is our group, we let S be G itself, or all cosets of some subgroup H, or various other suitably selected sets of subsets of G. The reader interested in finite groups and in seeing some harder applications of groups acting on sets should enjoy this section. The reader who wants to proceed directly to other parts of modern algebra should go on to Chap. 6 or Chap. 5.

The material in this section is more difficult than the preceding material in the book.

Proposition 1   (the strong Cayley theorem) Let H be a subgroup of G, and let $S = \{H, xH, yH, \ldots\}$ be the set of all left cosets of H. Let G act on S by left multiplication, so that $xH \xrightarrow{g} gxH$. This is an action, and the kernel of the action homomorphism is $\bar{H} = \bigcap_{x \in G} xHx^{-1}$.

PROOF   By definition, $\theta(g) = \begin{pmatrix} xH & yH & \cdots \\ gxH & gyH & \cdots \end{pmatrix}$. Obviously, $\theta(e) = I$, the identity permutation.

$\theta(gh)$

$$= \begin{pmatrix} xH & yH & \cdots \\ ghxH & ghyH & \cdots \end{pmatrix} = \begin{pmatrix} xH & yH & \cdots & hxH & \cdots \\ gxH & gyH & \cdots & ghxH & \cdots \end{pmatrix} \begin{pmatrix} xH & yH & \cdots \\ hxH & hyH & \cdots \end{pmatrix}$$

$$= \theta(g)\theta(h)$$

and, indeed, G acts on S via θ.

$$g \text{ is in } \ker(\theta) \text{ if } I = \begin{pmatrix} xH & yH & \cdots & zH & \cdots \\ gxH & gyH & \cdots & gzH & \cdots \end{pmatrix}.$$

This is equivalent to $xH = gxH$ $\forall x \in G$, or $H = x^{-1}gxH$ $\forall x \in G$, or $x^{-1}gx \in H$ $\forall x \in G$, or $g \in xHx^{-1}$ $\forall x \in G$, or $g \in \bigcap_{x \in G} xHx^{-1} = \bar{H}$.   ////

If H is of index n in G, and θ is the action described above, we note that θ gives a homomorphism from G into Sym (n).

Exercise

1   If H is a subgroup of G, show that $\bar{H} = \bigcap_{x \in G} x^{-1}Hx$ is the largest normal subgroup contained in H.

REMARK   If $H = \{e\}$, the identity subgroup, we have Cayley's theorem. Here $\bar{H}$ must also be equal to $\{e\}$ and, since $\ker(\theta) = \bar{H}$, θ is one-to-one.

Cayley's theorem   Any group $(G, \circ)$ is isomorphic to a subgroup of Sym (G). If G is finite, then G is isomorphic to a subgroup of Sym $(|G|)$.

Philosophically, Cayley's theorem is very interesting, since it says that any group is isomorphic to a subgroup of a symmetric group. Thus, in a sense, group theory is just a part of the theory of the symmetric groups and their subgroups. In practice this does not work as well, since, if we wanted to look at groups of order 12, we would have to locate them in Sym (12), which has order 479,001,600. The strong Cayley theorem often allows us to embed the group we are considering into a small enough symmetric group so that something can be learned. For instance, let's look at all groups of order 6. They all contain an element and thus a subgroup H of order 2, by Exercise 4 in Sec. 1.3.

H is either normal or not. If not, then $\bar{H} = \bigcap_{x \in G} x^{-1}Hx = \{e\} = \ker(\theta)$, and we can use the strong Cayley theorem to see that G is isomorphic to a subgroup of Sym (3), since it has three cosets. Since $|G| = 6$, G is isomorphic to Sym (3). Otherwise H is normal and, by Exercise 4 in Sec. 1.13, $H \lhd Z(G)$.

Thus, by Proposition 3 of Sec. 4.2, $Z(G) = G$ and G is abelian. This can only occur if G is cyclic.

Therefore, there are exactly two nonisomorphic groups of order 6, Sym (3) and $\mathbb{Z}_6$.

The following exercises depend on Sec. 15 of Chap. 1.

Exercises

2 *If a group G has order $2m$, where m is odd, show that it must have a normal subgroup of order m.

3 If a group G of order 12 has a nonnormal subgroup of order 3, show that $G \simeq$ Alt (4). You may use the fact that Sym (4) has only one subgroup of order 12.

4 (a) Given that Alt (n) has no proper normal subgroups for $n \geq 5$, show that Alt (5) has no subgroups of order 15, 20, or 30.

   (b) Show that Alt (n) has no subgroups of index 2, 3, ..., $n - 1$ if $n \geq 5$.

Using the strong Cayley theorem, we can obtain direct results saying that the existence of a large subgroup guarantees the existence of a reasonably large normal subgroup. Conversely, if normal subgroups are sparse, then so are large subgroups. One such result is the following:

Proposition 2 If a group G has a subgroup H of finite index greater than 1, then G also has a normal subgroup of finite index greater than 1.

   PROOF Since G has a subgroup H of index n, there is a homomorphism θ from G into Sym (n). Im (θ), the image of θ, is a subgroup, and the fundamental theorem of homomorphisms says that $G/\ker(\theta) \simeq$ Im (θ), where $|\text{Im}(\theta)| \leq n!$. Thus, $\ker(\theta)$ is a normal subgroup of finite index. Also, $\ker(\theta) = \bar{H}$ is contained in H, so $G \neq \bar{H}$. ////

Exercises

5 Show that, if a group G has order 10,000, then G cannot be simple. (The first Sylow theorem can be used here, two pages ahead of time, to assert that G has a subgroup of order 5^4.)

The next two problems are from recent issues of the *American Mathematical Monthly* and are easy if set up with the proper group or subgroup acting on the correct set.

6 *If G is of order $p^n m$, where $m < 2p$ and p is prime, then G has a normal subgroup of order p^n or p^{n-1}. (Again the first Sylow theorem gives the existence of a subgroup of order p^n.)

7 *If G is a torsion group, and H a subgroup of finite index m such that each nonidentity element of H has order $\geq m$, show that H is normal. (It is convenient to consider m prime and composite separately.) A *torsion group* is one with all elements of finite order. A torsion group itself may be finite or infinite.

Next we develop a short proof of the Sylow theorems using virtually no group theory. The standard proof due to Frobenius [25] can be found in many books such as Paley and Weichsel [68], as well as in MacDonald [59]. We start with some elementary results from ring theory. Our approach is due in part to Wielandt [89]. First we repeat some exercises from Chap. 2.

8 Prove that the binomial theorem holds in any commutative ring.
9 If p is a prime, show that

(a) $p \left| \begin{pmatrix} p \\ k \end{pmatrix} \right.$ for $k = 1, 2, \ldots, p - 1$, where $\begin{pmatrix} p \\ k \end{pmatrix} = p!/[k!(p-k)!]$ is the number of subsets of k elements each that can be selected from a set with p elements.

(b) $(a + b)^p = a^p + b^p$ in any commutative ring of characteristic p.

(c) The *Frobenius map* $a \xrightarrow{\theta} a^p$ is a ring homomorphism in any commutative ring of characteristic p.

(d) $a \xrightarrow{\theta^k} a^{p^k}$ is a ring homomorphism in any commutative ring of characteristic p.

One application of Exercise 9c is the following second proof of Fermat's little theorem.

10 (a) Show that $a^p = a$ for all $a \in \mathbb{Z}_p$, the integers modulo p.
    (b) Equivalently, show that $a^p \equiv a \pmod{p}$ for $a \in \mathbb{Z}$.

Lemma 1

$$\begin{pmatrix} p^a m \\ p^a \end{pmatrix} \equiv m \pmod{p}$$

We can obtain a stronger version of the Sylow theorems by doing a little extra work. This extra work and the stronger versions are put in brackets and can be omitted at a first reading.

Lemma 2 $\begin{pmatrix} p^a m \\ p^b \end{pmatrix} = p^{a-b} m \alpha$, where $\alpha \equiv 1 \pmod{p}$ and $0 \leq b \leq a$.

PROOF Expand $\begin{pmatrix} p^a m \\ p^b \end{pmatrix}$ carefully. ////

PROOF OF LEMMA 1   Since $\mathbb{Z}_p[x]$, the ring of polynomials over $\mathbb{Z}_p$, is a commutative ring with 1, we can use Exercise 9d. Thus, $\alpha(f(x)) = (f(x))^{p^a}$ is a ring homomorphism. It follows that $(1 + x)^{p^a} = \alpha(1 + x) = \alpha(1) + \alpha(x) = 1^{p^a} + x^{p^a} = 1 + x^{p^a}$. Thus, $(1 + x)^{p^a m} = (1 + x^{p^a})^m$; using the binomial theorem to compare coefficients of x^{p^a} on both sides, we have

$$\binom{p^a m}{p^a} = m \qquad\qquad \text{in } \mathbb{Z}_p$$

or, equivalently,

$$\binom{p^a m}{p^a} \equiv m \;(\text{mod } p) \qquad \text{in } \mathbb{Z} \qquad ////$$

Exercise

11 Try this out for some low values of p^a and m.

The first Sylow theorem   Let G be a finite group of order $p^a m$, where p is a prime not dividing m. Then G has a subgroup of order p^a. Any such subgroup of order p^a is called a *p-Sylow subgroup*.

[**Stronger version** G has subgroups of order $1, p, p^2, \ldots, p^a$.   ]

PROOF   Let S be the set of all subsets of G containing p^a elements. Then $|S| = \binom{|G|}{p^a} = \binom{p^a m}{p^a} \equiv m \;(\text{mod } p)$ by Lemma 1. Let G act on S by right multiplication. If $T \in S$ so that T is a subset of p^a elements, then $Te = T$ and $(Ta)b = Tab$. So this is an action of G on S. Since $p \nmid m$, at least one orbit of G has length not divisible by p. Call this orbit $\mathcal{O}$, and let T be an element (i.e., a subset of p^a elements) of $\mathcal{O}$.

Since $|G| = |G_T||\mathcal{O}|$ and $p \nmid |\mathcal{O}|$, we see that $p^a | |G_T|$. Now consider T as a set of p^a elements, and let $H = G_T$ act on T by right multiplication. If $t \in T$, then $tg = t$ only if $g = e$ or, put differently, $H_t = \{e\}$. Thus, let $\mathcal{O}_1$ be a orbit of H acting on T, and we have $|H| = |\mathcal{O}_1||H_t| = |\mathcal{O}_1| \leq |T| = p^a$.

Thus $p^a = |H| = |G_T|$, and we have our desired subgroup. Before going on, note that $T = \mathcal{O}_1 = \{th \mid h \in H\} = tH$, so if $p \nmid |\mathcal{O}|$, then every $T \in \mathcal{O}$ is a left coset of a p-sylow subgroup H. Going back further, we see that if $p \nmid |\mathcal{O}|$, then $|\mathcal{O}| = m$. We will need these remarks shortly.

For the stronger version, let S be all subsets of order p^b, use Lemma 2 to show there exists an orbit $\mathcal{O}$ whose length is not divisible by p^{a-b+1}, and proceed as above. Make the same note at the end.   ////

Informally, this proof can be summarized as follows. G acts on S, where S is the set of all subsets of order p^a. At least one orbit is not divisible by p, and this is the one to look at. If T is in this orbit, G_T must have order p^a or bigger. Letting G_T act on T again by right multiplication gives an upper bound of p^a, and we are done.

Before proceeding, let us see what we have accomplished. For example, if we are examining Sym(5), we know it has order $120 = 2^3 \cdot 3 \cdot 5$. The first Sylow theorem then says it must have a 2-sylow subgroup of order 8, a 3-sylow subgroup of order 3, and a 5-sylow subgroup of order 5.

[The stronger version tells us that we also have subgroups of order 2 and 4.]

It works out that Sym(5) also has subgroups of order 60, 24, 20, 12, 10, and 6, but none of order 15, 30, or 40. Thus, Sylow's theorem can be viewed as a partial converse of Lagrange's theorem.

A second application goes as follows: Prove that every group G of order 500 has a proper normal subgroup. By the first Sylow theorem, G has a subgroup H of order $125 = 5^3$. The strong Cayley theorem with respect to H says there is a nontrivial homomorphism θ from G into Sym(4), since H has 4 cosets. ker (θ) is our desired normal subgroup.

We now recall the terminology of the last proof and proceed to the second Sylow theorem.

The second Sylow theorem   A group G of order $p^a m$, where $p \nmid m$, has $1 + kp$ p-Sylow subgroups.

[In fact, the number of subgroups of order p^b is $1 + k'p$ for all $b \le a$.   ]

PROOF   We have two kinds of orbits in S—those of length not divisible by p, which we call $\mathcal{O}_1, \mathcal{O}_2, \ldots, \mathcal{O}_l$, and of those order divisible by p, which we call $\mathcal{N}_1, \mathcal{N}_2, \ldots, \mathcal{N}_k$. Partitioning S by orbits, we get

$$S = \mathcal{O}_1 \cup \mathcal{O}_2 \cup \cdots \cup \mathcal{O}_l \cup \mathcal{N}_1 \cup \mathcal{O}_2 \cup \cdots \cup \mathcal{N}_k$$

so

$$|S| = |\mathcal{O}_1| + |\mathcal{O}_2| + \cdots + |\mathcal{O}_l| + |\mathcal{N}_1| + |\mathcal{N}_2| + \cdots + |\mathcal{N}_k|$$

and

$$|S| = lm + \sum_{i=1}^{k} |\mathcal{N}_i|$$

If xP, a left coset of a p-Sylow subgroup, is in $\mathcal{N}_i$, then $\mathcal{N}_i = \{xPg \mid g \in G\}$, so $xPx^{-1} \in \mathcal{N}_i$, and $xPx^{-1} = Q$ is a p-sylow subgroup of G. Thus $\mathcal{N}_i = \{Qg \mid g \in G\}$, $|\mathcal{N}_i| = m$, and $p \nmid |\mathcal{N}_i|$, a contradiction. Hence no $\mathcal{N}_i$ contains any left coset of any p-Sylow subgroup.

Since each $\mathcal{O}_i$ is comprised of left cosets of various p-Sylow subgroups (see last proof), and since each p-Sylow subgroup has m cosets all contained in $\bigcup_{i=1}^{l} \mathcal{O}_i$, we see that l is the number of p-Sylow subgroups of G. But

$$m \equiv |S| \equiv lm \ (\text{mod } p)$$

so

$$l \equiv 1 \ (\text{mod } p) \quad \text{and} \quad l = 1 + kp$$

The proof of the stronger version results from replacing P with p^{a-b+1} and "p-Sylow subgroup" with "subgroup of order p^b," and using Lemma 2 at the end while being careful.   ////

The third Sylow theorem   If G is a finite group, then all the p-Sylow subgroups of G are conjugate.

PROOF   Let S now be the set of all p-Sylow subgroups, and let G act on S by conjugation. Assume S has two distinct orbits, $\mathcal{O}_1$ and $\mathcal{O}_2$. By the second Sylow theorem we can assume that at least one orbit has length not divisible by p, and we may as well call this orbit $\mathcal{O}_1$. Let P act on $\mathcal{O}_1$ by conjugation, where P is a p-Sylow subgroup not in $\mathcal{O}_1$. By the p-group fixed-point theorem there is an S in $\mathcal{O}_1$ fixed by P so that $g^{-1}Sg = S$ for all g in P. If $p' \in P - S$ we see that $\langle S, p' \rangle$ is a subgroup of G whose order is divisible by p^{a+1}, contradicting Lagrange's theorem. Thus there is but one orbit, and all the p-Sylow subgroups are in it and thus conjugate to one another.   ////

Exercises

12   If G is a finite group, and Q is a subgroup of G of order p^b, show (by letting Q act by conjugation on S) that Q is contained in some p-Sylow subgroup P.

13   *Show that the smallest symmetric group which contains a subgroup isomorphic to the quaternions is Sym(8).

Letting G act on this set of conjugate p-Sylow groups now tells us that $|G| = (1 + kp)|G_p|$, since $|\mathcal{O}| = 1 + kp$ and $G_p = \{g | g \in G \text{ and } g^{-1}Pg = P\}$, where P is a p-sylow subgroup. In particular,

$$P \subseteq G_p \quad \text{so} \quad 1 + kp \left| \frac{|G|}{|P|} \right.$$

Also, $G_p = N_G(P)$, the normalizer of P in G, so

$$|N_G(P)| = \frac{|G|}{1 + kp} \quad \text{and} \quad 1 + kp = \frac{|G|}{|N_G(P)|}$$

We have proved the following corollary:

Corollary The number of p-Sylow subgroups of a finite group is $|G|/|N_G(P)|$, where P is any p-Sylow subgroup. This implies the useful fact that $|N_G(P)| = |G|/(1 + kp)$, where $1 + kp$ is the number of p-Sylow subgroups.

EXAMPLE A   Every group G of order 350 has a proper normal subgroup. To show this, we first note that $350 = 2 \cdot 5^2 \cdot 7$. G has subgroups of order 25; in fact, by the second Sylow theorem, the number of 5-Sylow subgroups is on the list $1, 6, 11, 16, \ldots$. Also, this number $1 + kp$ must divide $|G|/|P| = 350/25 = 14$. Of the numbers on the list, only 1 divides 14, so $|G|/|N_G(P)| = 1$ or $G = N_G(P)$, and P is normal in G.

EXAMPLE B   Every group of order 60 has either one or six subgroups of order 5. To prove this, we examine the 5-Sylow subgroups of G, where G has order $60 = 2^2 \cdot 3 \cdot 5$. We see that the 5-Sylow subgroups have order 5. There are either 1, or 6, or 11, or 16, . . . of them. This number also must divide $|G|/|P| = 60/5 = 12$, so the only choices are 1 and 6.

EXAMPLE C   Up to isomorphism there is only one group of order 35. There is a cyclic group, $(\mathbb{Z}_{35}, \oplus)$, of this order, and all cyclic groups of the same order are isomorphic, so we have only to show that every group of order 35 is cyclic. First, using a procedure very similar to Example A, we can show that a group G of order 35 has normal subgroups H and K of order 5 and 7. Since 5 and 7 are primes, H and K are cyclic with generators h and k. h and k commute since $H \cap K = \{e\}$, and H and K are both normal. Thus, $(hk)^n = h^n k^n$, and hk has order 35, showing that hk is generator for G. Thus G is cyclic, and we are done.

If, in studying a group G, we can find a proper normal subgroup, then often we can find useful information by looking at the smaller group G/N. As a result much of the research in finite groups has been devoted to producing normal subgroups from knowledge of the order of the group. Example A is a very typical example, as are Exercises 14 and 16. In more intricate cases the Sylow theorems plus one more trick often will suffice. The modified Cayley theorem (See Exercise 5) or Burnside's transfer theorem (see Scott [78] or Hall [32]) or just cleverness (Example D) will often suffice.

EXAMPLE D   Every group G of order 56 has a proper normal subgroup. Since $56 = 2^3 \cdot 7$, we shall look at the 7-sylow subgroups. There are $1 + 7k$ such, and $1 + 7k$ must divide $|G|/7 = 8$. Thus there are either one or eight subgroups of order 7. If there is but one, it is normal and we are done. If not, we can assume there are eight subgroups, each containing six nonidentity elements (each of order 7). Thus there are 48 elements of order 7, leaving only eight other elements in the group. But G must have a 2-Sylow subgroup T, of order 8, and this takes care of the remaining elements. There cannot be any more 2-Sylow subgroups, so $T \triangleleft G$.

Exercises

14 Show that every group of order 200 or 1,975 has a proper normal subgroup.

15 Show that every group of order pq, where p and q are primes, has a proper normal subgroup.

16 Show that every group of order $5^2 \cdot 7^2$ is isomorphic to a direct product of two smaller groups.

17 Show that every group of order 992 has a proper normal subgroup. Is the same true for groups of order 1,000?

5

FURTHER RESULTS ON GROUPS

5.1 SOLVABLE GROUPS

In this chapter we discuss solvable groups and finitely generated abelian groups. The term *solvable group* arose in the study of polynomials, and this connection is discussed in Chap. 10 on Galois theory. It is a rather remarkable transition between the groups discussed here and the solving of polynomials. The material up through Proposition 3 is used in Chap. 10. The rest of the section treats some more advanced material, mostly on finite groups.

Definition A **normal series** $\{N_i\}_{i=1}^k$ for a group G is a finite set of subgroups N_i such that

$$G = N_1 \supseteq N_2 \supseteq \cdots \supseteq N_k = \{e\}$$

and

$$N_{i+1} \lhd N_i \qquad i = 1, 2, \ldots, k-1$$

Since $\{e\} \lhd G$, we always have a normal series $\{N_i\}_{i=1}^2$ with $N_1 = G$, $N_2 = \{e\}$.

Definition If $\{N_i\}_{i=1}^k$ is a normal series for G, then the groups N_i/N_{i+1}, for $i = 1, \ldots, k - 1$, are called the **factors** of the series. If H is a subgroup of G such that $H = N_i$ for some normal series, then H is **subnormal** in G.

If $H \lhd G$ then, taking $N_1 = G$, $N_2 = H$, $N_3 = \{e\}$, we see that any normal subgroup is subnormal.

EXAMPLE A Let us look at the subgroup lattice of D_8, given by a, b with

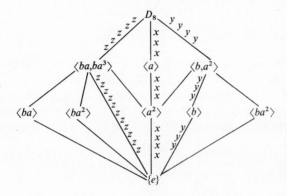

$a^4 = b^2 = e$, $ba = a^{-1}b$. The x's in the figure mark one normal series, the y's a second, and the z's a third. More specifically, three normal series for D_8 are

$$N_1 = D_8, \quad N_2 = \langle a \rangle, \qquad N_3 = \langle a^2 \rangle, N_4 = \{e\}$$
$$M_1 = D_8, \quad M_2 = \langle b, a^2 \rangle, \quad M_3 = \langle b \rangle, M_4 = \{e\}$$
$$L_1 = D_8, \quad L_2 = \langle ba, ba^3 \rangle, \quad L_3 = \{e\}$$

Another normal series would be $P_1 = P_2 = P_3 = D_8$, $P_4 = P_5 = \{e\}$.

Definition A group is **solvable** if it has a normal series $\{N_i\}_{i=1}^k$ such that all its composition factors N_i/N_{i+1} are abelian. Such a normal series is called a **solvable series.**

By taking $N_1 = A$ and $N_2 = \{e\}$, we see that any abelian group is solvable since there is only one factor, $N_1/N_2 = A/\{e\}$, and it is isomorphic to A. Another example of solvable groups is the dihedral groups D_{2n}. Let us say that a is a rotation one nth of the way around the n-gon, and b is some reflection of the n-gon. Thus $D_{2n} = \{a^i b^j \mid 0 \leq i \leq n - 1, j = 0, 1, b^{-1}ab = a^{-1}\}$. Let $N_1 = D_{2n}$, $N_2 = \langle a \rangle \simeq \mathbb{Z}_n$, $N_3 = \{e\}$. Since N_2 has index 2 in D_{2n}, we have $N_2 \lhd N_1$; since N_1/N_2 has order 2, it is cyclic and thus abelian. $N_2/N_3 \simeq \mathbb{Z}_n/\{0\} \simeq \mathbb{Z}_n$ is cyclic and abelian also. Thus D_{2n} is solvable.

By recalling the subgroup lattice of Alt (4), we can quickly show that it is also solvable.

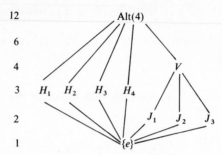

Since V is the only subgroup of order 4, it is normal in Alt (4). $V = \{e, (12)(34), (13)(24), (14)(23)\} \simeq \mathbb{Z}_2 \times \mathbb{Z}_2$ and is abelian. Let $N_1 = $ Alt (4), $N_2 = V$, and $N_3 = \{e\}$. Examining the factors of $\{N_i\}_{i=1}^3$, we see that $\{N_1/N_2\}$ has three elements, so $N_1/N_2 \simeq \mathbb{Z}_3$ and $N_2/N_3 \simeq V/\{e\} \simeq V$. Since both $\mathbb{Z}_3$ and V are abelian, Alt (4) is solvable. We also have enough information to show that

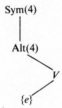

Sym (4) is solvable. Let $M_1 = $ Sym (4), $M_2 = $ Alt (4), $M_3 = V$, and $M_4 = \{e\}$. Since $M_1/M_2 \simeq $ Sym (4)/Alt (4) $\simeq \mathbb{Z}_2$, $M_2/M_3 \simeq \mathbb{Z}_3$, $M_3/M_4 \simeq V$, and $\mathbb{Z}_2$, $\mathbb{Z}_3$, and V are all abelian, Sym (4) is solvable.

Using the fact that Alt (n) has no proper normal subgroups for $n \geq 5$, we can show that Alt (n) is not solvable for $n \geq 5$. If $\{N_i\}_{i=1}^k$ is a normal series, we may as well assume that $N_i \supsetneqq N_{i+1}$, because $N_i = N_{i+1}$ means N_{i+1} is extraneous. Thus, let Alt (n) $= N_1$. If $N_2 \lhd N_1 = $ Alt (n), and $N_2 \neq N_1$, we see that $N_2 = \{e\}$ and then $N_1/N_2 \simeq $ Alt (n)/$\{e\} \simeq $ Alt (n), which is not abelian. Thus Alt (n), for $n \geq 5$, is not solvable. In fact, the same reasoning shows that any nonabelian simple group is not solvable. Feit and Thompson have shown in 1963 in an extremely important paper [22] that any group of odd order is solvable. This was conjectured by Burnside [14] and had been an open question for over 50 years. The Feit-Thompson result runs to 254 pages and is nontrivial.

The next group we would like to examine is Sym (n), for $n \geq 5$. If we assume that $N_1 = $ Sym (n) and $N_2 = $ Alt (n), then we can find no N_3 such that N_2/N_3 is abelian [unless $N_3 = $ Alt (n), which doesn't help]. Sym (n), for $n \geq 5$, could be shown not to be solvable at this point, but we postpone this until we have developed some theory. We now give an example of an infinite solvable nonabelian group. $G = \left\{ \begin{pmatrix} a & b \\ 0 & 1 \end{pmatrix} \middle| a \neq 0, \, a,b \in \mathbb{R} \right\}$ is an infinite and nonabelian

group. Let $N_1 = G$, $N_2 = \left\{ \begin{pmatrix} 1 & b \\ 0 & 1 \end{pmatrix} \middle| b \in \mathbb{R} \right\}$, and $N_3 = \left\{ \begin{pmatrix} 1 & 0 \\ 0 & 1 \end{pmatrix} \right\}$. Then $N_1/N_2 \simeq$

$(\mathbb{R} - (0), \cdot)$; $N_2/N_3 \simeq N_2 \simeq (\mathbb{R}, +)$.

Exercises

1 Show that, if G and H are solvable groups, then so is the group $G \times H$.

2 Show that $G = \left\{ \begin{pmatrix} 1 & a & b \\ 0 & 1 & c \\ 0 & 0 & 1 \end{pmatrix} \middle| \text{all entries in } \mathbb{Z}_3 \right\}$ is solvable. Note that every

element in this group has order 3, but the group is nonabelian. (Compare with Exercise 6 of Sec. 1.3.)

Proposition 1 If G is a solvable group and H is a subgroup of G, then H is solvable.

PROOF   Let $G = N_1 \supseteq N_2 \supseteq N_3 \supseteq \cdots \supseteq N_k = \{e\}$ be a normal series for G with N_i/N_{i+1} abelian, and look at

$$H = H \cap N_1 \supseteq H \cap N_2 \supseteq \cdots \supseteq H \cap N_k = \{e\}$$

The series $H \cap N$ starts and ends at the right groups, and we shall show that H is solvable if we can show two things: $H \cap N_{i+1} \lhd H \cap N_i$ so that $\{H \cap N_i\}_{i=1}^k$ is a normal series, and $H \cap N_i/H \cap N_{i+1}$ is abelian for all i. Let $x \in H \cap N_i$. Then

$$\begin{aligned} x^{-1}(H \cap N_{i+1})x &= x^{-1}Hx \cap x^{-1}N_{i+1}x \\ &= H \cap x^{-1}N_{i+1}x \qquad \text{since } x \in H \\ &= H \cap N_{i+1} \qquad \text{since } x \in N_i \end{aligned}$$

Thus

$$H \cap N_{i+1} \lhd H \cap N_i$$

and we can legitimately examine the group $H \cap N_i/H \cap N_{i+1}$.

Our first real application of the second isomorphism theorem shows that $H \cap N_i/H \cap N_{i+1} \simeq N_{i+1}(H \cap N_i)/N_{i+1} \subseteq N_i/N_{i+1}$, which is abelian. (The second isomorphism theorem says if $B \lhd AB$ then $AB/B \simeq A/A \cap B$. Here we set $A = H \cap N_i$ and $B = N_{i+1}$. See Exercise 17 of Sec. 1.13.) /////

Recently, Thompson [84] has classified all simple groups which have only solvable proper subgroups. This is another extremely important and extremely deep theorem.

We would like now to prove that a factor group of a solvable group is solvable.

Lemma   If $\bar{G}$ is a homomorphic image of G, and G is solvable, then so is $\bar{G}$.

PROOF   Let θ be a homomorphism from G onto $\bar{G}$, and let $G = N_1 \supseteq N_2 \supseteq \cdots \supseteq N_k = \{e\}$ be a normal series for G. We make the obvious choice and look at the series

$$\bar{G} = \theta(G) = \theta(N_1) \supseteq \theta(N_2) \supseteq \cdots \supseteq \theta(N_k) = \{\theta(e)\} = \{e_{\bar{G}}\}$$

By the correspondence theorem, θ takes normal subgroups to normal subgroups, so if $N_{i+1} \triangleleft N_i$, then $\theta(N_{i+1}) \triangleleft \theta(N_i)$. Thus the $\theta(N_i)$ do form a normal series. If A/B is abelian, then $a_1 B a_2 B = a_2 B a_1 B$ for all a_i in A. Thus, $\theta(a_1)\theta(B)\theta(a_2)\theta(B) = \theta(a_2)\theta(B)\theta(a_1)\theta(B)$, and $\theta(A)/\theta(B)$ is also abelian. In particular, each of the $\theta(N_i)/\theta(N_{i+1})$ is abelian, and $\theta(G) = \bar{G}$ is solvable.   ////

Proposition 2   If N is a normal subgroup of G, and G is solvable, then G/N is solvable.

PROOF   Let $\theta: G \to G/N$ be the canonical homomorphism of G onto G/N, and apply the lemma.   ////

Proposition 3   Let G be a group with a normal subgroup N such that both N and G/N are solvable. Then G itself is solvable.

PROOF   Let $N = N_1 \supseteq N_2 \supseteq \cdots \supseteq N_k = \{e\}$ and $G/N = M_1 \supseteq M_2 \supseteq \cdots \supseteq M_l = N/N = $ the identity of G/N be solvable series for N and G/N. If M_i is a subgroup of G/N, then $M_i = M_i'/N$, where M_i' is a subgroup of G containing N.

Thus, the second series can be written as

$$\frac{G}{N} = \frac{M_1'}{N} \supseteq \frac{M_2'}{N} \supseteq \cdots \supseteq \frac{M_l'}{N} = \frac{N}{N}$$

If we try

$$G = M_1' \supseteq \cdots \supseteq M_l' = N \supseteq N_2 \supseteq \cdots \supseteq N_k = \{e\} \qquad (*)$$

we see that it is a normal series (Exercise 3).

$$\frac{N_i}{N_{i+1}}$$

is abelian, since $\{N_i\}_{i=1}^k$ is a solvable series for N.

$$\frac{M_i'}{M_{i+1}'} \simeq \frac{M_i'/N}{M_{i+1}'/N}$$

by the first isomorphism theorem (Exercise 24 of Sec. 1.13)

$$= \frac{M_i}{M_{i+1}}$$

which is abelian since the $\{M_i\}_{i=1}^l$ are a solvable series for G/N. Thus (*) is a solvable series.   ////

Exercise

3   Show that series (*) is a normal series in this last proof.

The last two propositions can be looked at in the following way. Examine

$$N \xrightarrow{\alpha} G \xrightarrow{\theta} \frac{G}{N}$$

where $N \lhd G$, θ is the canonical homomorphism, and α is the inclusion homomorphism from N into G. Then the outside terms N and G/N are solvable if and only if the inside term is solvable.

Exercise

4   Use Proposition 1 and the fact that Alt (5) is not solvable to show Sym (5) is not solvable. This also shows Sym (n) is not solvable for $n \geq 5$.

Our second approach to solvable groups is via commutator subgroups. We recall from Sec. 1.14 that if G is a group then its commutator subgroup is the subgroup generated by all elements of the form $a^{-1}b^{-1}ab$. The main proposition says that if G' is the commutator subgroup of G, then $G' \lhd G$, G/G' is abelian, and if G/N is also abelian then $G' \subseteq N$. Now G' is itself a group, so it in turn has a commutator subgroup which we shall call G'' or $G^{(2)}$. Continuing, we let $G^{(k+1)}$ be the commutator subgroup of $G^{(k)}$, and we call $G^{(k)}$ the kth *commutator subgroup* or kth *derived group* of G.

Proposition 4   A group G is solvable if and only if $G^{(k)} = \{e\}$ for some positive integer k.

PROOF   If $G^{(k)} = \{e\}$, then

$$G \supseteq G' = G^{(1)} \supseteq G^{(2)} \supseteq \cdots \supseteq G^{(k)} = \{e\}$$

is a normal series with abelian factors. Both these assertions follow from the definition of $G^{(i+1)}$ as the commutator subgroup of $G^{(i)}$.

Conversely, if G is solvable and $\{N_i\}_{i=0}^l$ is a solvable series for G, then

$$G = N_0 \supseteq N_1 \supseteq \cdots \supseteq N_l = \{e\}$$

We shall show by induction that $G^{(i)} \subseteq N_i$, and this will give us $G^{(l)} = \{e\}$. Since G/N_1 is abelian, we know that $N_1 \supseteq G' = G^{(1)}$, which starts the induction.

Assume $G^{(j)} \subseteq N_j$; then, since N_j/N_{j+1} is abelian, $N_{j+1} \supseteq N_j'$. But since $N_j \supseteq G^{(j)}$, we must have $N_j' \supseteq G^{(j)'} = G^{(j+1)}$; this yields the desired $N_{j+1} \supseteq G^{(j+1)}$.   ////

Exercises

5 Show that if $G = \text{Sym}(4)$, then $G' = \text{Alt}(4)$, $G'' = V$, and $G^{(3)} = \{e\}$.
6 Find the commutator series for D_8 and D_{2n}. (There will be two cases, depending on whether n is even or odd.)
7 If G is a group of order pq, where p and q are primes, show G to be solvable. (This exercise probably requires Sylow's theorems in Chap. 4 and fore-shadows Burnsides' theorem in Chap. 9).
8 If $G = \left\{ \begin{pmatrix} a & b \\ 0 & c \end{pmatrix} \middle| a,b,c \in \mathbb{R},\ ac \neq 0 \right\}$, what are G' and G''?

If $\alpha: G \to G$ is an automorphism, then we find that $\alpha(G') = G'$. This follow, since

$$\alpha(x^{-1}y^{-1}xy) = \alpha(x^{-1})\alpha(y^{-1})\alpha(x)\alpha(y)$$
$$= \alpha(x)^{-1}\alpha(y)^{-1}\alpha(x)\alpha(y) \in G'$$

Since any inner automorphism f_g given by $f_g(x) = g^{-1}xg$ is an automorphism, it follows immediately that G' is normal. In fact, G' is even more than normal.

Definition   If H is a subgroup of G such that $\alpha(H) \subseteq H$ for all automorphisms α of G, then H is **characteristic** in G. This is denoted H char G.

The next exercises show why this property of being characteristic, though harder to establish than being normal, is often more useful. We have seen already that H char G implies $H \lhd G$.

Exercises

9 (The transitive property for characteristic subgroups) If H char G and K char H, show that K char G.
10 Show that G', $G^{(2)}$, $G^{(3)}$, ... are all characteristic in G and therefore normal in G.
11 (The lack of transitivity for normal subgroups) In D_8, find H and K such that $H \lhd D_8$ and $K \lhd H$ but K is not normal in D_8.
12 If $H \lhd G$ and K char H, show that $K \lhd G$.
13 Show that $Z(G)$ is characteristic in G, where $Z(G)$ is the center of G.
14 All subgroups of Klein's four group V are normal. Why? Which of them are characteristic? [If $V = \{e,a,b,c\}$, it might be noted that $\alpha(a) = b$, $\alpha(b) = c$, $\alpha(c) = a$, $\alpha(e) = e$ yields an automorphism α of V.]
15 If G is solvable, show tht G contains a normal, abelian, nonidentity subgroup.

The last set of exercises in this section treats p-groups and shows that they are solvable. The key fact used is that $Z(P) \neq \{e\}$, and the method used is the correspondence theorem and induction.

Proposition 5   If H is a proper subgroup of a finite p-group P, then $N_P(H) \supsetneq H$. Here, $N_P(H)$ is the normalizer of H in P.

   PROOF   If $H \subsetneq Z(P)$ then, since $N_P(H) \supseteq Z(P)$, we would have $N_P(H) \supsetneq H$. Thus we may safely assume $H \supseteq Z(P)$. Since $Z(P) \lhd P$, we have

$$\bar{P} = \frac{P}{Z(P)} \supseteq \bar{H} = \frac{H}{Z(P)} \supseteq \{\bar{e}\} = \frac{Z(P)}{Z(P)}$$

$|\bar{P}| < |P|$, and thus we apply the induction hypothesis to $\bar{P}$ and assume $\bar{N} = N_{\bar{P}}(\bar{H}) \supsetneq \bar{H}$. The second part of the fundamental homomorphism theorem finishes the proof.   ////

Corollary   If H is a subgroup of P of index p, then $H \lhd P$.

Exercise

16   Prove the corollary.

Proposition 6   If P is a finite p-group, then P is solvable.

   PROOF   We proceed by induction on the order of P. Since $Z(P) \neq \{e\}$, $P/Z(P)$ has smaller order than P and, by the induction hypothesis, is solvable. However, $Z(P)$, being abelian, is solvable. Proposition 3 now says P itself must be solvable.   ////

Exercises

17   If H is a subgroup of P, show that there exists a solvable series with H as one of the series. (*Hint:* Proposition 1 will suffice to get from H down to $\{e\}$, and Proposition 5 does the rest.)

18   If $|P| = p^n$ and $0 \leq m \leq n$, then there exists a normal subgroup of P of order p^m. [*Outline:* If $m = 0$, this is trivial. The result is true for abelian groups, which will be proved in the next section. Thus $Z(P)$ contains a subgroup N of order p. Show that P/N contains a normal subgroup of order p^{m-1} by the induction hypothesis. Go back to P.]

19   If P is a nontrivial finite p-group, show that $P \neq P'$.

20   **Let P be a finite p-group with $p \neq 2$, and let P contain only one subgroup of order p^m for some $p^m \neq 1$ or $|P|$. Show that P is cyclic.

21   (This requires the Sylow theorems as well as Propositions 3 and 6.) Show that any group of order 200 is solvable.

We now give some additional exercises on automorphisms and automorphism groups.

22   If $x \in G$ and we define $f_x : G \to G$ by $f_x(g) = x^{-1} g x$, show that f_x is a automorphism of G. (f_x is called an *inner automorphism* or the *inner automorphism induced by x.*)

23   Show that the automorphisms of G form a group [which we denote Aut (G)]. Show that the inner automorphisms form a normal subgroup [which we denote Inn (G)].

24   Show that $G/Z(G) \simeq$ Inn (G). [*Hint*: Let G act on itself by conjugation. Then $\theta(x) = f_x$ is the action homomorphism. What is ker (θ)?]

25   If α is automorphism of G show that
    (a) $\alpha(g)$ and g have the same order
    (b) $N \lhd G \Rightarrow \alpha(N) \lhd G$
    (c) $\alpha(g)\alpha(h) = \alpha(h)\alpha(g) \Leftrightarrow gh = hg$
    (d) $\alpha(Z(G)) = Z(G)$

26   Find Aut $(\mathbb{Z})$ and Aut $(\mathbb{Z}_5, \oplus)$. [*Hint*: g is a generator $\Leftrightarrow \alpha(g)$ is a generator.]

27   Show that Inn $(\text{Sym}(n)) \simeq$ Sym (n) for all $n \geqslant 3$. [Actually, Inn $(\text{Sym}(n)) =$ Aut $(\text{Sym}(n)) \simeq$ Sym (n) for all $n \geqslant 3$ except for $n = 6$. This is an old theorem of Hölder. See Burnside [14] or Passman [69] for a proof.]

5.2   FINITELY GENERATED ABELIAN GROUPS

The third major classification theorem we come to in this book is the fundamental theorem of finitely generated abelian groups. It says that any finitely generated abelian group is the direct sum of cyclic groups. Thus it depends on our first major classification theorem, which says all cyclic groups of the same order are isomorphic. It also strongly resembles the classification of finite-dimensional vector spaces by dimension. This material is used in Chap. 9 on group representations.

The theorem was first proved in the finite case by Kronecker in 1870. That the finite cyclic factors could be taken to be of prime power order was proved by Frobenius and Stickelberger [26] in 1878, although special cases were known previously. It has since been rephrased in terms of finitely generated modules over a principal ideal domain.

First let us recall how the decomposition of G as a direct product goes. G is isomorphic to $H_1 \times H_2$ if $H_1 \lhd G$, $H_2 \lhd G$, $H_1 \cap H_2 = \{e\}$, and $G = H_1 H_2$. Here, since G will be abelian, we shall write everything in additive notation, so this will become $H_1 \cap H_2 = \{0\}$ and $G = H_1 + H_2$. Since G is abelian, the normality of the subgroups is automatic.

If this G were a direct sum of H_1, H_2, ..., H_k, then we would have $H_i \cap (H_1 + \cdots + H_{i-1} + H_{i+1} + \cdots + H_k) = \{0\}$ for $i = 1, 2, \ldots, k$, and $G = H_1 + H_2 + \cdots + H_k$.

Definition An abelian group G is said to be **finitely generated** if there exists a finite subset $\{g_1, g_2, \ldots, g_k\}$ of G such that each element of G can be written as $m_1 g_1 + m_2 g_2 + \cdots + m_k g_k$ for some choice of integers m_i. The subset $\{g_i\}_{i=1}^k$ is then said to be a **set of generators** of G. If k is the smallest integer so that some set of k generators can be found, then k is the **rank** of G. A cyclic group is precisely a finitely generated abelian group of rank 1.

The fundamental theorem of abelian groups If A is a finitely generated abelian group, then A is isomorphic to a direct sum of cyclic groups. In fact, if A has rank r, then it is isomorphic to a direct sum of r cyclic groups.

OUTLINE OF PROOF Assume A is finitely generated and thus has some rank r. Also assume A is written additively, so if $\{g_i\}_{i=1}^r$ is a set of generators, then each element of A can be written as $n_1 g_1 + n_2 g_2 + \cdots + n_r g_r$. We want to find a new set of generators $\{h_i\}_i^r$ that has the additional property that

$$0 = n_1 h_1 + n_2 h_2 + \cdots + n_r h_r \text{ implies each } n_i h_i = 0$$

This second property is equivalent to the condition that

$$\langle h_i \rangle \cap (\langle h_1 \rangle + \langle h_2 \rangle + \cdots + \langle h_{i-1} \rangle + \langle h_{i+1} \rangle + \cdots + \langle h_r \rangle) = \{0\}$$

Since the $\{h_i\}_{i=1}^r$ also generate A, this will suffice to prove $A = \langle h_1 \rangle \oplus \langle h_2 \rangle \oplus \cdots \oplus \langle h_r \rangle$.

One point here differs from what one might expect. We specify $n_i h_i = 0$ and not $n_i = 0$. This modification is necessary if we want to handle finite groups. For instance, if a generates a cyclic group of order 5, it is certainly true that $0 \cdot a = 5 \cdot a = 10 \cdot a = 0$.

The proof proceeds by induction on the rank. Assume it is true for all abelian groups of rank $k - 1$. If, for some generating set $\{g_i\}_{i=1}^h$, we have that $0 = n_1 g_1 + n_2 g_2 + \cdots + n_k g_k$ implies all $n_i g_i = 0$, then we are done. So we may assume each such generating set and each such relationship has some of the $n_i g_i$ not equal to 0. Let m_1 be the smallest positive integer such that $m_1 g_1' \neq 0$ among all the m_i in all generating sets $\{g_i'\}_{i=1}^k$, where $\sum_{i=1}^k m_i g_i' = 0$. We can relabel so that m_1 is this smallest integer.

It can be shown (Exercise 2) that m_1 divides all p_1, where $p_1 g_1' + p_2 g_2 + \cdots + p_k g_k' = 0$. It can then be shown (Exercise 3) that m_1 divides $m_2, m_3, \ldots, m_k$ for any relationship

$$m_1 g_1' + m_2 g_2' + \cdots + m_k g_k' = 0$$

Thus $m_i = a_i m_1$ for some $a_i \in \mathbb{Z}$, for $i = 2, 3, \ldots, k$.

Now we construct a new set of generators,

$$g_1'' = g_1' + a_2 g_2' + a_3 g_3' + \cdots + a_k g_k'$$
$$g_2'' = g_2' \qquad g_3'' = g_3' \qquad \cdots \qquad g_k'' = g_k'$$

We now can conclude that $\langle g_1'' \rangle \cap \langle g_2'', g_3'', \ldots, g_k'' \rangle = 0$ and apply induction to $\langle g_2', g_3', \ldots, g_k' \rangle$, which has rank $k - 1$.

Rewriting $\langle g_2'', g_3'', \ldots, g_k'' \rangle$ as a direct sum of $k - 1$ cyclic groups will, together with $\langle g_1'' \rangle$, give us A as a direct sum of k cyclic groups.

////

The following exercises refer to the last proof.

Exercises

1   Show that the theorem is true if the rank of A is 1.

2   Show that if m_1 is minimal with respect to $m_1 g_1' + m_2 g_2' + \cdots + m_k g_k' = 0$, and $m_1 g_1' \neq 0$, then $m_1 | p_1$ for any p_1 satisfying $p_1 g_1' + p_2 g_2' + \cdots + p_k g_k' = 0$.

3   Using the same notation, show that $m_1 | m_2$, $m_1 | m_3$, $\ldots$, $m_1 | m_k$ whenever $m_1 g_1' + m_2 g_2' + \cdots + m_k g_k' = 0$.

4   Writing $m_i = a_i m_1$ for $i = 2, 3, \ldots, k$, show that $g_1'', g_2'', \ldots, g_k''$ is also a set of generators, where

$$g_1'' = g_1' + a_2 g_2' + a_3 g_3' + \cdots + a_k g_k'$$
$$g_2'' = g_2' \qquad g_3'' = g_3' \qquad \cdots \qquad g_k'' = g_k'$$

(It is obvious that each g_i'' is a sum of g_i', so the converse is all that need be shown.)

5   Show $\langle g_1'' \rangle \cap \langle g_2'', \ldots, g_k'' \rangle = \{0\}$. (Otherwise, show that g_1' is in $\langle g_2', g_3', \ldots, g_k' \rangle$, which would lower the rank of A.)

Definition   If $A \simeq \underbrace{\mathbb{Z} \oplus \mathbb{Z} \oplus \mathbb{Z} \oplus \cdots \oplus \mathbb{Z}}_{k \text{ copies}} \oplus F$ and F is a finite abelian group, then from algebraic topology we call k the **Betti number** of A.

6   If A is any abelian group, then let $T(A)$ be those elements of A of finite order. Show that $T(A)$ is a subgroup of A. (This subgroup is called the *torsion subgroup* of A.)

7   If A has no element of finite order other than the identity, then A is said to be *torsion-free*. Show that for abelian group A, the group $\bar{A} = A/T(A)$ is torsion-free.

8   In the group $\mathbb{Z} \oplus \mathbb{Z}_2$, show that the elements of infinite order together with $(0, \bar{0})$, do not form a group. What is the Betti number of this group?

9   If A is a finite abelian group and p is prime, show that A contains the following two subgroups: $A_{p'}$, those elements of A whose order is a power of p, and $A_{p'}$, those elements of A whose order is relatively prime to p. Show that $A = A_p \oplus A_{p'}$.

We want to look more closely at the finite abelian groups. It's easy to see that any finite abelian group is finitely generated, since we could take the whole group as a set of generators. Using either Exercise 9 or the first Sylow theorem, we see that we can examine one prime at a time.

What are the possibilities for a finite abelian p-group? Up to isomorphism, the only such groups of order p^4 are $\mathbb{Z}_{p^4}$, $\mathbb{Z}_{p^3} \oplus \mathbb{Z}_p$, $\mathbb{Z}_{p^2} \oplus \mathbb{Z}_p \oplus \mathbb{Z}_p$, $\mathbb{Z}_{p^2} \oplus \mathbb{Z}_{p^2}$, and $\mathbb{Z}_p \oplus \mathbb{Z}_p \oplus \mathbb{Z}_p \oplus \mathbb{Z}_p$.

We note that the ways to partition 4 as a sum of positive integers are

$$4 = 3 + 1 = 2 + 1 + 1 = 2 + 2 = 1 + 1 + 1 + 1$$

Let $\rho(n)$ be the *number of partitions* of the positive integer n into a sum of positive integers. Then the number of nonisomorphic abelian groups of order p^n is $\rho(n)$. Accepting this without proof for a minute, we can do the following exercises.

Exercises

10 Find all seven nonisomorphic abelian groups of order p^5. How many nonisomorphic abelian groups of order p^6 are there?

11 Find all the nonisomorphic abelian groups of order 90,000.

12 (($\Phi(16)$, $\cdot$) is an abelian group of order 8 consisting of 1, 3, $\ldots$, 13, 15 together with multiplication modulo 16. Is this group isomorphic to $\mathbb{Z}_8$, $\mathbb{Z}_4 \oplus \mathbb{Z}_2$, or $\mathbb{Z}_2 \oplus \mathbb{Z}_2 \oplus \mathbb{Z}_2$? How can this be generalized?

To show that each partition of n gives a distinct abelian p-group, assume $n = n_1 + n_2 + \cdots + n_k = m_1 + m_2 + \cdots m + _l$, where $n_1 \geq n_2 \geq \cdots \geq n_k$ and $m_1 \geq m_2 \geq \cdots \geq m_l$. We want to show that

$$A = \mathbb{Z}_{p^{n_1}} \oplus \mathbb{Z}_{p^{n_2}} \oplus \cdots \oplus \mathbb{Z}_{p^{n_k}} \quad \text{and} \quad B = \mathbb{Z}_{p^{m_1}} \oplus \mathbb{Z}_{p^{m_2}} \oplus \cdots \oplus \mathbb{Z}_{p^{m_l}},$$

are nonisomorphic. We do this by induction on n, leaving $n = 1$ to the reader.

If, say, $n_1 > m_1$, then A has elements of order p^{n_1} but B does not, so we can assume $n_1 = m_1$. Thus $A \simeq B$ would imply $\bar{A} = A/\mathbb{Z}_{p^{n_1}} \simeq B/\mathbb{Z}_{p^{m_1}} = \bar{B}$. We can now apply the induction hypothesis to $\bar{A}$ and $\bar{B}$, which represent different partitions of $n - n_1$, to obtain the contradiction. Thus A and B must be nonisomorphic.

It is straightforward to check that this gives a complete list of abelian p-groups. A finite abelian group A is finitely generated; therefore A is isomorphic to a finite direct sum of cyclic groups. Each of these cyclic groups must be finite for A to be finite. If, in addition, A is a p-group, then each of the cyclic groups must be of order p^{n_i}. Thus,

$$p^n = |A| = |\mathbb{Z}_{p^{n_1}} \oplus \mathbb{Z}_{p^{n_2}} \oplus \cdots \oplus \mathbb{Z}_{p^{n_k}}|$$

and $n = n_1 + n_2 + \cdots + n_k$ is a partition of n into positive integers.

In one sense we have complete information about finite abelian groups. In another sense it just means that we ask harder questions. The next few exercises are a few of these harder questions.

Exercises

13 (a) If $A = \mathbb{Z}_2 \oplus \mathbb{Z}_2 \oplus \mathbb{Z}_2$, show that A has seven subgroups of order 2.

(b) *How many subgroups of order 4 are there? [One way to do this is to count ordered pairs (T,F), where $|T| = 2$, $|F| = 4$, and T is a subgroup of F. Count these in two ways.]

14 *Find all subgroups of $\mathbb{Z}_2 \times \mathbb{Z}_m$.

15 *Find the automorphism group of $\mathbb{Z}_m$.

16 Let $E_{p^n} = \mathbb{Z}_p \oplus \mathbb{Z}_p \oplus \cdots \oplus \mathbb{Z}_p$, which is called an *elementary abelian p-group*.

(a) Show that any homomorphism ϕ from $(E_{p^n}, +)$ to $(E_{p^n}, +)$ satisfies $\phi a = a\phi$ for any $a \in \mathbb{Z}_p$. Thus, any homomorphism from E_{p^n} to itself is a linear transformation over $\mathbb{Z}_p$. Here p denotes a prime.

(b) Show that Aut $(E_{p^n}) \simeq$ GL $(n, \mathbb{Z}_p)$, where Aut (E_{p^n}) denotes the automorphism group of E_{p^n}.

17 *Find conditions on a finite abelian group A so that Aut (A) will also have to be abelian.

Turning to some easier questions, we can use the fundamental theorem to show the following:

18 If A is a finite abelian group and m is a positive integer dividing $|A|$, show that A has a subgroup of order m. Thus, for abelian groups, Lagrange's theorem has a complete converse.

19 (a) Let A be a finite abelian group and $|A| = p_1^{m_1} p_2^{m_2} \cdots p_k^{m_k}$, where the p_i's are distinct primes. Show that

$$A \simeq A_1 \oplus A_2 \oplus \cdots \oplus A_k \qquad \text{where} \qquad |A_i| = p_i^{m_i}$$

(b) If $n = p_1^{m_1} p_2^{m_2} \cdots p_k^{m_k}$, where the p_i's are distinct primes, then the number of nonisomorphic abelian groups of order n is $\rho(m_1)\rho(m_2) \cdots \rho(m_k)$. Here ρ is the partition function.

20 Show that an abelian group of order 108 has but one subgroup of order 27. *Show that the number of subgroups of order 3 is either 1, 4, or 13.

21 *Any subgroup of a finitely generated abelian group of rank r is a finitely generated abelian group of rank at most r. Prove this statement.

22 *Find two abelian groups A_1 and A_2 such that A_1 is isomorphic to a subgroup of A_2, A_2 is isomorphic to a subgroup of A_1, but A_1 and A_2 are not isomorphic. (Obviously, the A_i are not finite; using Exercise 21, the A_i cannot even be finitely generated.)

23 (a) Show that $\mathbb{Z}_2 \oplus \mathbb{Z}_3 \simeq \mathbb{Z}_6$.

(b) Show that $\mathbb{Z}_{m_1} \oplus \mathbb{Z}_{m_2} \simeq \mathbb{Z}_{m_1 m_2}$ if and only if gcd $(m_1, m_2) = 1$.

(c)   Show that $\mathbb{Z}_8 \oplus \mathbb{Z}_4 \oplus \mathbb{Z}_9 \oplus \mathbb{Z}_3 \simeq \mathbb{Z}_{72} \oplus \mathbb{Z}_{12}$.

(d)   *Show that any finite abelian group is isomorphic to $\mathbb{Z}_{m_1} \oplus \mathbb{Z}_{m_2} \oplus \cdots \oplus \mathbb{Z}_{m_k}$, where $m_k \mid m_{k-1}$, $m_{k-1} \mid m_{k-2}$, ..., $m_2 \mid m_1$. (These m_i satisfying $m_i \mid m_{i-1}$ are called the *torsion coefficients* of the abelian group.)

(e)   Compose some examples, and find their torsion coefficients.

There are four directions which the reader might want to pursue. First, the decomposition of a finitely generated abelian group to a direct sum of cyclic groups can be done with matrices and can even be programed. Dean [20] and Schreier and Sperner [77] are the best references for this. This kind of computation is important in algebraic topology.

The second direction is the wide world of finite nonabelian groups. Recommended here are Rotman [73], MacDonald [59], Kurosh [50], Hall [32], and Scott [78].

The third topic of related interest is infinite abelian groups. Here the number of books is small, and they are all worthwhile. The complete list is Kaplansky [45], Griffith [31], and Fuchs [27].

The fourth topic is principal ideal domains and their finitely generated modules. Principal ideal domains are discussed in Chap. 6, but the most important example is $\mathbb{Z}$. An R-module, briefly, is a vector space over a ring R. Vector spaces are modules over fields, and $\mathbb{Z}$-modules turn out to be the same as abelian groups. Had we phrased things a little differently and checked two or three more details, we would have proved that any finitely generated module over a principal ideal domain is isomorphic to a direct sum of cyclic modules. References for this approach are Lang [53], Sah [74], and Fuchs [27].

6

INTEGRAL DOMAINS

6.1 DEFINITIONS AND QUOTIENT FIELDS

We saw earlier that the various classes of rings can be put in a lattice such as

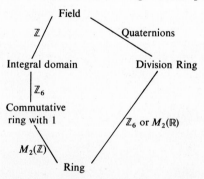

where separating examples have been placed along the edges.

In this chapter we want to discuss various sorts of integral domains; at the end, we shall have the following:

Field
/
/ $\mathbb{Z}$

Euclidean domain = ED

$\left| \mathbb{Z}\left(\dfrac{1+\sqrt{-19}}{2} \right) \right.$.

Principal Ideal Domain = PID

$\left| F(x,y) \right.$

Unique Factorization Domain = UFD

$\left| \mathbb{Z}(\sqrt{-5}) \right.$

Integral Domain

The name for integral domains around the turn of the century was "domains of integrity." As we shall see, this righteous nomenclature applies more to the commutative and cancellation laws than it does to unique factorization into primes. To save space, we shall sometimes refer to integral domains as domains, and R^* will denote the nonzero elements of a ring R.

The applications of this material are primarily in number theory and in algebraic geometry. Some topics in number theory utilizing unique factorization are discussed in the next chapter. Some other uses for integral domains are illustrated in the exercises.

In the process of constructing the real numbers, one crucial step is to use the integers to construct the rationals. The same basic technique can be used to construct a field of quotients from any integral domain.

If we leave out details, the basic construction goes as follows: Let D be an integral domain and let $D^* = D - \{0\}$.

1 Form ordered pairs (a,b), where a, $b \in D$ and $b \in D^*$. (These are to correspond to the fraction a/b.)

2 Let $\sim$ be defined by $(a,b) \sim (c,d)$ if $ad = bc$, and show that $\sim$ is an equivalence relation. (This says, for instance, that $\frac{3}{4} \sim \frac{15}{20}$.)

3 Denote the equivalence class containing (a,b) by $[a,b]$, and define addition and multiplication as follows:

$$[a,b] + [c,d] = [ad + bc, bd] \qquad [a,b] \cdot [c,d] = [ac,bd]$$

Show that if $[a,b] = [a',b']$ we have $[a,b] + [c,d] = [a',b'] + [c,d]$ and $[a,b]$ $[c, d] = [a',b'][c,d]$. These definitions correspond to $a/b + c/d = (ad + bc)/bd$ and $(a/b)(c/d) = ac/bd$.

4 Verify that the $\{[a,b]\}$, with the given addition and multiplication, form a field F. Note that $[1,1] = [a,a]$ is the multiplicative identity; $[0,1]$ is the additive identity; and if $[a,b] \neq [0,1]$ so that a and b are both nonzero, then $[a,b]^{-1} = [b,a]$. Checking of the associative, commutative, and distributive laws is left as an exercise.

5　Show that the mapping $d \rightarrow [d,1]$ is a monomorphism of D into F, so D can be considered as a subring of F.

6　If D is a subring of any field G, then G contains a subfield isomorphic to F. (Thus F is the "smallest" field containing D.)

Exercises

1　Complete the details of the preceding construction.

We see that $\mathbb{Q}$ is the quotient field of $\mathbb{Z}$ since, if $\mathbb{Z}$ is contained in a field F, then all the ab^{-1}, where $a \in \mathbb{Z}$, $b \in \mathbb{Z}^*$, must also be in F. Thus $\mathbb{Q}$ must be a subring of F, and $\mathbb{Q}$ is thus the "smallest" field containing $\mathbb{Z}$.

2　What are the quotient fields of $\mathbb{Z}(\sqrt{2})$ and $\mathbb{Z}[i]$? What is the quotient field of $F[x]$?

3　Where does the construction fail if we only assume D is a commutative ring with 1?

The next exercise is stated entirely in terms of domains but the straightforward proof needs the existence of quotient fields.

4　If d and d' are in a domain D, and if $d^m = d'^m$ and $d^n = d'^n$ for relatively prime integers m and n, show that $d = d'$. (*Warning*: In the ring $\mathbb{Z}_9$, we have $3^2 = 0^2$ and $3^3 = 0^3$ but $0 \neq 3$.)

5　What is the field of quotients of $D = \{n/2^k \mid n, k \in \mathbb{Z}\}$? Can this be generalized?

6　Find the error in the following alternative "proof" of the existence of a quotient field F of a domain D.

(a)　$D \circ D$ is a commutative ring with unity, where the elements of $D \circ D$ are ordered pairs (a,b), for $a \in D$ and $b \in D^*$. Addition is given by $(a,b) + (c,d) = (ad + bc, bd)$, and multiplication is given by $(a,b) \cdot (c,d) = (ac,bd)$.

(b)　$M = \{(0,d) \mid d \in D^*\}$ is a maximal ideal in $D \circ D$.

(c)　$F = (D \circ D)/M$ is a field.

(d)　The mapping $d \mapsto (d,1) + M$ is a monomorphism from D into F.

In a commutative ring R we say that r *divides* s or $r \mid s$ if there is an element r' such that $rr' = s$. An element u is *invertible* in a commutative ring with 1 if $u \mid 1$. (See Sec. 2.1.) Two elements s and t in R are *associates* if $s = tu$, where u is invertible. An invertible element is often called a *unit*. (A unit should not be confused with the multiplicative unity, 1.)

7　In a commutative ring with identity, let $s \sim t$ if s and t are associates. Show that $\sim$ is an equivalence relation.

8 If R is a ring with 1, and $\mathscr{U}(R)$ is the group of units or group of invertible elements, show that $\mathscr{U}(R)$ acts on R by left multiplication. Show that the orbits of this action are the equivalence classes of the preceding exercise.

9 Describe all the units and classes of associated elements in $\mathbb{Z}[i]$.

Definition A nonzero element a of a commutative ring with 1 is **irreducible** if a is not a unit and $a = bc$ implies that b or c is a unit.*

10 What are the units and irreducibles of $\mathbb{Z}_{12}$?

11 What are the units and irreducibles of $\mathbb{Z}$?

12 What are the units and irreducibles in a field F?

13 **What are the units and irreducibles in $\mathbb{Z}[i] = \{a + bi \mid a, b \in \mathbb{Z}\}$? [*Hint*: Watch out, since $29 = (5 + 2i)(5 - 2i)$.]

14 (*a*) In any integral domain D, let the group of units, $\mathscr{U}(D)$, act on D by left multiplication. Show that all the orbits not containing 0 have the same length.

    (*b*) *Find two examples of commutative rings with 1 that are not integral domains but which have this equal-orbit property. That is, all classes of associates have the same length if we exclude 0.

15 (J. Joseph) How close is $\mathbb{Z}$ to being the only commutative ring with 1 such that each nonzero class of associates has two elements?

6.2 EUCLIDEAN DOMAINS AND PRINCIPAL IDEAL DOMAINS

An integral domain D is called a *euclidean domain* if it is close enough to being the integers so that the euclidean algorithm will work. Thus, if $a \in D$ and $b \in D^*$, we need to be able to find c and r such that $a = bc + r$, where r is in some sense smaller than b. The absolute-value function works in $\mathbb{Z}$ (see Sec. 1.3) to give us the concept of smaller. The formal definition is as follows:

Definition An integral domain E together with a function $\delta\colon E^* \to \mathbb{Z}^+ \cup (0)$ is a **euclidean domain** if

    (*a*) $\delta(a) \leq \delta(ab)$ for all a, b in E^*.

    (*b*) For any a in E and b in E^*, there exist c and r in E such that

$$a = bc + r \quad \text{and} \quad \delta(r) < \delta(b) \quad \text{or} \quad r = 0$$

It would be more precise, although more cumbersome, to say (E, δ) is a

---

* The two most familiar rings are $\mathbb{Z}$ and $F[x]$. Our definition of "irreducible" corresponds to the primes in $\mathbb{Z}$ and to the irreducible polynomials in $F[x]$. Thus, both prime and irreducible would be logical choices for the term we have defined. The slightly more common term nowadays is irreducible, with "prime" being used for the generator of a prime ideal.

euclidean domain. Certainly you have to have an appropriate δ function in mind, as a domain all by itself cannot be considered to be euclidean. The premier example of a euclidean domain, is of course, $\mathbb{Z}$ [or $(\mathbb{Z}, |\ |)$]. After we have established a few elementary properties of euclidean domains, we shall show that $F[x]$, the ring of polynomials over a field, and $\mathbb{Z}[i]$ are euclidean.

We first note that $\delta(1)$ is a minimal element in the range of δ, since $\delta(1) \le \delta(1 \cdot a) = \delta(a)$ for all a in E^*. If u is a unit, then

$$\delta(u) \le \delta(uu^{-1}) = \delta(1)$$

so $\delta(u) = \delta(1)$.

Conversely, if $\delta(v) = \delta(1)$, then, using condition b, we can find c and r such that $1 = vc + r$ and either $\delta(r) < \delta(v) = \delta(1)$ or $r = 0$. The first possibility is eliminated by the minimality of $\delta(1)$, so we must have $1 = vc + 0 = vc$, and v is a unit. We have thus proved the following:

Proposition 1 If E together with δ is a euclidean domain, then

(a) $\delta(1)$ is a minimal element in the range of δ.
(b) $\delta(a) = \delta(1)$ if and only if a is a unit in E.

Exercises

1   Show that every field is a euclidean domain.
2   Use the previous proposition on $\mathbb{Z}$ together with $|\ |$ to find (once again) the units of $\mathbb{Z}$.
3   (a)   Let $E_p = \{a/b \mid p \nmid b$ and $a, b \in \mathbb{Z}, p$ a fixed prime$\}$. Show that E_p is an integral domain.
      (b)   Find the units of E_p.
      (c)   Define a δ such that (E_p, δ) is a euclidean domain. Note first that a/b in E_p can be written as $(a'/b)p^k$, where $p \nmid a'$, $p \nmid b$, and $k = 0, 1, 2, \ldots$.

Let F be a field, and let $F[x]$ denote the ring of all polynomials in one variable with coefficients in F. We have shown earlier that this is a commutative ring with unity, and now we want to show that it is an integral domain. It seems easier to show there are no zero divisors rather than to check a cancellation law. So assume $a = \sum_{i=0}^{n} a_i x^i$ and $b = \sum_{i=0}^{m} b_i x^i$ are nonzero polynomials with $a_n \ne 0$ and $b_m \ne 0$, and that $(\sum_{i=0}^{n} a_i x^i)(\sum_{i=0}^{m} b_i x^i) = 0$. Multiplying out, we have

$$\sum_{k=1}^{m+n} c_k x^k = 0 \qquad \text{where} \qquad c_k = \sum_{i=0}^{k} a_{k-i} b_i$$

In particular, the highest term $c_{mn} = a_n b_m = 0$, which is a contradiction since $a_n \ne 0$ and $b_m \ne 0$, and F is a field.

Exercise

4 (*a*) If D is an integral domain, show that $D[x]$ is also an integral domain.

(*b*) Show that $D[x,y] = (D[x])[y]$ is also an integral domain.

(*c*) Show that $D[x_1, x_2,...,x_k]$ is an integral domain.

We can show that $F[x]$ is a euclidean domain by defining $\delta(\sum_{i=0}^{n} a_i x^i) = n$ (asuming, of course, that $a_n \neq 0$). First, we note that δ is a map from $F(x)$ to the nonnegative integers. Second, we do not define δ for the zero polynomial. All the other constant polynomials are mapped by δ to 0; the linear polynomials $ax + b$ are mapped to 1; $a_2 x^2 + a_1 x + a_0$ to 2; and so on. We can easily verify condition *a*, since

$$\delta\left(\sum_{i=0}^{n} a_i x^i\right)\left(\sum_{i=0}^{m} b_i x^i\right) = \delta\left(\sum_{i=0}^{m+n} c_k x^k\right) = m + n \geq n = \delta\left(\sum_{j=0}^{n} a_i x^i\right)$$

It remains only to verify condition *b*, and this turns out to be just long division by polynomials, a topic from high-school mathematics. If $a = \sum_{i=0}^{m} a_i x^i$ and $b = \sum_{i=0}^{m} b_i x^i \neq 0$, then by dividing out we have

$$\frac{\sum_{i=0}^{n} a_i x^i}{\sum_{i=0}^{m} b_i x^i} = \sum_{i=0}^{k} c_i x^i + \frac{\sum_{i=0}^{l} r_i x^i}{\sum_{i=0}^{m} b_i x^i} \quad \text{and} \quad l < m$$

Since we haven't defined division in this ring, the rigorous way to write this is $\sum_{i=0}^{n} a_i x^i = \sum_{i=0}^{k} c_i x^i \sum_{i=0}^{m} b_i x^i + \sum_{i=0}^{l} r_i x^i$. Since $l < m$, we see that condition *b* is satisfied for δ. This shows that $F[x]$ is a euclidean domain. Two points are to be noted here. First, for

$$\sum_{i=0}^{n} a_i x^i = \sum_{i=0}^{k} c_i x^i \cdot \sum_{i=0}^{m} b_i x^i + \sum_{i=0}^{l} r_i x^i \quad l < m$$

we must have $c_k = a_n(b_m^{-1})$, and this is where we need the fact that F is a field. Second, we are using induction on n, and $\sum_{i=0}^{n} a_i x^i - c_k x^k \sum_{i=0}^{m} b_i x^i$ has degree $n - 1$; thus, the induction hypothesis does apply to it.

Our next example of a euclidean domain is the ring of *Gaussian integers*, $\mathbb{Z}[i] = \{a + bi \mid a, b \in \mathbb{Z}\}$. This is a subring of $\mathbb{C}$, the complex numbers, and so is commutative, has an identity, has no zero divisors, and consequently is an integral domain. We define $\delta: \mathbb{C}^* \rightarrow \mathbb{R}^+$ by $\delta(z) = z\bar{z}$. This is equivalent to $\delta(a_1 + a_2 i) = a_1^2 + a_2^2$, where $a_1, a_2 \in \mathbb{R}$.

We first observe that δ takes $\mathbb{Z}[i]^*$ into $\mathbb{Z}^*$ and then that

$$\delta(a_1 + a_2 i)\delta(b_1 + b_2 i) = (a_1^2 + a_2^2)(b_1^2 + b_2^2)$$
$$= (a_1 b_1 - a_2 b_2)^2 + (a_1 b_2 + a_2 b_1)^2$$
$$= \delta((a_1 b_1 - a_2 b_2) + (a_1 b_2 + a_2 b_1)i)$$
$$= \delta[(a_1 + a_2 i)(b_1 + b_2)]$$

Thus $\delta(z_1)\delta(z_2) = \delta(z_1 z_2)$ for all *complex numbers* z_1, z_2. If, in addition, z_1, $z_2 \in \mathbb{Z}[i]$, we must have $\delta(z_1) \leq \delta(z_1 z_2) = \delta(z_1)\delta(z_2)$, which is condition a.

Considering the gaussian integers as the square lattice in the plane, it is obvious that for any complex number $z = x_1 + x_2 i$ we can find a gaussian integer $c_1 + c_2 i$ such that $|x_1 - c_1| \leq \frac{1}{2}$ and $|x_2 - c_2| \leq \frac{1}{2}$. If $a_1 + a_2 i \in \mathbb{Z}[i]$ and $b_1 + b_2 i \in \mathbb{Z}[i]^*$, then let

$$\frac{a_1 + a_2 i}{b_1 + b_2 i} = x_1 + x_2 i = c_1 + c_2 i + s_1 + s_2 i$$

so that $s_1 = x_1 - c_1$, $s_2 = x_2 - c_2$, and $|s_1|$, $|s_2|$ are $\leq \frac{1}{2}$. Then

$$\delta(s_1 + s_2)i = s_1{}^2 + s_2{}^2 \leq \tfrac{1}{4} + \tfrac{1}{4} = \tfrac{1}{2}$$

So, multiplying out, we have

$$a_1 + a_2 i = (b_1 + b_2 i)(c_1 + c_2 i) + (b_1 + b_2 i)(s_1 + s_2 i)$$
$$= (b_1 + b_2 i)(c_1 + c_2 i) + (r_1 + r_2 i)$$

$r_1 + r_2 i \in \mathbb{Z}[i]$, since all the other elements are in $\mathbb{Z}[i]$, and

$$\delta(r_1 + r_2 i) = \delta(b_1 + b_2 i)\delta(s_1 + s_2 i) \leq \delta(b_1 + b_2 i)\tfrac{1}{2} < \delta(b_1 + b_2 i)$$

This completes the proof of condition b, and $\mathbb{Z}[i]$ is a euclidean domain. This is an interesting proof since, to solve a problem in $\mathbb{Z}[i]$ we go to $\mathbb{C}$, which at first sight seems to be a needless complication.

Exercises

5 If $a = 5 + 2i$ and $b = 3 + 3i$ are elements of $\mathbb{Z}[i]$, find c and r in $\mathbb{Z}[i]$ that will give $a = bc + r$, where $\delta(r) < \delta(b) = 18$.

6 If, in $\mathbb{Q}[x]$, $a = x^3 + 5x^2 + 3x + 2$ and $b = 3x^2 + 3x + 3$, what c and r will give $a = bc + r$, where the degree of r is less than 2?

As with groups and vector spaces, we let $\langle S \rangle$ denote the smallest subring containing S, where S is the subset of a ring. We say then that S *generates* $\langle S \rangle$.

Definition A **principal ideal domain** (abbreviated PID) is an integral domain where every ideal is generated by one element. That is, if A is a PID, and J is an ideal in A, then there is some element j in A such that $J = \langle j \rangle = jA = \{ja \,|\, a \in A\}$. Such an element j is the **generator** of $jA = J$.

Our next proposition will supply us with a number of examples of PID's.

Proposition 2 Every euclidean domain is a principal ideal domain.

PROOF Let E together with δ be a euclidean domain, and let I be any ideal in E. $\delta(I^*)$ is a subset of the nonnegative integers and, as such, has a minimal element $\delta(i)$. If i is an element of I^* such that $\delta(i)$ is minimal, then $I \supseteq \langle i \rangle = \{ie \,|\, e \in E\}$ by the ideal property. To prove the reverse inclusion, let $x \in I$. Since E is euclidean, $x = ci + r$ and $\delta(r) < \delta(i)$ or $r = 0$.

Since x and ci are in I, r is in I, but then the minimality of $\delta(i)$ excludes the possibility $\delta(r) < \delta(i)$. Therefore $r = 0$ and $x = ci$, so $x \in \langle i \rangle$ and $\langle i \rangle \supseteq I$. Thus $\langle i \rangle = I$ and, as I was an arbitrary ideal, E is a PID.

$////$

Thus we have that $\mathbb{Z}$, $\mathbb{Z}[i]$, any field F, and $F[x]$ and E_p from Exercise 3 are all PID's. For an example of an integral domain that is not a PID, consider the domain $\mathbb{Z}[x]$. It has an ideal $I = \langle 2, x \rangle =$ all polynomials with integer coefficients whose constant term is even. If I were principal, I would be equal to $\langle p(x) \rangle$, and there would have to be two other polynomials q and r so that $p(x)q(x) = 2$ and $p(x)r(x) = x$. This is impossible unless $p(x) = 1$, so $\mathbb{Z}[x]$ is not a PID. By the last proposition, $\mathbb{Z}[x]$ is not euclidean either.

There are examples of domains that are PID's but are not euclidean. One of these is $\mathbb{Z}[(1 + \sqrt{-19})/2]$. See the paper by Motzkin [ ] for a proof of this. Wilson [ ] has a more detailed and accessible presentation of this same example.

Exercises

7   If R is a ring with 1, u a unit of R, I an ideal in R, and $u \in I$, show that $I = R$.

8   Show that $F[x, y]$, the ring of all polynomials in two variables over a field F, is not euclidean.

9   This exercise presents a result in one-dimensional algebraic geometry. Let K be a field, and let $V \subseteq K$ be a *variety* if every point of V is a root of every polynomial in I, where I is some ideal of $K[x]$.

   (a)   Show that the only varieties possible in this setup are K, ϕ, and finite subsets of K. (If you have had topology, how can this result be rephrased?)

   (b)   Show that if $K = \mathbb{C}$ then every proper ideal I yields a proper variety. (*Hint*: Any polynomial in $\mathbb{C}[x]$ factors into linear factors.)

6.3   UNIQUE FACTORIZATION DOMAINS AND APPLICATIONS

In this section we are going to examine domains in which every element can be factored uniquely into a finite product of irreducible elements. First let us examine some examples in which this phenomenon does not occur. We look at $\bar{A} = K[y_1, y_2, y_3, \ldots]$, where $y_1 = y_2{}^2$, $y_2 = y_3{}^2, \ldots, y_i = y_{i+1}{}^2$, and we try to factor the monomial y_1. It factors easily enough, since $y_1 = y_2{}^2 = y_3{}^4 = \cdots = y_n{}^{2^{(n-1)}} = \cdots$, and none of the y_i are units. However, y_1 certainly doesn't factor, as a finite number of primes. We haven't shown that such a ring exists or that it is a domain, but have simply demonstrated the phenomenon of infinite factorization. See Exercise 1 for a construction of this ring.

A second example is $\mathbb{Z}(\sqrt{-5})$, where $9 = 3 \cdot 3 = (2 - \sqrt{-5}) \cdot (2 + (\sqrt{-5})$. It can be shown that, although $3, 2 - \sqrt{-5}$, and $2 + \sqrt{-5}$ are primes, 3 is not an associate of $2 - \sqrt{-5}$ or $2 + \sqrt{-5}$. We shall discuss this example later in this section and in the next section.

Exercise

1 Let $A = F[x_1, x_2, x_3, \ldots]$ be the ring of all polynomials in an infinite number of variables over a field F. To set up the infinite factorization we want for $F[y_1, y_2, y_3, \ldots]$, we use the principle that "to make something zero in the quotient, place it in the kernel." We want $y_i = y_{i+1}{}^2$ or $y_i - y_{i+1}{}^2 = 0$. Thus put $x_i{}^2 - x_{i+1}{}^2$ in some ideal I. More precisely, define $I = \langle x_1 - x_2{}^2, x_2 - x_3{}^2, x_3 - x_4{}^2, \ldots \rangle = $ the smallest ideal containing all $x_i - x_{i+1}{}^2$. Now let θ be the canonical map from A to A/I, and let $y_i = \theta(x_i)$. Show that this works.

We now want to find some reasonable conditions to guarantee factorization and then unique factorization. The ring $\bar{A} = K[y_1, y_2, y_3, \ldots]$ given above can be eliminated by the following finiteness condition.

Definition   An **ascending chain** $\{C_i\}_{i=0}^n$ or $\{C_i\}_{i=0}^\infty$ of subsets of a set is a set of subsets such that $C_0 \subseteq C_1 \subseteq C_2 \subseteq \cdots C_i \subseteq C_{i+1} \subseteq \cdots$. A **decending chain** satisfies $C_0 \supseteq C_1 \supseteq C_2 \supseteq \cdots \supseteq C_i \supseteq C_{i+1} \supseteq \cdots$.

Definition   A ring R satisfies the **ascending-chain condition** (acc) if no infinite ascending chain of ideals exists in R, each properly containing the previous ideal. The **descending-chain condition** has a similar definition and is denoted dcc.

Proposition 1   Let R be a commutative ring with 1. Then $\langle a \rangle = aR$ contains $\langle b \rangle = bR$ if and only if $a|b$. This proposition is sometimes summarized by the mnemonic "to divide is to contain."

PROOF   If $a|b$ then there exists c such that $ac = b$, so $b \in \langle a \rangle$. Since $\langle a \rangle$ is an ideal, $\langle b \rangle \subseteq \langle a \rangle$.
If $\langle a \rangle \supseteq \langle b \rangle$ then $b \in \langle a \rangle$, so $b = ca$ and $a|b$.   ////

We can now see that $\mathbb{Z}$ does not satisfy the dcc since $\langle 2 \rangle \supset \langle 4 \rangle \supset \langle 8 \rangle \supset \langle 16 \rangle \supset \langle 32 \rangle \cdots$. However, if $\langle n \rangle$ is any ideal in $\mathbb{Z}$, then n has only a finite number of proper divisors. So any chain starting $\langle n \rangle \subset \langle$proper divisor of $n \rangle \subset \cdots$ can have only a finite number of terms if all the inclusions are proper. Thus $\mathbb{Z}$ does satisfy the acc.

Exercises

2 (a) Find all the divisors of 60.

   (b) Find all ideals in $\mathbb{Z}$ containing $\langle 60 \rangle$.

   (c) Write down all strictly ascending chains in $\mathbb{Z}$ starting with $\langle 60 \rangle$. What is the maximum number of terms possible? How can this be generalized?

3 Show that $\mathbb{Z}(\sqrt{-3}) = \{a + b\sqrt{-3} \mid a, b \in \mathbb{Z}\}$ satisfies the acc but not the dcc. (*Hint*: If $a + b\sqrt{-3}$ is in I, so is the integer $a^2 + 3b^2$.)

4 Show that any finite ring and any field satisfy both chain conditions.

Proposition 2 If D is a domain satisfying the acc, then any nonzero element in D factors into a finite product of primes.

PROOF If $d \in D^*$ and d is prime, it can be considered as a product of just one prime. Otherwise, $d = d_1' d_2$, where neither d_1' nor d_2 is a unit. If d_1' and d_2 are both primes, we are done. Otherwise, say d_2 is not a prime and factors properly as $d_2 = d_2' d_3$. Again, if d_1', d_2', and d_3 are all primes we are done, so we may assume, say, that $d_3 = d_3' d_3$. If the process doesn't terminate, we obtain $d_4 = d_4' d_5, \ldots, d_i = d_i' d_{i+1}$, etc. This leads to the chain of ideals $\langle d \rangle \subsetneqq \langle d_2 \rangle \subsetneqq \langle d_3 \rangle \subsetneqq \langle d_4 \rangle \subsetneqq \cdots$, where all inclusions are proper. This violates the acc, which is a contradiction. Thus the factorization process must halt after a finite number of steps, and the proposition is proved.   ////

Definition A domain is a **factorization domain** if every nonzero element is either a unit or a product of a finite number of primes.

Our next propositions will guarantee us a supply of examples of unique factorization domains, whereas the proposition just proved gives us one condition that at least guarantees factorization into a finite number of primes.

Definition A domain D is a **unique factorization domain** (abbreviated UFD) if every element in D factors into a finite number of primes in essentially a unique way. That is, if $d = up_1 p_2 \cdots p_k = vq_1 q_2 \cdots q_l$, where u, v are units and the p_i's and q_j's are primes, then $k = l$, and after a relabeling of the subscripts (if necessary) each p_i is associated with q_i.

For an example, $\mathbb{Z}$ is a UFD, and the factorization $10 = (-2)(-5) = 5 \cdot 2$ is essentially unique since 2 and -2 are associates as are 5 and -5.

Proposition 3 Any principal ideal domain D satisfies the ascending-chain condition and thus is a factorization domain.

PROOF Let $\langle d \rangle \subsetneqq \langle d_2 \rangle \subsetneqq \langle d_3 \rangle \subsetneqq \langle d_4 \rangle \cdots$ be a strictly ascending chain of ideals. We let $I = \bigcup_{i=1}^{\infty} \langle d_i \rangle$ and can show that I is an ideal.

Since I is an ideal in a **PID**, we have $I = \langle f \rangle$ for some f in D. However, $f \in \bigcup_{i=1}^{\infty} \langle d_i \rangle$, so $f \in \langle d_j \rangle$ for some integer j, and thus $\langle d_j \rangle = I = \langle d_{j+1} \rangle = \langle d_{j+2} \rangle = \cdots$. We have just shown that D satisfies the acc. By the previous proposition every element of D factors into a finite number of primes. $////$

Exercise

5 Show that I above is indeed an ideal. (For instance, if a and b are in I then a is in some $\langle d_j \rangle$, and b is in $\langle d_k \rangle$. Either j or k is larger. If, say, k is larger, then a and b both are in $\langle d_k \rangle$, as is $a + b$. Thus $a + b$ is in I.)

Before showing that factorization is unique in a PID, several equivalent formulations of unique factorization will be developed. We start with some straightforward definitions.

Definition In a domain, a **greatest common divisor** of two elements a and b is an element d such that $d \mid a$, $d \mid b$, and any other element d' also dividing a and b must divide d. Similarly, a **least common multiple** m of a and b is divisible by a and b and divides any other element divisible by both a and b. As with $\mathbb{Z}$, the abbreviations gcd (a, b) and lcm(a, b) are used. As in $\mathbb{Z}$, it is easily seen that the gcd's and lcm's are not unique but are unique up to associates.

Definition A **principal ideal** in a ring R with 1 is an ideal I such that $I = \langle a \rangle = \{ ar \mid r \in R \}$ for some a in I.

As a cautionary prelude to our theorem, we examine a domain where many reasonable things go awry.

Examine $\mathbb{Z}(\sqrt{-5})$, and observe that $9 = 3 \cdot 3 = (2 + \sqrt{-5}) \cdot (2 - \sqrt{-5})$. Accepting that $3, 2 + \sqrt{-5}$, and $2 - \sqrt{-5}$ are all primes, we see the following:

1 If $a = 3$ and $b = 2 + \sqrt{-5}$, then lcm(a, b) doesn't exist.

2 If $c = 9$ and $d = 3(2 + \sqrt{5})$, then gcd (c, d) doesn't exist.

3 $3 \mid (2 - \sqrt{-5})(2 + \sqrt{-5})$, but $3 \nmid (2 - \sqrt{-5})$ and $3 \nmid (2 + \sqrt{-5})$.

4 If $I = \langle 3 \rangle$ and $J = \langle 2 + \sqrt{-5} \rangle$ are principal ideals, then $I \cap J$ contains 9 and $3(2 + \sqrt{-5})$. Thus, using 1 above, we see $I \cap J$ is not principal.

For instance, for item 1 above, we easily see that 9 and $3(2 + \sqrt{-5})$ are both multiples of 3 and $2 + \sqrt{-5}$. If a lcm existed, it would have to divide both 9 and $3(2 + \sqrt{-5})$. But $9 = 3 \cdot 3 = (2 + \sqrt{-5})(2 - \sqrt{-5})$ are essentially the only factorizations of 9, so no multiple of 3 and $2 + \sqrt{-5}$ divides 9 except 9 itself. And $9 \nmid 3(2 + \sqrt{-5})$, so 9 is not a lcm. Thus no lcm exists.

Items 2, 3, and 4 are established by similar reasoning.

NOTATION If a and b are in a domain D and their greatest common divisor exists, we denote it by (a,b). Their least common multiple is denoted $[a,b]$ if it exists.

Lemma 1 If D is an integral domain, and greatest common divisors exist for any two elements in D, then the following statements are true:

(a) $(a,b)c \sim (ac,bc)$. In particular, $(c,bc) \sim c$.

(b) If $(a,b) \sim 1$ and $(a,c) \sim 1$, then $(a,bc) \sim 1$.

(c) If $(a,b) \sim d$, then $(a/d,b/d) \sim 1$.

PROOF (a) Since $(a,b)|a$ and $(a,b)|b$, we have $(a,b)|ac$ and $(a,b)c|bc$, so by the definition of greatest common divisors we must have $(a,b)c|(ac,bc)$. Let $(a,b)cg = (ac,bc)$. Thus $(a,b)cg$ divides ac and bc. Unless $c = 0$, which reduces the statement to $0 = 0$, we can cancel c to obtain that $(a,b)g$ divides a and b. Thus $(a,b)g|(a,b)$; this implies g is a unit. So $(a,b)c \sim (ac,bc)$. Taking $a = 1$ yields $(c,bc) \sim (1,b)c \sim 1c = c$

(b) $1 \sim (a,b)$ and $1 \sim (a,c)$ yield $1 \sim (a,b) \sim (a,(a,c)b) \sim (a,(ab,cb))$ $\sim ((a,ab),cb) \sim (a,cb)$. [It is easily checked that $(x,(y,z)) = ((x,y),z)$.]

(c) If $(a/d,b/d) \sim d' \nsim 1$, then (a) gives us $(a,b) \sim d(a/d,b/d) \sim dd' = (a,b)d'$, and this is impossible. ////

Exercise

6 *If for a and b in an integral domain D both (a,b) and $[a,b]$ exist, show that $(a,b)[a,b] \sim ab$.

We are now set to state an important theorem giving equivalent characterizations of UFD's. An immediate corollary of this theorem is the remainder of the proof that every PID is a UFD. The proof of the theorem will be given as a series of lemmas (which we recommend as exercises rather than reading matter).

Theorem 1 If D is an integral domain in which every element factors into a finite product of primes, then the following statements are equivalent:

(a) D is a UFD.

(b) If p is a prime in D and $p|ab$, then $p|a$ or $p|b$.

(c) For any two elements in D, there exists a greatest common divisor.

(d) For any two elements in D, there exists a least common multiple.

(e) The intersection of any two principal ideals in D is a principal ideal.

PROOF (a) $\Leftrightarrow$ (b) Lemma 2
         (a) $\Rightarrow$ $(c), (d)$ Lemma 3
         (c) $\Rightarrow$ (d) Lemma 4
         (d) $\Rightarrow$ (c) Exercise 7
         (d) $\Leftrightarrow$ (e) Lemma 5
         (c) $\Rightarrow$ (b) Lemma 6   ////

Before the lemmas, let us prove the corollary.

Corollary Every principal ideal domain is a unique factorization domain.

PROOF We have already shown (a few pages ago) that a PID satisfies the ascending-chain-condition and that this implies that every element factors as a finite product of primes. To finish, use part e of the theorem.   ////

Lemma 2 If D is a factorization domain, then the following are equivalent:

(a) D is a UFD.
(b) $p \mid ab \Rightarrow p \mid a$ or $p \mid b$ for all primes p in D.

PROOF If D is a UFD and $p \mid ab$, then $px = ab$ for some element x in D. Writing a and b as products of primes, we see by the uniqueness of factorization that p is associated with one of the prime factors of a or of b. Thus either $p \mid a$ or $p \mid b$ (or possibly both).

Conversely, if $p \mid ab \Rightarrow p \mid a$ or $p \mid b$ and D is not a UFD, we arrive at a contradiction. Let $up_1 p_2 \cdots p_l = vq_1 q_2 \cdots q_m$ be two essentially different factorizations of some element in D, where u and v are units, and the p_i's and q_j's are primes. Assume also that we have picked in such a way that l is the smallest possible integer allowing this nonunique factorization. We may assume p_1 is not associated with any q_i, for if it were we could cancel and arrive at a shorter equality.

Then $p_1 \mid vq_1 q_2 q_3 \cdots q_m$ and, letting $a = vq_1$ and $b = q_2 \cdots q_m$, we see that either $p_1 \mid vq_1$ or $p_1 \mid q_2 \cdots q_m$. The first is impossible, and the second is the situation we started the paragraph with, but with one fewer prime factor. Continuing (i.e., using an appropriate induction hypothesis), we arrive at a contradiction. Thus, D must be a UFD.   ////

Lemma 3 Suppose D is a unique factorization domain. Then every two elements of D have a greatest common divisor and a least common divisor.

PROOF   Select one irreducible element from each equivalence class of associated irreducible elements, and call it p_i. Then if we select any two elements a and b from D, we see that

$$a = up_1^{m_1} p_2^{m_2} \cdots p_k^{m_k} \quad \text{and} \quad b = vp_1^{n_1} p_2^{n_2} \cdots p_n^{n_k}$$

where u and v are units. (We allow some m_i or n_i equal to 0 so as to be able to use the same p_i's for both decompositions.)

Now $c = \gcd(a,b) = p_1^{d_1} p_2^{d_2} \cdots p_k^{d_k}$, where $d_i = \min(m_i, n_i)$. It is clear that $c \mid a$ and $c \mid b$, and equally clear that any element dividing a and b, when written out as a product of primes, will have exponents at most equal to the d_i.

Similarly, $\operatorname{lcm}(a,b) = p_1^{l_1} p_2^{l_2} \cdots p_k^{l_k}$, where $l_i = \max(m_i, n_i)$.   ////

Lemma 4   If D is a domain in which gcd's exist for all pairs of elements, then lcm's exist for all pairs of elements.

PROOF   Let a and b be an arbitrary pair of elements in D. We may assume both are nonzero, as $[0,b] = 0$. If a and b are nonzero, so is (a,b). We contend that $x = ab/(a,b)$ is a lcm for a and b. Since $a \mid x = a[b/(a,b)]$ and $b \mid x = [a/(a,b)]b$, x is a multiple of a and b. If m is another multiple of a and b, look at (x,m). Since $a \mid x$ and $a \mid m$, by the definition of greatest common divisor, $a \mid (x,m)$. Similarly, $b \mid (x,m)$. Since $(x,m) \mid x$, we have $(x,m)f = x = ab/(a,b)$ for some element f in D.

Thus $[(x,m)/a] f = b/(a,b)$, so $f \mid b/(a, b)$ and, by similar reasoning, $f \mid a/(a,b)$. Again by the definition of gcd, $f \mid (a/(a,b), b/(a,b)) \sim 1$ by Lemma 1c. So f is a unit, $(x,m) \sim x$, $x \mid m$, and x is a least common multiple a and b.   ////

Exercise

7   If D is a domain in which every two elements have a least common multiple, show that every two elements in D have a greatest common divisor.

Lemma 5   If $A = \langle a \rangle$ and $B = \langle b \rangle$ are principal ideals in a domain D, then $A \cap B$ is also a principal ideal if and only if $[a,b]$, the least common multiple of a and b, exists.

PROOF   If $[a,b]$ exists, then $a \mid [a,b]$ and $b \mid [a,b]$, so $[a,b] \in A$ and $[a,b] \in B$; thus, $[a,b]$ is in $A \cap B$. If $c \in A \cap B$, then c is in A and B, $a \mid c$ and $b \mid c$. Thus $[a,b] \mid c$, and thus $A \cap B = \langle [a,b] \rangle$.

Conversely, if $A \cap B = \langle d \rangle$ is principal, then $d \in A$ and $d \in B$, so $a \mid d$ and $b \mid d$. If d' is any element divisible by both a and b, then $d' \in A \cap B = \langle d \rangle$, so $d \mid d'$. Thus d is a lcm of a and b.   ///

Lemma 6 If, in a domain D, there exist greatest common divisors for any two elements, then for any prime p in D, $p|ab$ implies $p|a$ or $p|b$.

PROOF p is prime, so if $p \nmid a$ we have $(p,a) \sim 1$; similarly $p \nmid b$ gives $(p,b) \sim 1$. Using Lemma 1b, we see that this implies $(p,ab) \sim 1$, so $p \nmid ab$. Thus, if gcd's always exist, the conditions $p|ab$, $p \nmid a$, $p \nmid b$ cannot hold simultaneously, and so $p|ab$ implies $p|a$ or $p|b$.   ////

This completes all the parts of Theorem 1 giving five equivalent characterizations of unique factorization domains.

We have now also completed the proof that a PID is a UFD. As a corollary, we obtain the fundamental theorem of arithmetic, which says that every integer factors into primes in an essentially unique fashion.

Corollary (fundamental theorem of arithmetic) $\mathbb{Z}$ is a unique factorization domain

PROOF $\mathbb{Z}$ is a PID and therefore a UFD.   ////

Corollary (the Bezout identity)   Suppose that D is a unique factorization domain, and that $aR + bR = dR$ for some d in R. Then $d = \gcd(a,b)$, and thus there exist r_1 and r_2 in R such that $d = ar_1 + br_2$.

PROOF   Since a and b are both in $\langle d \rangle = dR$, we have $d|a$ and $d|b$. Conversely, if $c|a$ and $c|b$, then $aR \subseteq \langle c \rangle$ and $bR \subseteq \langle c \rangle$, so that $\langle d \rangle = aR + bR \subseteq \langle c \rangle$ and $c|d$. Thus $d = \gcd(a,b)$.   ////

Exercises

8   Make up several examples to check the Bezout identity in various domains.

9   Let P be a prime ideal in a PID. Suppose $\langle x \rangle = P$ and $x|ab$. Show that either $x|a$ or $x|b$. (This motivates defining an element y to be *prime* if $y|cd \Rightarrow y|c$ or $y|d$.)

10   What are five distinct examples of UFD's?

11   (Determination of the pythagorean triples) Let x, y, z be three integers such that $x^2 + y^2 = z^2$ and such that x, y, and z are pairwise relatively prime. Factoring in $\mathbb{Z}[i]$, we have $(x + iy)(x - iy) = z^2$. Show first that if $p = a + bi$ is a prime in $\mathbb{Z}[i]$ and $p|x + iy$, then $p \nmid x - iy$, and vice versa (unless $p = 1 + i$). Using unique factorization in $\mathbb{Z}[i]$, show that we must have $x + iy = (u + vi)^2$. Therefore, $x + iy = u^2 - v^2 + 2uvi$. Show finally that $x = u^2 - v^2$, $y = 2uv$, and $z = u^2 + v^2$.

12   Find all integer solutions to $x^2 + 2y^2 = z^2$, where x, y, and z, are coprime. [*Hint:* $\mathbb{Z}(\sqrt{-2})$ is a UFD.]

13 For the integral domain $\mathbb{Z}(\sqrt{m}) = \{a + b\sqrt{m} \mid a, b, m \in \mathbb{Z}\}$, let $\theta: \mathbb{Z}(\sqrt{m}) \to \mathbb{Z}$
be defined by $\theta(a + b\sqrt{m}) = a^2 - mb^2$. In general θ is not a euclidean norm,
but show that it does have the following properties:
   (a) $\theta(ab) = \theta(a)\theta(b)$.
   (b) u is a unit if and only if $\theta(u) = 1$.
   (c) If p is a prime in $\mathbb{Z}$ and $\theta(a + b\sqrt{m}) = \pm p$, then $a + b\sqrt{m}$ is a prime
        in $\mathbb{Z}(\sqrt{m})$.
14 In $\mathbb{Z}(\sqrt{3})$ show that $5 + 3\sqrt{3}, 4 + \sqrt{3}, 19 + 11\sqrt{3}$, and $5 - 2\sqrt{3}$ are primes.
(The last exercise might be useful here.) Why isn't $(5 + 3\sqrt{3})(4 + \sqrt{3}) = 29 + 17\sqrt{3} = (19 + 11\sqrt{3})(5 - 2\sqrt{3})$ a contradiction to the fact that $\mathbb{Z}(\sqrt{3})$
is a UFD?
15 If $a^2 + b^2 = c^2$ in $\mathbb{Z}$, then $(a/b)^2 + (b/c)^2 = 1$ in $\mathbb{Q}$, and $(a/b, b/c)$ is said to
be a *rational point* on the circle. Similarly, if $d^2 + f^2 = 1$ for d and f in $\mathbb{Q}$,
then one can find corresponding pythagorean triples. Show that the
rational points on the circle form a group (it might be helpful to look at
the plane as the complex plane), and thus that the pythagorean triples (or
classes thereof) can be made into a group.
16 In a PID show that every nonzero prime ideal is maximal. (*Hint*: Let
$\langle x \rangle = P \neq 0$, where $D \supsetneq N \supsetneq P$ so that P is nonmaximal. Since $N = \langle n \rangle$,
what can we say about the relation between x and n? See Exercise 9.)

6.4 ODDS AND ENDS

In this section we mention a few more results on integral domains, including
some unusual examples. Anyone interested in the question of when an integral
domain is a unique factorization domain should see the survey articles by Samuel
[15] and Cohn [17] or the lecture notes by Samuel [76].

Let

$$\mathbb{Y}[\sqrt{m}] = \{a + b\sqrt{m} \mid a, b, m \in \mathbb{Z}\} \qquad \text{for } m \equiv 2, 3 \text{ (mod 4)}$$

and

$$\mathbb{Y}[\sqrt{m}] = \left\{ \frac{a}{2} + \frac{b}{2}\sqrt{m} \mid a, b, m \in \mathbb{Z}, a \equiv b \text{ (mod 2)} \right\} \qquad \text{if } m \equiv 1 \text{ (mod 4)}$$

($\mathbb{Y}[\sqrt{m}]$ is the ring of algebraic integers of a quadratic extension of $\mathbb{Q}$.) We
also assume m has no proper divisors which are squares.

Then $\mathbb{Y}[\sqrt{m}]$ is a unique factorization domain if $m = -1, -2, -3, -7,$
$-11, -19, -43, -67,$ or -163. These nine negative numbers were known by
Gauss, and in 1967 it was proved that these are the only 9 such numbers (Stark
[82]). $\mathbb{Y}[\sqrt{m}]$ is also a UFD for $m = 2, 3, 5, 6, 7, 11, 13, 17, 19, 21, 22, 23, 29,$

33, 37, 38, 41, 43, 46, 47, ... Extensive tables of such m have been tabulated, but it is still only a conjecture that there are an infinite number of such m.

If we define $\delta(c + d\sqrt{m}) = c^2 - md^2$, δ is a euclidean norm for $\mathbb{Y}[\sqrt{m}]$ if and only if $m = -11, -7, -3, -2, -1, 2, 3, 5, 6, 7, 11, 13, 17, 19, 21, 29,$ 33, 37, 41, 57, or 73. Other of the $\mathbb{Y}[\sqrt{m}]$ may be euclidean under some other norm map, but no such example has yet been produced. Bolker [10] contains an excellent presentation of the $\mathbb{Y}[\sqrt{m}]$ and their applications.

Exercise

1 Give an example of a UFD that contains a subdomain that is not a UFD.

> **Proposition 1** If D is a UFD, then $D[x]$, the ring of polynomials in one variable over D, is also a UFD. The proof of this is not hard and can be found in a number of texts, for instance, Herstein [38], Dean [20], and Fraleigh [23].

If F is a field, then $F[x_1]$ is euclidean and thus a UFD. Then $F[x_1,x_2] = (F[x_1])[x_2]$ is also a UFD by the previous proposition and, by induction, so are $F[x_1,x_2,x_3]$ and $F[x_1,x_2,...,x_n]$. We state this separately.

> **Corollary** If F is a field, then $F[x_1,x_2,...,x_n]$, the ring of all polynomials in n variables over F, is a unique factorization domain.

By this proposition both $F[x,y]$ and $\mathbb{Z}[x]$ are UFD's. Since they are not PID's, we have examples separating the two concepts. Two more examples are as follows.

EXAMPLE A   Let F be any field so that -1 is not a square in F. (For instance $F = \mathbb{Q}$, $\mathbb{R}$, any field in between or, as we shall see in the next chapter, $\mathbb{Z}_p$, where $p = 4k + 3$.) Let I be the ideal generated by the polynomial $x^2 + y^2 + z^2 - 1$ in $F[x,y,z]$. Then $D = F[x,y,z]/I$ is a UFD.

EXAMPLE B   Let F be any field, and let I be the ideal generated by the polynomial $x^s + y^r + z^t$, where $\gcd(s,r) = \gcd(s,t) = \gcd(r,t) = 1$. Again $F[x,y,z]/I$ is a UFD. (See Samuel [76].)

We want now to return to an earlier example, $R = \mathbb{Z}(\sqrt{-5}) = \{a + b\sqrt{-5}$ $a,\ b \in \mathbb{Z}\}$. We asserted then that R is a domain in which factorization is not unique, since $9 = 3 \cdot 3 = (2 + \sqrt{-5})(2 - \sqrt{-5})$. First let $\theta(a + b\sqrt{-5}) = a^2 + 5b^2$. θ here is not a euclidean norm, but it is a map into the nonnegative integers, and it is easily seen that

> *1* $\theta(r)\theta(s) = \theta(rs)$ for all r, s in R.

2 $\theta(r) = 0$ iff $r = 0$ (since $a^2 + 5b^2 = 0 \Leftrightarrow a = b = 0$).

3 $\theta(r) \neq 3$ (since $a^2 + 5b^2 = 3$ has no integer solutions).

4 $\theta(r) = 1$ iff r is a unit.

Item 1 follows from a straightforward computation, and item 4 follows from item 1. An immediate corollary of item 4 is that $\mathscr{U}(R) = \{1, -1\}$.

We now show that 3 is prime [note that $29 = (2 + \sqrt{-5})(2 - \sqrt{-5})$, so there is something to be proved here]. If $3 = (a + b\sqrt{-5})(c + d\sqrt{-5})$, then we can save some energy by applying θ to both sides:

$$9 = \theta(3) = \theta((a + b\sqrt{-5})(c + d\sqrt{-5})) = \theta(a + b\sqrt{-5})\theta(c + d\sqrt{-5})$$
$$= (a^2 + 5b^2)(c^2 + 5d^2)$$

If either $a^2 + 5b^2$ or $c^2 + 5d^2$ is 1, then $a + b\sqrt{-5}$ or $c + d\sqrt{-5}$ is a unit by item 4 above, and 3 has not been genuinely factored. But $a^2 + 5b^2 = c^2 + 5d^2 = 3$ is an impossibility. Thus there is no genuine factorization of 3 in R. Almost identical computations show that $2 + \sqrt{-5}$ and $2 - \sqrt{-5}$ are prime. Since $\mathscr{U}(R) = \{1, -1\}$, the only associates of 3 are 3 and -3, so 3 is not an associate of either $2 + \sqrt{-5}$ or $2 - \sqrt{-5}$, and we have nonunique factorization.

Exercises

2 Show that $\mathbb{Z}(\sqrt{-6})$ is not a UFD. (*Hint:* Try factoring 10.)

3 Show that $\mathbb{Z}(\sqrt{15})$ is not a UFD.

4 *Find the group of units of $\mathbb{Z}(\sqrt{2})$.

Using the fact that $\mathbb{Z}_p[x]$ is a UFD, we can give slick proofs to several theorems from Sec. 2.6 on polynomial rings.

5 (Another proof of Eisenstein's criterion)

(a) Let $\phi: \mathbb{Z} \to \mathbb{Z}_p$ be the usual ring homomorphism, so that $\phi(a) = \bar{r}$, where $a = c \cdot p + r$, for $0 \leq r \leq p - 1$. Define a new homomorphism $\phi^*: \mathbb{Z}[x] \to \mathbb{Z}_p[x]$.

(b) Let $f(x) = a_0 + a_1 x + a_2 x^2 + \cdots + a_n x^n \in \mathbb{Z}[x]$. Show that if $f(x)$ is reducible, then so is

$$\phi^*(f(x)) = \bar{a}_0 + \bar{a}_1 x + \bar{a}_2 x + \cdots + \bar{a}_n x^n \in \mathbb{Z}_p[x]$$

(c) Assume $p \mid a_0, a_1, a_2, \ldots, a_{n-1}, p^2 \nmid a_0, p \nmid a_n$. Then $\phi^*(f(x)) = \bar{a}_n x^n$. If $\phi^*(f(x)) = \bar{a}_n x^n$ factors, it must be into $\bar{b}_m x^m \cdot \bar{c}_k x^k$, where

$$a_0 + a_1 x + \cdots + a_n x^n = (b_0 + b_1 x + \cdots + b_m x^m)(c_0 + c_1 x + \cdots + c_k x^k)$$

Why are no other factorizations possible?

(d) Except in the trivial cases $n = m$ and $n = k$, show that $p \mid b_0$, $p \mid c_0$, and thus that $p^2 \mid a_0$, contradicting our assumption.

6 (Gauss's theorem) Continue the notation of Exercise 5. Let

$$b(x) = b_0 + b_1 x + \cdots + b_m x^m \qquad \text{and} \qquad c(x) = c_0 + c_1 x + \cdots + c_k x^k$$

be primitive [that is, gcd $(b_0, b_1, \ldots, b_m) = 1 = \gcd(c_0, c_1, \ldots, c_k)$], and assume that $a(x) = b(x)c(x)$ is not primitive, so that some p exists with $p \mid a_0$, $p \mid a_1, \ldots, p \mid a_n$. Show that in $\mathbb{Z}_p[x]$ this leads to $0 = \phi^*(b(x))\phi^*(c(x))$. Why is this contradictory?

NUMBER THEORY

7.1 BASIC RESULTS

In this chapter we want to demonstrate that number theory and modern algebra are closely linked subjects. Number theory is a very attractive subject, and one of its charms is that many of its results can be understood by a nonmathematical audience. It also has been a somewhat neglected subject in the last 20 years, and that is one motivation for including slightly more than the minimum necessary amount.

Definition 1 (sophisticated) For the ring $\mathbb{Z}_n$ we define $\Phi(n)$ as the group of units of $\mathbb{Z}_n$, and $\phi(n) = |\Phi(n)|$. The function ϕ is called **Euler's phi function**.

Definition 2 (naive) For any positive integer n, $\phi(n)$ is the number of positive integers less than n and relatively prime to n.

For example, if $n = 15$, then $\Phi(n) = \{\bar{1},\ \bar{2},\ \bar{4},\ \bar{7},\ \bar{8},\ \bar{11},\ \bar{13},\ \bar{14}\}$ so that $\phi(n) = |\Phi(n)| = 8$. A quick observation says that 1, 2, 4, 7, 8, 11, 13, and 14 are those numbers less than 15 and relatively prime to 15, so the second definition yields the same answer. That this happens in general is the content of the next proposition.

Proposition 1   If n is a positive integer, then $\Phi(n) = \{\bar{k} \mid \gcd(n,k) = 1\}$. Thus, by picking k between 1 and n, we see that the two definitions of $\phi(n)$ coincide.

PROOF   We know that $\gcd(n,k) = 1$ implies the existence of integers a and b such that $an + bk = 1$. Reducing modulo n yields $\bar{a} \cdot \bar{0} + \bar{b} \cdot \bar{k} = \bar{1}$, and thus $\bar{b} = \bar{k}^{-1}$ and $\bar{k} \in \Phi(n)$, as $\bar{k}$ is a unit. This process is reversible since $an + bk = 1$ certainly implies that $\gcd(n,k) = 1$. Finally, since $\bar{k}$ is the coset $\{k,\ k + n,\ k - n,\ k + 2n,\ k - 2n,\ \ldots\}$, we can pick some element of $\bar{k}$ between 1 and n.   ////

Exercise

1   Use the method of this proof, starting with Euclid's algorithm, to find $\bar{3}^{-1}$ in Z_{10}, $\overline{31}^{-1}$ in Z_{67}, and $\overline{288}^{-1}$ in Z_{401}.

We would now like to apply several of the results of the previous chapters to some problems in number theory. Let us start by recalling some notation and then proving the Euler-Fermat theorem.

The following statements are equivalent:

$$n \mid a - b \qquad\qquad\qquad\qquad\qquad \text{where } n,a,b \in Z$$
$$a \equiv b \ (\text{mod } n) \qquad \text{or} \qquad a - b \equiv 0 \ (\text{mod } n) \qquad \text{where } n,a,b \in Z$$

or

$$\bar{a} = \bar{b} \text{ in } \mathbb{Z}_n$$

where $\bar{a}$, $\bar{b}$ are cosets of the subgroup $n\mathbb{Z} = \bar{0}$ of $\mathbb{Z}$, and $a \in \bar{a}$, $b \in \bar{b}$.

Computations modulo n are referred to as *modular arithmetic*. This modular arithmetic is in some ways easier than regular arithmetic, but in some senses less informative. We have just shown that $(\Phi(n), \cdot)$ is a group of order $\phi(n)$, where ϕ is Euler's phi function. Thus, if $\bar{a} \in \Phi(n)$, then, by Lagrange's theorem, the order of $\bar{a}$ divides $\phi(n)$, and we must have $\bar{a}^{\phi(n)} = \bar{1}$ or $a^{\phi(n)} \equiv 1$ (mod n) or $n \mid a^{\phi(n)} - 1$. Thus we have most of the following:

Euler-Fermat theorem   If $(a,n) = 1$, then $n \mid a^{\phi(n)} - 1$.

PROOF $(a,n) = 1$ is equivalent to $\bar{a} \in \Phi(n)$, which completes the proof.   ////

The next exercise gives a result thath as been known a long time and may have been a motivation for the development of modular arithmetic.

Exercise

2 Compute the decimal expansion of $\frac{1}{7}$ to show $7\,|\,10^6 - 1$.

3 Show that $\equiv_n$ is an equivalence relation on $\mathbb{Z}$, where $a \equiv_n b$ is defined by $n\,|\,a - b$.

As an illustration of how computations can be simpler in modular arithmetic, let us compute 2^{40} modulo 9. The laborious approach would be to compute 2^{40}, divide by 9, and write down the remainder. One quicker approach is to note that $2^5 \equiv 5 \pmod{9}$.

Thus $2^{10} \equiv (2^5)^2 = 5^2 \equiv -2 \pmod{9}$. Now $2^{40} = (2^{10})^4 \equiv (-2)^4 = 16 \equiv 7 \pmod{9}$. Using the Euler-Fermat theorem we have $2^6 \equiv 1 \pmod{9}$, since $\phi(9) = 6$. Then, again, $2^{40} = (2^6)^6 2^4 \equiv 1^6 \cdot 2^4 \equiv 7 \pmod{9}$.

Exercises

4 (a) Compute 3^{1000} modulo 10 (alternatively, $\overline{3}^{1000}$ in $\mathbb{Z}_{10}$).
   (b) Compute $\overline{1} \cdot \overline{2} \cdot \overline{3} \cdot \overline{4} \cdot \cdots \cdot \overline{11} \cdot \overline{12} =$ in $\mathbb{Z}_{13}$.
   (c) Compute $1^{10} + 2^{10} + 3^{10} + 4^{10}$ in $\mathbb{Z}_5$.

As another demonstration of the utility of modular arithmetic, we prove that $641\,|\,2^{2^5} + 1$. This fact is historically important since Fermat conjectured that all numbers of the form $2^{2^n} + 1$ are prime. In Chap. 10 we discuss the geometric importance of these Fermat numbers. We don't care here how many times 641 divides $2^{2^5} + 1$, and this saves us the onerous task of computing $2^{32} + 1$. Since

$$641 = 5 \cdot 2^7 + 1$$

we have $$5 \cdot 2^7 \equiv -1 \pmod{641}$$

Thus $$5^4 \cdot 2^{28} \equiv (-1)^4 \equiv 1 \pmod{641}$$

However, $$5^4 \equiv 625 \equiv -16 \equiv -2^4 \pmod{641}$$

so $$-2^4 \cdot 2^{28} \equiv -2^{32} \equiv 1 \pmod{641}$$

Thus $$2^{32} + 1 \equiv 0 \pmod{641}$$

so $$641\,|\,2^{32} + 1$$

Exercise

5 (The Chinese hypothesis) Another once-famous mathematical conjecture is the following: $n\,|\,2^n - 2$ if and only if n is prime.

(a)   Show that if p is prime then $p|2^p - 2$.

(b)   Show that $31 \cdot 11 = 341|2^{341} - 2$ (unfortunately).

Corollary 1 (Fermat)   If p is a prime and $p \nmid a$, then $p|a^{p-1} - 1$.

PROOF   If p is prime, $\phi(p) = p - 1$, and $p \nmid a$ is equivalent to $(a,p) = 1$.   ////

Corollary 2   If p is prime, $p|a^p - a$ or $a^p \equiv a \pmod{p}$.

PROOF   If $p \nmid a$, then $a^{p-1} \equiv 1 \pmod{p}$, and we multiply both sides by a. If $p|a$, then $a^p \equiv 0 \equiv a \pmod{p}$ is true.   ////

This is our second proof of Corollary 2, having given a ring-theory proof in Sec. 4.4. Historically, a reverse process was followed. Corollary 1 was proved by Gauss [28] by listing the elements of $\Phi(p)$ as follows:

$$a, \quad a^2, \quad a^3, \ldots, \quad a^m = 1 \qquad \text{where } m \text{ is the order of } a$$
$$xa, \quad xa^2, \quad xa^3, \ldots, \quad xa^m$$
$$ya, \quad ya^2, \quad ya^3, \ldots, \quad ya^m$$

Then it is easy to show that each row contains m distinct elements, that the rows contain disjoint elements, and thus that $m|p - 1$. This was all done before groups were defined. However, if the word "row" is changed to "coset," this is essentially our first proof of Lagrange's theorem. Miller [61] gives an interesting account of the early history of group theory.

One result that we use several times in this chapter is that a polynomial of degree n over a field F has at most n roots. This is proved by first showing that if a is root of $f(x)$ then $x - a$ divides $f(x)$ and then by using induction on n and unique factorization in $F[x]$. The details are written out Sec. 10.3.

We now want to prove that $\Phi(p)$ is cyclic; this is a corollary of a stronger result of independent interest.

Theorem   If M is a finite subgroup under multiplication of the non-zero elements in a field, then M is cyclic.

PROOF   We compare M with a cyclic group C, also of order $|M|$. For every positive integer k, $x^k = 1$ has at most k solutions in M, since M is in a field. In $C = \{e, c^2, c^3, \ldots, c^{m-1}\}$, $x^k = e$ has e, $c^{|M|/k}$, $c^{2|M|/k}$, $\ldots$, $c^{(k-1)|M|/k}$ as its k solutions.

Thus, for every possible proper divisor of $|M|$, we find that C has as many or more elements of that order than has M. But the two groups both have $|M|$ elements. Thus, since C has an element of order $|M|$, so must M. (M cannot have more elements of lower order.) Since M has an element of order $|M|$, this element is a generator, and M is cyclic.   ////

Observe that $\{1, i, -1, -i\}$ is indeed a cyclic subgroup of $\mathbb{C}$.

Corollary 1 $(F - (0), \cdot)$ is cyclic for any finite field F.

Corollary 2 $\Phi(p)$ is cyclic.

Definition   A generator of the cyclic group $\Phi(p)$ is called a **primitive root** modulo p.

EXAMPLE $\Phi(7)$ has order 6, and the powers of $\bar{3}$ are

$$\bar{3}^2 = \bar{2} \qquad \bar{3}^3 = \bar{6} = -\bar{1} \qquad \bar{3}^4 = \bar{4} \qquad \bar{3}^5 = \bar{5} \qquad \bar{3}^6 = \bar{1}$$

Thus $\bar{3}$ is a primitive root of 7 (as is $\bar{5}$).

Exercises

6   Find primitive roots for $p = 3$, 5, 11, and 17.
7   Show that $\Phi(p)$ has $\phi(p - 1)$ primitive roots for any prime p.

In general, finding primitive roots is very difficult.   For any particular prime a primitive root can be found by trial and error.   The following is a famous conjecture.

ARTIN'S CONJECTURE   Any fixed positive nonsquare integer k is a primitive root for infinitely many primes.

In some later exercises we shall produce some primitive roots for some special kinds of primes.

Proposition 2 $\Phi(n)$ is cyclic if and only if $n = 2$, 4, p^a, or $2p^a$, where p is an odd prime.

In fact, the groups $\Phi(n)$ have been completely analyzed as finite abelian groups.   See, for instance, Bolker [10], Ireland and Rosen [41], or Shanks [81].

Exercises

8   Write the following as direct products of cyclic groups: $\Phi(8)$, $\Phi(12)$, and $\Phi(18)$.   This can be done, since each of these is a finite abelian group.   See Sec. 5.2.
9   Write down a complete subgroup lattice for $\Phi(13)$.
10   If p is a prime other than 2 or 5, then look at the set of remainders R obtained when writing $1/p$ in decimal form.   For instance, if $p = 13$, $\frac{1}{13}$ yields $R = \{10,9,12,3,4,1\}$.   Show that R is a subgroup of $\Phi(p)$.   What can be said about the remainders arising from a/p?   When does $R = \Phi(p)$?

7.2 QUADRATIC RESIDUES

Having shown that $\Phi(p)$ is cyclic, we now want to see which elements in $\Phi(p)$ are squares of other elements. If $y \in \Phi(p)$, and $y = x^2$ for some other element x in $\Phi(p)$, then x is called a *quadratic residue* or, more briefly, a *residue*.

Proposition 1 If p is an odd prime, then the homomorphism $\bar{x} \xrightarrow{\theta} \bar{x}^2$ from $\Phi(p)$ to $\Phi(p)$ has $(p-1)/2$ elements in its image (these are the *residues*), and $(p-1)/2$ elements not in its image (the *nonresidues*). Let us call the subgroup of residues R.

> PROOF θ is a homomorphism, and $\ker(\theta) = \{\bar{x} \mid \bar{x}^2 = \theta(\bar{x}) = \bar{1}\}$. However, $\mathbb{Z}_p$ is a field, so a second-degree polynomial over $\mathbb{Z}_p$ has at most two roots. Thus $x^2 - 1$ has at most two roots over $\mathbb{Z}_p$, and these are $\bar{1}$ and $-\bar{1} = \overline{p-1}$. Thus $|\ker(\theta)| = 2$, $|\Phi(p)| = p-1$, and $|R| = |\text{Image}(\theta)| = (p-1)/2$ by the homomorphism theorem.
>
> $p - 1 - (p-1)/2 = (p-1)/2 =$ the number of nonresidues. In fact R, the residues, are a subgroup of index 2, and the nonresidues are the other coset. ////

If $p = 11$, then $\bar{1}^2 = \overline{10}^2 = \bar{1}$, $\bar{2}^2 = \bar{9}^2 = \bar{4}$, $\bar{3}^2 = \bar{8}^2 = \bar{9}$, $\bar{4}^2 = \bar{7}^2 = \bar{5}$, and $\bar{5}^2 = \bar{6}^2 = \bar{3}$, so $\{\bar{1}, \bar{3}, \bar{4}, \bar{5}, \bar{9}\} = R$ are the residues, and $\bar{2}$, $\bar{6}$, $\bar{7}$, $\bar{8}$, and $\overline{10}$ are nonresidues. An element n in $\mathbb{Z}$ is said to be a residue or nonresidue (mod p) according to whether $\bar{n}$ is a residue or nonresidue.

Exercises

1 Find all the residues and nonresidues (mod 17). What happens in $\Phi(8)$?
2 See if -1 and 2 are residues for $p = 3, 5, 7, 11, 13, 17, 23, 29, 31, 37$, and 41. Can you conjecture general results here?

If, for a given prime p (for instance, 7), we know a primitive root g (for instance, $\bar{3}$), then since $\Phi(p) = \{g, g^2, g^3, \ldots, g^{p-1} = 1\}$, the residues are just $\{g^2, (g^2)^2, (g^3)^2, \ldots, [g^{(p-1)/2}]^2\}$ (for instance, $\bar{3}^2 = \bar{2}$, $\bar{3}^4 = \bar{4}$, $\bar{3}^6 = \bar{1}$). In general, however, finding a primitive root is difficult whereas, after the machinery has been developed, finding out whether a specific number is a residue is easy.

Since $\Phi(p)/R$ is a group with two elements, we often identify it with the group $(\{1, -1\}, \cdot)$. This gives us a homomorphism $l: \Phi(p) \rightarrow \{1, -1\}$ with each residue mapped to 1 and each nonresidue to -1. If we extend this map a bit, we get a function called the *Legendre symbol*, which is defined as follows:

$$\left(\frac{a}{p}\right) = \begin{cases} l(\bar{a}) & \text{if } p \nmid a \\ 0 & \text{if } p \mid a \end{cases} = \begin{cases} +1 & \text{if } a \text{ is a residue (mod } p) \\ -1 & \text{if } a \text{ is a nonresidue (mod } p) \\ 0, & \text{if } p \mid a \end{cases}$$

Proposition 1

(a) $\left(\dfrac{a}{p}\right)\left(\dfrac{b}{p}\right) = \left(\dfrac{ab}{p}\right).$

(b) $\left(\dfrac{a^2}{p}\right) = 1$ if $p \nmid a$ so, in particular, $\left(\dfrac{1}{p}\right) = 1.$

(c) $\left(\dfrac{a + kp}{p}\right) = \left(\dfrac{a}{p}\right)$

PROOF (a) If $p \mid a$ or $p \mid b$ the statement reduces to $0 = 0$, so we may assume $p \nmid a$ and $p \nmid b$. In this case we have

$$\left(\frac{ab}{p}\right) = l(ab) = l(a)l(b) = \left(\frac{a}{p}\right)\left(\frac{b}{p}\right).$$

(b) $\left(\dfrac{a^2}{p}\right) = \left(\dfrac{a}{p}\right)\left(\dfrac{a}{p}\right) = (\pm 1)^2 = 1.$

(c) a and $a + kp$ are in the same coset $\bar{a}$ in $\mathbb{Z}_p \approx \mathbb{Z}/p\mathbb{Z}$, so

$$\left(\frac{a}{p}\right) = l(\bar{a}) = \left(\frac{a + kp}{p}\right). \qquad\qquad ////$$

EXAMPLE We have seen earlier that $(\frac{1}{7}) = (\frac{4}{7}) = (\frac{2}{7}) = 1$, while $(\frac{3}{7}) = (\frac{5}{7}) = (\frac{6}{7}) = -1$. Using the various parts of the previous proposition, we can show that

$$\left(\frac{2{,}004}{7}\right) = \left(\frac{2 \cdot 2 \cdot 501}{7}\right) = \left(\frac{2^2}{7}\right)\left(\frac{501}{7}\right) = \left(\frac{501}{7}\right) = \left(\frac{4}{7}\right) = 1$$

and

$$\left(\frac{2{,}400}{7}\right) = \left(\frac{2^5 \cdot 3 \cdot 5^2}{7}\right) = \left(\frac{2^2}{7}\right)\left(\frac{2^2}{7}\right)\left(\frac{5^2}{7}\right)\left(\frac{6}{7}\right) = \left(\frac{6}{7}\right) = -1$$

Each of these computations could be done in a variety of ways.

We now state a result that greatly extends our ability to compute quadratic residues.

Proposition 2 If p is an odd prime, then

(a) $\left(\dfrac{-1}{p}\right) = (-1)^{(p-1)/2}$; alternatively, $\left(\dfrac{-1}{p}\right) = \begin{cases} 1 & \text{if } p = 4k+1 \\ -1 & \text{if } p = 4k+3 \end{cases}$

(b) $\left(\dfrac{2}{p}\right) = (-1)^{(p^2-1)/8}$; alternatively, $\left(\dfrac{2}{p}\right) = \begin{cases} 1 & \text{if } p = 8k+1 \text{ or } 8k+7 \\ -1 & \text{if } p = 8k+3 \text{ or } 8k+5 \end{cases}$

(c) If p and q are both odd primes, then

$$\left(\frac{p}{q}\right)\left(\frac{q}{p}\right) = (-1)^{[(p-1)/2][(q-1)/2]}$$

This is the *law of quadratic reciprocity*. An alternative statement is that $\left(\dfrac{p}{q}\right) = \left(\dfrac{q}{p}\right)$ unless both p and q are of the form $4k + 3$, when $\left(\dfrac{p}{q}\right) = -\left(\dfrac{q}{p}\right)$.

PROOF (*a*) We want to see when $x^2 = -1$ has a solution in $\mathbb{Z}_p$. Squaring, we obtain $x^4 = (x^2)^2 = (-1)^2 = 1$, so we want to know when $x^4 - 1$ has four solutions in $\mathbb{Z}_p$. If $x^4 - 1$ has but two solutions, these are 1 and -1, which are no help. Since $\Phi(p)$ is cyclic, so are all its subgroups, and we want conditions on $\Phi(p)$ so that it will have a subgroup of order 4. By Lagrange's theorem this requires that $4 \,|\, |\Phi(p)| = p - 1$, which is also sufficient, since a finite cyclic group has subgroups of all orders dividing the order of the group. Thus $4 \,|\, p - 1$ or $p \equiv 1 \pmod 4$ or $p = 4k + 1$ is a necessary and sufficient condition.

For proofs of parts *b* and *c*, see almost any number-theory text. For instance, Bolker [10] contains a pleasant and coherent geometric proof of part *c*, as does Archibald [3]. The latter also has several other proofs and some history.   ////

Exercise

3 Show that $x^4 + 1 = 0$ is solvable in $\mathbb{Z}_p$ if and only if $p \equiv 1 \pmod 8$, where p is an odd prime.

The remainder of this section is devoted to applications of quadratic residues.

APPLICATION 1   All odd prime divisors of $n^2 + 1$ must have the form $4k + 1$. It is probably worth considering some numerical values first. For $n = 1,000$ it is far from obvious even how to factor $n^2 + 1$. However, if $p \,|\, n^2 + 1$, then $-1 \equiv n^2 \pmod p$, so -1 is a residue $\pmod p$ and, by Proposition 2*a*, p has the form $4k + 1$.

APPLICATION 2   Is 379 a quadratic residue modulo 401 (401 is prime)? Using the various parts of Propositions 1 and 2, we have

$$\left(\frac{379}{401}\right) = \left(\frac{401}{379}\right) = \left(\frac{22}{379}\right) = \left(\frac{2}{379}\right)\left(\frac{11}{379}\right) = (-1)\left(\frac{11}{379}\right)$$

$$= (-1)\left(\frac{379}{11}\right)(-1) = \left(\frac{379}{11}\right) = \left(\frac{5}{11}\right) = \left(\frac{11}{5}\right)$$

$$= \left(\frac{1}{5}\right) = 1$$

Exercise

4 Compute the following: $\left(\dfrac{2,005}{103}\right)$, $\left(\dfrac{2,009}{101}\right)$, and $\left(\dfrac{2,020}{401}\right)$. (101, 103, and 401 are primes.)

APPLICATION 3   For what odd primes p is 3 a quadratic residue? Rephrasing, we want to solve $\left(\dfrac{3}{p}\right) = 1$. Using quadratic reciprocity, we have that

$$\left(\frac{3}{p}\right) = (-1)^{[(3-1)/2][(p-1)/2]}\left(\frac{p}{3}\right) = (-1)^{(p-1)/2}\left(\frac{p}{3}\right)$$

p must have the form $3k + 1$ or $3k + 2$, so we treat two cases. If $p = 3k + 1$, we have

$$\left(\frac{3}{p}\right) = (-1)^{(p-1)/2}\left(\frac{3k+1}{3}\right) = (-1)^{(p-1)/2}$$

so $\left(\dfrac{3}{p}\right) = 1$ if p also is of the form $4k + 1$. To satisfy both conditions, p must be of the form $12k + 1$.

If $p = 3k + 2$, then

$$\left(\frac{3}{p}\right) = (-1)^{(p-1)/2}\left(\frac{3k+2}{3}\right) = (-1)^{(p-1)/2}\left(\frac{2}{3}\right) = (-1)^{(p-1)/2}(-1) = (-1)^{(p+1)/2}$$

so we need $p = 4k + 3$ if we want $\left(\dfrac{3}{p}\right) = 1$.

Combining the conditions $p = 4k + 3$ and $p = 3k + 2$ tells us that $p = 12k + 11$. This can be seen by looking at the intersection of the sets $\{3,7,11,15, 19,23,...\}$ and $\{2,5,8,11,14,17,20,23,...\}$. So our result is that 3 is a quadratic residue of all primes p of the form $12k + 1$ or $12k + 11$.

Exercises

5 Show that if an odd prime p divides a number of the form $n^2 - 3$, then p cannot have the form $12k + 5$ or $12k + 7$. Formulate a similar result about numbers of the form $n^2 + 5$ and their odd divisors.
6 For which odd primes p is 7 a quadratic residue?

APPLICATION 4   Factor 9,997. A quick examination reveals that neither 3, 5, 7, nor 11 divides 9,997 evenly, so it is time to think. If 9,997 is not prime, at least one factor is less than $\sqrt{9,997} < 100$. So the primes 3, 5, 7, 11, 13, ..., 97 are candidates. We can thin out this list by noting that if $p\,|\,9,997 = (100)^2 - 3$, then $+3 \equiv (100)^2 \pmod{p}$, so 3 must be a residue $\pmod{p}$. By application 3 we see that p must have the form $12k + 1$ or $12k + 11$. This gives us a smaller list consisting of 11, 13, 23, 37, 47, 59, 61, 71, 73, 83, and 97. Now it is time to divide. 11 doesn't work, but $13 \cdot 769 = 9,997$. Since $\sqrt{769} < 30$, it remains only to see if 11, 13, or 23 divides 769. They do not, so a complete factorization of 9,997 is $13 \cdot 769$. This particularly convenient example is from Niven and Zuckerman [65].

Exercises

7   Show that 9,993 is prime.

8   (a)   Is $x^2 + 6x + 3$ solvable in $\mathbb{Z}_{97}$? In $\mathbb{Z}_{107}$? (*Hint:* Complete the square.)
    (b)   For what primes p is $x^2 + 6x + 3$ solvable?
    (c)   Factor 10,603.

9   Show that in $\mathbb{Z}_p$ every primitive root is a nonresidue, where p is an odd prime.

As a last application we present two cases in which primitive roots are easily found.

Definition   A prime f of the form $2^m + 1$ is called a *Fermat prime*.

Exercise

10   Show that any nonresidue $\pmod{f}$ is a primitive root. Show that if $f \geq 5$ then 3 is a primitive root $\pmod{f}$. (*Hint:* Show that $3 \mid 2^m + 1$ if and only if m is odd.)

Fermat primes also show up in geometry. A regular n-gon is constructible by straightedge and compass if and only if $n = 2^q f_1 f_2 \cdots f_k$, where the f_i are distinct Fermat primes. This will be discussed further in Chap. 10. They also have an interesting history (see, for instance, Shanks [81]) since Fermat conjectured that all the numbers $2^{2^n} + 1$, for $n = 0, 1, 2, 3, \ldots$, were prime. This has turned out to be far from true.

Proposition 3 $\left(\dfrac{a}{p}\right) = a^{(p-1)/2} \pmod{p}$.

PROOF   If $p \mid a$ this just says $0 = 0$. We can thus assume $p \nmid a$. Since $[a^{(p-1)/2}]^2 = a^{p-1} = 1 \pmod{p}$, we must have $a^{(p-1)/2} = \pm 1$. If $1 = \left(\dfrac{a}{p}\right)$, then $a = b^2$ for some b in $\Phi(p)$, so $a^{(p-1)/2} = b^{p-1} = 1$. Since there are $(p-1)/2$ distinct residues in $\mathbb{Z}_p$, they are all the solutions of $x^{(p-1)/2} - 1$.

Since $x^{p-1} - 1$ has $p - 1$ solutions and

$$x^{p-1} - 1 = [(x^{(p-1)/2} + 1)][x^{(p-1)/2} - 1)]$$

the $(p - 1)/2$ nonresidues are the solutions of $x^{(p-1)/2} + 1$. So if n is a nonresidue, $n^{(p-1)/2} + 1 = 0$ or $n^{(p-1)/2} = -1 = \left(\dfrac{n}{p}\right)$.   ////

Exercise

11 If $p = 4q + 1$, where p and q are primes, show that 2 is a primitive root of p. [*Hint:* Write down a subgroup lattice for $\Phi(p)$, use the last proposition to identify the residues, and show that any nonresidue is either a primitive root or has order 4.]

7.3   THE TWO-SQUARE THEOREM OF FERMAT

We devote this section to proving the famous "two-square theorem" due originally to Fermat. This says:

Theorem An odd prime is respresentable as a sum of two squares if and only if the prime is of the form $4k + 1$.

PROOF Note first that if $p = x^2 + y^2$, for p an odd prime, and $x, y \in Z$, then $p \nmid x$ and $p \nmid y$. Thus,

$$x^2 + y^2 \equiv 0 \ (\text{mod } p)$$

or $\bar{x}^2 + \bar{y}^2 = 0$ in $\mathbb{Z}_p$ for $\bar{x}, \bar{y} \neq 0$

Hence, $\bar{x}^2 = -\bar{y}^2$ in $\mathbb{Z}_p$

and $(\bar{x}\bar{y}^{-1})^2 = -1$ in $\mathbb{Z}_p$

So -1 is a residue (mod p), which can take place if and only if p has the form $4k + 1$.

If we try to reverse the argument, we obtain the following: p has the form $4k + 1$ implies -1 is a residue (mod p). Equivalently, there is a $\bar{z}$ in $\mathbb{Z}_p$ such that $\bar{z}^2 = -1$ in $\mathbb{Z}_p$. Thus, $p \mid z^2 + 1$ for some z in $\mathbb{Z}$ or $pm = z^2 + 1$. Factoring $z^2 + 1$ in $\mathbb{Z}[i]$ yields $pm = (z + i)(z - i)$. If p is prime in $\mathbb{Z}[i]$, then $p \mid z + i$ or $z - i$. Say $p \mid z + i$, so $p(a + bi) = z + i$, and $a + bi = z/p + i/p$ is in $\mathbb{Z}[i]$, which is a contradiction. Similarly, $p \nmid z - i$, so p is not a prime in $\mathbb{Z}[i]$.

Let $u + iv$ be a proper divisor of p. Then there exists an $a + bi$ such that

$$(u + iv)(a + bi) = p$$

Since $x + iy \xrightarrow{\theta} x - iy$ is a ring automorphism, of $\mathbb{Z}[i]$, we have $(u - iv)(a - bi) = \theta((u + iv)(a + bi)) = \theta(p) = p$. Multiplying, we have $(u^2 + v^2)(a^2 + b^2) = p^2$, so $u^2 + v^2 = 1$, p, or p^2.

$u^2 + v^2 \neq 1$, since this would make $u + iv$ a unit. $u^2 + v^2 \neq p^2$, since this would make $a + ib$ a unit. Thus, $u^2 + v^2 = p$, and the theorem is complete.   ////

Exercises

1   Show that -2 is a residue (mod p) for an odd prime p if and only if p is of the form $8k + 1$ or 3.

2   If p is an odd prime, show that p can be written as $x^2 + 2y^2$, where $x, y \in \mathbb{Z}$, if and only if p is of the form $8k + 1$ or 3. (*Hint:* $\mathbb{Z}[\sqrt{-2}]$ is a UFD.)

3   If p is an odd prime, show that p can be written as $z^2 + 6w^2 - 4zw$ if and only if p has the form $8k + 1$ or 3.

4   Show that the equation $7y^2 - 1 = x^2$ has no solutions with $x, y \in \mathbb{Z}$.

5   By factoring in $\mathbb{Z}[i]$, show that if m and n both can be written as the sum of two squares in $\mathbb{Z}$, then so can mn.

6   If m is an integer, show that m^2 has the form $4k$ or $4k + 1$. If $x^2 + y^2 = p$ is an odd prime, show that x and y have opposite parity. Thus show that p has the form $4k + 1$.

A similar theorem was conjectured by Fermat and proved by Lagrange. It states that every integer can be written as a sum of four squares. A brief check shows that this seems reasonable. For instance,

$$1 = 1^2 + 0^2 + 0^2 + 0^2 \qquad 11 = 3^2 + 1^2 + 1^2$$
$$2 = 1^2 + 1^2 \qquad 12 = 2^2 + 2^2 + 2^2$$
$$3 = 1^2 + 1^2 + 1^2 \qquad 13 = 3^2 + 2^2$$
$$4 = 2^2 \qquad 14 = 3^2 + 2^2 + 1^2$$
$$5 = 2^2 + 1^2 \qquad 15 = 3^2 + 2^2 + 1^2 + 1^2$$
$$6 = 2^2 + 1^2 + 1^2 \qquad 16 = 4^2$$
$$7 = 2^2 + 1^2 + 1^2 + 1^2 \qquad 17 = 4^2 + 1^2$$
$$8 = 2^2 + 2^2 \qquad 18 = 3^2 + 3^2$$
$$9 = 3^2 \qquad 19 = 3^2 + 3^2 + 1^2$$
$$10 = 3^2 + 1^2 \qquad 20 = 4^2 + 2^2$$

An algebraic proof of this theorem is in Herstein [38], while an *ad hoc* proof is in Archibald [3]. See also Exercise 19 in Chap. 8.

We have treated a very few topics that come up in number theory, and none of those in any generality. One of the charms of number theory is that many of the results can be understood by nonmathematicians. For instance "an odd prime $p = x^2 + y^2$ if and only if p has remainder 1 when divided by 4" can be understood by almost anyone, although the proof might be more difficult. The early proofs, it must be admitted, did not use cyclic groups, quadratic residues, and unique factorization domains, as these terms were not in the mathematical vocabulary until at least 200 years later. Fermat used the method of infinite descent, which is a backward induction method. See Archibald [3] for a different proof of this result, not using groups or unique factorization domains.

Number theory also has an attraction for good writers, so there are many fine books on various aspects of the subject. The following are good texts on elementary number theory: Archibald [3], Davenport [19], Hardy and Wright [35], Niven and Zuckerman [65], and Shanks [81]. Two recent books treating elementary number theory, assuming some background in modern algebra, are Bolker [10] and Ireland and Rosen [41]. Both are well written. Other recommended books are Borevich-Shafarevich [11], Mordell [63], and Ore [67]; there are many other good texts on the subject (as well as some bad ones). "Disquisitiones Arithmeticae," the classic by Gauss [28], is also recommended.

The next few exercises concern several other topics in number theory and conclude this chapter.

Exercises

7   Let A be a finite abelian group where $A = \{a_1, a_2, \ldots, a_n\}$. Show that $\prod_{i=1}^{n} a_i = a_1 a_2 \cdots a_n = e$, unless A contains exactly one element t of order 2. In that case, $\prod_{i=1}^{e} a_i = t$.

8   In Z_p show that $\bar{1} \cdot \bar{2} \cdot \bar{3} \cdots \overline{p-1} = -\bar{1}$. Alternatively, $p \mid 1 \cdot 2 \cdot 3 \cdots (p-1) + 1$. We are again using p to represent a prime. This is *Wilson's theorem*.

9   Every element of $\mathbb{Z}_p$ satisfies $x^{p-1} - 1$. Since $\mathbb{Z}_p[x]$ is a UFD (why?), we must have $x^{p-1} - 1 = (x - \bar{1})(x - \bar{2}) \cdots (x - \overline{(p-1)})$. Why? Setting $x = \bar{0}$ gives a second proof of Wilson's theorem.

10   If p is a prime of the form $4k + 1$, show that $-g$ is a primitive root if and only if g is a primitive root.

11   Use the fact that 2 is a primitive root modulo 29 to do the following:
    (a)   Find four solutions of $x^4 = 1$ in Z_{29}.
    (b)   Solve $1 + x + x^2 + x^3 + x^4 + x^5 + x^6$ in Z_{29}.

12   If R is the subgroup of residues in $\Phi(p)$, show that $\prod_{r \in R} r = (-1)^{(p+1)/2}$. Here p is an odd prime.

8

THE WEDDERBURN THEOREMS (A SURVEY)

In this and in the following chapters several more advanced topics will be presented. I was first exposed to them informally, often incorrectly, as gossip in the graduate lounge. I will try to present them in the same spirit but with less misinformation. The first of these topics is the Wedderburn theorems.

In this chapter we shall assume all rings have a multiplicative identity 1, unless we specify otherwise.

Our first definition is that of an artinian ring. Emil Artin (1898–1962) was an extraordinary algebraist and teacher. Among his many accomplishments was the generalizing of Wedderburn's theorems from algebras to rings. In the last chapter we mentioned Artin's conjecture that any positive nonsquare integer is a primitive root for infinitely many primes. Coincidentally, both Wedderburn and (later) Artin were at Princeton for many years. There is an article by Artin [4] that discusses the importance of Wedderburn's work to modern algebra.

Definition A ring is **left artinian** if it satisfies the descending-chain condition on left ideals. That is, no infinite set $\{L_i\}$ of left ideals exists in R, where

$$L_1 \supsetneq L_2 \supsetneq L_3 \cdots$$

Recall that a subring L of a ring R is a left ideal of R if $RL \subseteq L$. For example, $\mathbb{Z}$ is not artinian since $\mathbb{Z} \supsetneq 2\mathbb{Z} \supsetneq 4\mathbb{Z} \supsetneq 8\mathbb{Z} \supsetneq \cdots$ is an infinite chain of ideals each properly containing the next. Obviously, we could go through the whole theory using right ideals and right-artinian rings. We shall, however, use "artinian" to mean "left artinian."

As examples of artinian rings we have division rings (including fields), $M_2(\mathbb{R})$, and finite rings (including all $\mathbb{Z}_n$).

Exercise

1 Prove that the rings mentioned above are all artinian.

Definition A ring R without 1 is **nilpotent** if, for some positive integer n, $R^{(n)} = \{0\}$, where

$$R^{(n)} = \left\{ \sum_{\text{finite}} r_1 r_2 \cdots r_n \,\middle|\, r_i \in R \right\}$$

To establish nilpotence, it suffices to show that each $r_1 r_2 \cdots r_n = 0$.

Two standard examples of nilpotent rings are

$$R_1 = \left\{ \begin{pmatrix} 0 & a \\ 0 & 0 \end{pmatrix} \,\middle|\, a \in \mathbb{R} \right\} \quad \text{and} \quad R_2 = \left\{ \begin{pmatrix} 0 & a & b \\ 0 & 0 & c \\ 0 & 0 & 0 \end{pmatrix} \,\middle|\, a,b,c \in \mathbb{R} \right\}$$

Another class of examples results from taking any abelian group A, writing it additively, and defining the multiplication always to be 0. The resulting ring is called the *trivial* ring on the group $(A, +)$.

The following definition is standard, though not universal.

Definition An artinian ring is **semisimple** if it has no nonzero nilpotent left ideals.

Another definition in use gives better motivation for the term semisimple. Recall that a ring is simple if it has no proper two-sided ideals.

Definition An artinian ring is **semisimple** if it is a direct sum of a finite number of simple rings.

The first Wedderburn theorem reconciles the two definitions.

First Wedderburn theorem An artinian ring has no nonzero nilpotent ideals if and only if it is isomorphic to a direct sum of simple ideals.

Exercises

2 Show that $\mathbb{Z}_{10}$ is semisimple but that $\mathbb{Z}_9$ is not. Is $\mathbb{Z}$ semisimple?
3 Characterize those m such that $\mathbb{Z}_m$ is semisimple.

4 Is $M_2(\mathbb{R})$ semisimple?

5 For what n is $\mathbb{Z}_n$ a simple ring?

6 Show that $M_n(D)$ is simple, where D is a division ring, and $M_n(D)$ is the ring of all $n \times n$ matrices over D. [Start by letting E_{ij} be the matrix of all zeros except for a 1 where the ith row intersects the jth column. If A is a nonzero matrix in some ideal I, look at $E_{ir} A E_{sj}$, where A is assumed to have a nonzero entry in the (r,s) position.]

The next theorem says that Exercise 6 presents the whole story.

Second Wedderburn theorem An artinian simple ring is isomorphic to $M_n(D)$, the ring of $n \times n$ matrices over a division ring D.

We won't prove these theorems. Readable proofs can be found in many places. Among the available sources are Jans [43], McCoy [60], Paley and Weichsel [68], and Herstein [39]. A particularly short proof is given by Henderson [37]. There is a different definition of semisimple for nonartinian rings, but it simplifies our presentation to assume artinian; all the definitions coincide in the artinian case.

Combining the first and second Wedderburn theorems, we have a complete classification of semisimple artinian rings.

Corollary 1 If R is a semisimple artinian ring, then

$$R \simeq M_{n_1}(D_1) \oplus M_{n_2}(D_2) \oplus \cdots \oplus M_{n_k}(D_k)$$

where $M_n(D)$ is the ring of all $n \times n$ matrices with entries from D.

Corollary 2 If R is a commutative semisimple artinian ring, then R is a finite direct sum of fields.

PROOF R is a finite direct sum of simple rings, so each of these simple rings must also be commutative. Each division ring must then be a field, and a ring of all $n \times n$ matrices is commutative only if $n = 1$. Since any field (or any ring) is isomorphic to the 1×1 matrices over it, the proof is complete. ////

Let R be a finite ring such that $x^2 = x$ for all x in R. First we observe that $x = x^2 = (-x)^2 = -x$. Next we see that $x + y = (x + y)^2 = x^2 + xy + yx + y^2$, so $xy = -(yx) = yx$, and R is commutative. Since R is finite, it satisfies the dcc. Since $x^2 = x$ for all x, R has no nonzero nilpotent elements, and thus certainly no nonzero nilpotent ideals. Applying the Wedderburn theorems, we have that $R \simeq F_1 \oplus \cdots \oplus F_k$, where each F_i is a field. In F_i we must have the same property that $x^2 = x$ for all x in F_i. In a field a quadratic can have at most two solutions, and 0 and 1 are solutions for $x^2 = x$. Thus $F_i = \{0,1\} \simeq \mathbb{Z}_2$, and $R \simeq \mathbb{Z}_2 \oplus \mathbb{Z}_2 \oplus \cdots \oplus \mathbb{Z}_2$. A ring is called *Boolean* if $x^2 = x$ for all x in the ring. Thus we have just classified all finite Boolean rings. If S is a finite set, and $P(S)$ is its power set, and we define $A \cdot B = A \cap B$ and $A + B = (A \cup B) - (A \cap B)$, then $P(S)$ is such a Boolean ring.

As another application, assume a ring has only two proper left ideals, and neither contains the other. Obviously, the dcc on left ideals is satisfied. We call this ring R, and the proper left ideals L_1 and L_2. $L_1 \cap L_2$ is also a left ideal, and so must be $\{0\}$. In Exercise 7 we show that R has no proper nilpotent ideals.

Thus $R \simeq \sum_{i=0}^{k} M_{n_i}(D_i)$, but each of the $M_{n_i}(D_i)$ is isomorphic to an ideal in R, so

$$R \simeq M_n(D) \qquad \text{or} \qquad R \simeq M_{n_1}(D_1) \oplus M_{n_2}(D_2)$$

Since each $M_{n_i}(D_1)$ contains at least n_i left ideals of the form

$$\begin{pmatrix} 0 & \cdots & 0 & d_1 & 0 & \cdots & 0 \\ 0 & \cdots & 0 & d_2 & 0 & \cdots & 0 \\ & & & \cdots & & & \\ 0 & \cdots & 0 & d_{n_i} & 0 & \cdots & 0 \end{pmatrix}$$

we have, in the second case, that $R \simeq D_1 \oplus D_2$.

We can eliminate the first case entirely. For the first case to work, we would have to have $R \simeq M_2(D)$, since $M_1(D) \simeq D$ would yield no proper left ideals, and $M_3(D)$ would yield too many. Moreover, in $M_2(D)$

$$L_1 = \left\{ \begin{pmatrix} x & 0 \\ y & 0 \end{pmatrix} \right\} \qquad L_2 = \left\{ \begin{pmatrix} 0 & x \\ 0 & y \end{pmatrix} \right\} \qquad L_3 = \left\{ \begin{pmatrix} x & x \\ y & y \end{pmatrix} \right\}$$

are distinct proper left ideals, and this case is eliminated. Using the Wedderburn theorems, we have shown that any ring with two proper left (or right) ideals each not containing the other is isomorphic to $D_1 \oplus D_2$, where D_1 and D_2 are division rings. This would not have worked if we hadn't required 1 in R.

Exercises

7 (a) Let R be a ring with exactly two proper ideals, L_1 and L_2, neither containing the other. Show that $R \simeq L_1 \oplus L_2$. Thus $1 = e_1 + e_2$ with $e_1 \in L_1$ and $e_2 \in L_2$. Show that $e_1^2 = e_1$ and $e_2^2 = e_2$, and that $e_1 \neq 0 \neq e_2$. Finally, show that neither L_1 nor L_2 is nilpotent and, thus, that R is semisimple.

  (b) Let $\mathbb{Z}_p{}^0$ be the abelian group $(\mathbb{Z}_p, \oplus)$ made into a ring by defining $ab = 0$ for all a and b in $\mathbb{Z}_p$. Show that $R = D \oplus \mathbb{Z}_p{}^0$, where D is a division ring, has exactly two proper left ideals but no multiplicative identity.

8   Let R be a ring with 1 and with no proper left ideals. Show that R is a division ring. (If R doesn't have 1 in these circumstances, then $R \simeq \mathbb{Z}_p{}^0$.)

We recall from Chap. 2 that an element a of a ring is nilpotent if $a^n = 0$ for some positive integer n. Obviously, every element of a nilpotent ideal is nilpotent, since if all $r_1 r_2 \cdots r_n = 0$ we can just set each $r_i = a$.

9   Show that the nilpotent elements of the ring $\mathbb{Z}_{p^2} \oplus \mathbb{Z}_{p^3} \oplus \mathbb{Z}_{p^4} \oplus \cdots$ form a nonnilpotent ideal. Note also that this ring is not an artinian ring.

Proposition   In an artinian ring, any left ideal consisting of nilpotent elements is itself nilpotent.

PROOF   See Jans [43] or Paley and Weichsel [68].   /////

As another application of the Wedderburn structure theorems, we examine artinian integral domains. An integral domain has no nonzero zero divisors and thus no nonzero nilpotent elements. If we call our artinian domain D, we see that D is semisimple. Corollary 2 now applies, and $D \simeq F_1 \oplus F_2 \oplus \cdots F_k$. If $k > 1$, then $(1, 0, 0, \ldots) \cdot (0, 1, 0, 0, \ldots)$ yields zero divisors in D, which tells us that $k = 1$ and $D \simeq F_1$. We put this in the form of a proposition.

Proposition   An artinian integral domain is a field. In particular, no integral domain can have a finite number of proper ideals (except fields which have no proper ideals).

Exercises

10   Give an alternative proof of this proposition.

11   A ring is *strongly regular* if, for every a in the ring, there exists x such that $a = a^2 x$. Show that any finite strongly regular ring is a direct sum of fields. (Read the third Wedderburn theorem before starting.)

Assuming we can find nilpotent elements without undue difficulty, we need to know whether or not a ring is artinian. This is not always easy to establish, but is easy for one large class of examples.

Definition   An **algebra** over F is a ring R that is also a vector space over a field F. The multiplications must satisfy $\alpha(r_1 r_2) = (\alpha r_1) \cdot r_2 = r_1 \cdot (\alpha r_2)$ for all $r_i \in R$, $\alpha \in F$.

The following exercise will supply us with examples.

Exercises

12   Show that the following are examples of algebras:

(*a*)   Any field is an algebra over itself.

(*b*)   The ring $P(F)$ of all polynomials over a field F is an algebra over F.

(*c*)   If $E \supset F$, where E and F are fields, then E is an algebra over F.

(*d*) $M_n(F)$ is an algebra over F.

(e) $\left\{ \begin{pmatrix} x & y \\ 0 & z \end{pmatrix} \middle| x,y,z \in F \right\}$ is an algebra over F.

(f) Let $P_0(F)$ denote the set of all polynomials with constant term equal to 0. $P_0(F)$ is an algebra over F.

(g) The ring $C[0,1]$ is an algebra over $\mathbb{R}$. (See Chap. 2.)

Definition An **algebra ideal** of an algebra A is an ideal of A that is also a subspace. A similar definition holds for left or right ideals.

13 If A is an algebra with 1, show that every left ideal of A is an algebra ideal. (If $i \in L$, show that $\alpha i \in L$ by using the definition of algebra and the "absorption" laws.)

Using the last exercise, we can show the following.

Proposition If A is a finite-dimensional algebra with 1, then A is artinian.

PROOF   We have to show the dcc holds, so let

$$A \supseteq I_1 \supseteq I_2 \supseteq I_3 \supseteq \cdots$$

be a chain of left ideals. Each ideal is a subspace by the last exercise. Thus, $n = \dim(A) \geq \dim(I_1) \geq \dim(I_2) \geq \dim(I_3) \cdots \geq 0$. If $\dim(I_k) = \dim(I_{k+1})$, then $I_k = I_{k+1}$, so this chain can have at most $n+1$ distinct ideals, and the dcc holds.   ////

It was for this class of examples that Wedderburn proved his original theorem back in 1907. Going back to Exercise 9, we see that parts a, d, e, and sometimes c, yield examples of finite-dimensional algebras with 1. Another large set of examples, finite group algebras, are important for group representations.

Exercises

14 If $R = \left\{ \begin{pmatrix} x & y \\ 0 & z \end{pmatrix} \middle| x,y,z \in F \right\}$, then find the largest nilpotent ideal.

15 Show that $R = \left\{ \begin{pmatrix} x & 0 \\ y & 0 \end{pmatrix} \middle| x,y \in \mathbb{R} \right\}$ is an algebra without 1. Find an infinite descending chain of left ideals in R.

We conclude by listing some deeper applications of the Wedderburn structure theorems. The first is due to Wedderburn himself, in 1937. For Theorem 1 we remove the restriction that all rings have 1.

Theorem 1 Let A be a finite-dimensional algebra over a field F. Let A have a basis over F consisting of nilpotent elements. Then A is itself nilpotent.

Exercise

16 Let

$$R = \left\{ \begin{pmatrix} 0 & x & y \\ 0 & 0 & z \\ 0 & 0 & 0 \end{pmatrix} \,\middle|\, x,y,z \in \mathbb{Q} \right\}$$

$$b_1 = \begin{pmatrix} 0 & 1 & 0 \\ 0 & 0 & 0 \\ 0 & 0 & 0 \end{pmatrix} \qquad b_2 = \begin{pmatrix} 0 & 0 & 1 \\ 0 & 0 & 0 \\ 0 & 0 & 0 \end{pmatrix} \qquad b_3 = \begin{pmatrix} 0 & 0 & 0 \\ 0 & 0 & 1 \\ 0 & 0 & 0 \end{pmatrix}$$

Show that $\{b_1, b_2, b_3\}$ is a basis for R. Show also that each b_i is nilpotent. Conclude that R is nilpotent.

A group G is a *torsion group* if each element of G is of finite order. A stronger condition is local finiteness. A group is *locally finite* if $\langle x_1, x_2, \ldots, x_n \rangle$ is finite for any finite set of elements $\{x_1, x_2, \ldots, x_n\}$ in G. Recall that $\langle x_1, x_2, \ldots, x_n \rangle$ is the smallest subgroup of G containing all the x_i.

Burnside proved the following somewhat surprising theorem.

Theorem 2   If G is a torsion subgroup of $\mathrm{GL}(n, F)$, then G is locally finite.

A proof can be given using the Wedderburn theorems (see Herstein [39] or Kaplansky [46]).

Burnside posed the problem of establishing the same results without the context of matrices:

1   The general Burnside problem.   Is every torsion group locally finite?
2   The weak Burnside problem.   Is every torsion group, where $x^N = e$ for all x in G, locally finite? This sets up a separate Burnside problem for each positive number N.

A negative answer for Problem 1 has been established recently by Golod and Shafarevich [29]. Novikov has given a negative answer for Problem 2 where $N = 4,381$. If $N = 2, 3, 4,$ or 6, Problem 2 has a positive answer, but this problem becomes very difficult very fast. See Hall [32], Chap. 18, for a discussion of these cases.

Exercises

17   Establish the general Burnside problem affirmatively for abelian groups.
18   If $N = 2$, show G is abelian, establishing problem 1, and thus problem 2, for this case.

The other Wedderburn theorem is the simplest.

Third Wedderburn theorem   A finite division ring is a field.

For proofs of this see Herstein [38], Redei [72], Dean [20], or Paley and Weichsel [68]. Each of the many proofs of this result seems a bit harder than necessary, and a simpler proof may be possible. For instance, the proof that every finite integral domain is a field is a fairly routine exercise. One corollary of this result is that every finite projective plane which is Desaurgian is also Pappian. Hartshorne [36] gives an excellent account of this theory. Another corollary is the following:

Proposition Any finite semisimple ring is a direct sum of finitely many full matrix rings over finite fields.

Exercise

19 Let $\mathscr{Q}_p = \{a + bi + cj + dk \mid a,b,c,d \in \mathbb{Z}_p, \ p$ an odd prime$\}$, and $i^2 = j^2 = k^2 = ijk = -1$. Show that:

(a) $\mathscr{Q}_p$ is noncommutative.

(b) $\mathscr{Q}_p$ is not a division ring.

(c) $\mathscr{Q}_p$ is simple.

(d) $\mathscr{Q}_p \simeq M_2(\mathbb{Z}_p)$.

(e) $\mathscr{Q}_p$ has nonzero nilpotent elements.

(f) $p \mid a^2 + b^2 + c^2 + d^2$ for some choice of a, b, c, and d, each between 0 and $p - 1$ and not all 0.

To conclude this section, we prove the following result:

Proposition Any finite subgroup G of a division ring of characteristic p is cyclic.

PROOF We have already proved the same result for any field. If the division ring D has characteristic p, it contains $\mathbb{Z}_p = \{0,1,1 + 1,...\}$. Looking at $S = \{\sum_{i=1}^{k} z_i g_i \mid z_i \in \mathbb{Z}_p, \ g_i \in G\}$, we see that it forms a finite subring of D, and S is itself a division ring. By Wedderburn's theorem, S is a field; G, as a finite subgroup of a field, is cyclic. ////

Exercises

20 *Let R be an artinian ring such that $x^3 = x$ for all x in R. Show that R is commutative.

21 *Redo Exercise 20 without the assumption that the ring is artinian.

The monograph of Herstein [39] contains a superb account of such commutativity theorems, including the method of embedding a ring as a subdirect sum of primitive rings. The modest example of Exercise 20 has been greatly generalized in many directions.

9

GROUP REPRESENTATIONS (A SURVEY)

9.1 REPRESENTATIONS

The outline of this chapter is as follows: The first concept is that the irreducible representations are the building blocks from which all other representations can be constructed. We then find all irreducible respresentations of finite abelian groups.

Suppose next that we have a finite group G and that the inequivalent irreducible representations consist of $n_1 \times n_1$ matrices, $n_2 \times n_2$ matrices, ..., $n_k \times n_k$ matrices. Then $|G| = n_1^2 + n_2^2 + \cdots + n_k^2$ (the degree equation).

The next part of the chapter is devoted to the utilization of this equation. For instance, k equals the number of conjugate classes in G, and $|G/G'|$ is the number of n_i that equal 1. Using this equation, we can prove various small results about groups of order pq or p^n.

The next concept is rather mysterious, and it is that the trace of a representation is a valuable concept to us. The trace of a representation is called a *character*, and we put our information about traces into a $k \times k$ character table. The solution to the degree equation fills the first column of the character table. Various properties of this character table are given, such as the orthogonality properties. Some connections with algebraic numbers (another surprise) are made, and as a sample mysterious result we discuss Burnside's celebrated theorem that all groups of order $p^a q^b$ are solvable, where p and q are primes.

Definition   Let $GL(V)$ denote the group of all one-to-one onto linear transformations from a vector space V to itself. Let $GL(n,F)$ denote the group of all nonsingular $n \times n$ matrices with entries in F.

By picking an ordered basis for an n-dimensional vector space V over a field F, we set up an isomorphism between $GL(V)$ and $GL(n,F)$, and we use this isomorphism almost constantly in this chapter.

Definition   A **representation** of a group G is a homomorphism from G to $GL(V)$ or $GL(n,F)$.

Examples are easy to come by.

EXAMPLE A   Any group of nonsingular matrices over a field is a representation of itself.

EXAMPLE B   Any group can be mapped trivially to the element 1, which can be considered a 1×1 matrix over $\mathbb{C}$. This is called the *principal representation*.

EXAMPLE C   As we showed in Chap. 3, $\theta(a) = \begin{pmatrix} 0 & 1 \\ -1 & 0 \end{pmatrix} = M_1$ and $\theta(b) = \begin{pmatrix} 1 & 0 \\ 0 & -1 \end{pmatrix} = M_2$ will yield an isomorphism of the dihedral group D_8 into $GL(2,\mathbb{C})$. Actually, $\theta(a^i b^j) = M_1{}^i M_2{}^j$ defines the complete isomorphism.

EXAMPLE D   The homomorphism $\theta(r) = \begin{pmatrix} 1 & r \\ 0 & 1 \end{pmatrix}$ is a homomorphism from $(\mathbb{R},+)$ to $GL(2,\mathbb{R})$ or $GL(2,\mathbb{C})$.

EXAMPLE E   Let G be a group of permutations on $S = \{1,2,...n\}$. Pick an ordered basis $(b_1, b_2, ... b_n)$ for an n-dimensional vector space V over any field F. If $g = \begin{pmatrix} 1 & 2 & \cdots & n \\ g(1) & g(2) & \cdots & g(n) \end{pmatrix}$, then let $\theta(g)$ be the linear transformation taking each b_i to $b_{g(i)}$. (In fact, it suffices to let G be a group acting on S as in Chap. 4.)

EXAMPLE E′ Let $G = \text{Sym}(3)$, and $S = \{1,2,3\}$. Then, using the above procedure and going over to matrices yields

$$\begin{pmatrix} 1 & 2 & 3 \\ 1 & 2 & 3 \end{pmatrix} \longrightarrow \begin{pmatrix} 1 & 0 & 0 \\ 0 & 1 & 0 \\ 0 & 0 & 1 \end{pmatrix}$$

$$\begin{pmatrix} 1 & 2 & 3 \\ 2 & 1 & 3 \end{pmatrix} \longrightarrow \begin{pmatrix} 0 & 1 & 0 \\ 1 & 0 & 0 \\ 0 & 0 & 1 \end{pmatrix} \qquad \text{Discussed below}$$

$$\begin{pmatrix} 1 & 2 & 3 \\ 3 & 2 & 1 \end{pmatrix} \longrightarrow \begin{pmatrix} 0 & 0 & 1 \\ 0 & 1 & 0 \\ 1 & 0 & 0 \end{pmatrix}$$

$$\begin{pmatrix} 1 & 2 & 3 \\ 1 & 3 & 2 \end{pmatrix} \longrightarrow \begin{pmatrix} 1 & 0 & 0 \\ 0 & 0 & 1 \\ 0 & 1 & 0 \end{pmatrix}$$

$$\begin{pmatrix} 1 & 2 & 3 \\ 2 & 3 & 1 \end{pmatrix} \longrightarrow \begin{pmatrix} 0 & 1 & 0 \\ 0 & 0 & 1 \\ 1 & 0 & 0 \end{pmatrix}$$

$$\begin{pmatrix} 1 & 2 & 3 \\ 3 & 1 & 2 \end{pmatrix} \longrightarrow \begin{pmatrix} 0 & 0 & 1 \\ 1 & 0 & 0 \\ 0 & 1 & 0 \end{pmatrix}$$

For example, $\theta \begin{pmatrix} 1 & 2 & 3 \\ 2 & 1 & 3 \end{pmatrix}$ is the linear transformation taking b_1 to b_2, b_2 to b_1, and b_3 to b_3. Thus, $\left(\theta \begin{pmatrix} 1 & 2 & 3 \\ 2 & 1 & 3 \end{pmatrix}\right)(b_1) = 0 \cdot b_1 + 1 \cdot b_2 + 0 \cdot b_3$, and the coefficients $(0,1,0)$ are the first row of the corresponding matrix. Similarly,

$$\left(\theta \begin{pmatrix} 1 & 2 & 3 \\ 2 & 1 & 3 \end{pmatrix}\right)(b_2) = 1 \cdot b_1 + 0 \cdot b_2 + 0 \cdot b_3$$

and

$$\left(\theta \begin{pmatrix} 1 & 2 & 3 \\ 2 & 1 & 3 \end{pmatrix}\right)(b_3) = 0 \cdot b_1 + 0 \cdot b_2 + 1 \cdot b_3$$

yield the last two rows.

Exercises

1 Represent D_8 in a similar manner. This should result in 4×4 matrices with all zero entries except for exactly one 1 in each row and column.

2 Mechanize the above a bit by noting that each row is all zeros except for one entry equal to 1. If $g = \begin{pmatrix} 1 & 2 & \cdots & n \\ g(1) & g(2) & \cdots & g(n) \end{pmatrix}$, then which entry is the

nonzero entry in the first row of the corresponding matrix? In the second row? The ith row?

Definition A matrix is called a **permutation matrix** if it has all zero entries except for exactly one 1 in each row and column.

EXAMPLE E″ The symmetric group Sym(n) is representable as the group of all $n \times n$ permutation matrices. This takes place over any field.

EXAMPLE F To look at the symmetries of a cube, imagine the eight corners of the cube at $\{(\pm 1, \pm 1, \pm 1)\}$. Let G be the group of all 48 symmetries of the cube. This includes 24 rotations and 24 other motions, such as moving the top half to the bottom by taking (a,b,c) to $(a,b,-c)$. Taking a standard basis, we find that we have 48 matrices, each all zero except for exactly one entry of $+1$ or -1 in each row and column. G can be defined as all permutations of the eight vertices, always taking pairs of adjacent vertices to adjacent vertices.

EXAMPLE G We saw in Chap. 4 that any group $(G,*)$ is isomorphic to a group of permutations of $G = S$. This is the content of Cayley's theorem. If G has order n, this allows us to use Example E and represent G as a group of $n \times n$ permutation matrices. This is extremely important and is called the *regular representation* (due to C. S. Pierce in 1879; cf. Klein [47]).

EXAMPLE G′ Let $V = \{e,a,b,c\}$ be Klein's four group. First writing V as permutations, we have

$$e \leftrightarrow \begin{pmatrix} e & a & b & c \\ e & a & b & c \end{pmatrix}$$

$$a \leftrightarrow \begin{pmatrix} e & a & b & c \\ a & e & c & b \end{pmatrix}$$

$$b \leftrightarrow \begin{pmatrix} e & a & b & c \\ b & c & e & a \end{pmatrix} \quad \text{Explained below}$$

$$c \leftrightarrow \begin{pmatrix} e & a & b & c \\ c & b & a & e \end{pmatrix}$$

Letting (b_e, b_a, b_b, b_c) be a basis for V, we obtain

$$\theta(e) = \begin{pmatrix} 1 & 0 & 0 & 0 \\ 0 & 1 & 0 & 0 \\ 0 & 0 & 1 & 0 \\ 0 & 0 & 0 & 1 \end{pmatrix}$$

$$\theta(a) = \begin{pmatrix} 0 & 1 & 0 & 0 \\ 1 & 0 & 0 & 0 \\ 0 & 0 & 0 & 1 \\ 0 & 0 & 1 & 0 \end{pmatrix}$$

$$\theta(b) = \begin{pmatrix} 0 & 0 & 1 & 0 \\ 0 & 0 & 0 & 1 \\ 1 & 0 & 0 & 0 \\ 0 & 1 & 0 & 0 \end{pmatrix}$$

$$\theta(c) = \begin{pmatrix} 0 & 0 & 0 & 1 \\ 0 & 0 & 1 & 0 \\ 0 & 1 & 0 & 0 \\ 1 & 0 & 0 & 0 \end{pmatrix}$$

Recall that, say, b is represented by the permutation taking x to xb. Thus,

$$b \leftrightarrow \begin{pmatrix} e & a & b & c \\ eb & ab & bb & cb \end{pmatrix} = \begin{pmatrix} e & a & b & c \\ b & c & e & a \end{pmatrix}.$$

Exercises

3   Find the regular representations for $(\mathbb{Z}_4, +)$ and Sym(3).

4   Note that no matrix other than the one for e had other than zeros on the main diagonal in Example G' and in Exercise 3. Show that this is true for the regular representation of any finite group G.

Definition   If S is a set of linear transformations of a vector space V, and W is a subspace of V such that $g(W) \subseteq W$ for all g in S, then W is said to be S-**stable**. Let G be a group, θ a representation of G to GL(V), and $S = \theta(G)$. We shorten things a bit, since, if W is $\theta(G)$-stable, we say W is G-*stable*.

For instance, in Example E, where $S = \{1,2,3,...,n\}$, we let $w = 1 \cdot b_1 + 1 \cdot b_2 + \cdots + 1 \cdot b_n$. Then $(\theta(g))(w) = 1 \cdot b_{g(1)} + 1 \cdot b_{g(2)} + \cdots + 1 \cdot b_{g(n)} = w$. Thus, if W is the 1-dimensional subspace spanned by w, then W is G-stable.

Definition   If S is a set of linear transformations on V such that no subspace of V other than V and (0) is S-stable, then V is S-**irreducible**. Otherwise V is S-**reducible**. The space V is S-**decomposable** if two proper S-stable subspaces, W and W', can be found so that $V = W \oplus W'$. This means that $W \cap W' = (0)$ and that every v in V can be written as $w + w'$, with w in W and w' in W. If V is not S-decomposable, it is S-**indecomposable**.

Having set up these concepts in terms of linear transformations on V, we can now pick convenient bases and see how the matrices look. Let V be S-reducible so that W is an S-stable subspace. Let $(w_1, w_2, ..., w_k)$ be a basis, and extend this basis to a basis of V. Since W is S-stable, we have $g(w_i) = a_{i1}w_1 + a_{i2}w_2 + \cdots + a_{ik}w_k + 0$. If v_i is one the basis vectors not in W, then

$$g(v_i) = b_{i1}w_1 + b_{i2}w_2 + \cdots + b_{ik}w_k + c_{i,\,k+1}v_{i,\,k+1} + \cdots + c_{in}v_{in}$$

With this basis, g is represented by

$$\begin{pmatrix} a_{11} & a_{12} & \cdots & a_{1k} & 0 & \cdots & 0 \\ a_{21} & a_{22} & \cdots & a_{2k} & 0 & \cdots & 0 \\ \hdotsfor{7} \\ a_{k1} & a_{k2} & \cdots & a_{kk} & 0 & \cdots & 0 \\ b_{k+1,1} & b_{k+1,2} & \cdots & b_{k+1,k} & c_{k+1,k+1} & \cdots & c_{k+1,n} \\ \hdotsfor{7} \\ b_{n,1} & b_{n,2} & \cdots & b_{n,k} & c_{n,k+1} & \cdots & c_{n,n} \end{pmatrix}$$

If, as in our usual case, $S = \theta(G)$, then having an S-stable subspace will give us two smaller representations. One, on the subspace W, is obtained by restricting each linear transformation $\theta(g)$ to W. Here g would correspond to

$$\begin{pmatrix} a_{11} & a_{12} & \cdots & a_{1k} \\ a_{21} & a_{22} & \cdots & a_{2k} \\ \hdotsfor{4} \\ a_{k1} & a_{k2} & \cdots & a_{kk} \end{pmatrix}$$

Also, we would obtain a representation θ' on V/W by taking $\theta'(g) = \theta(g) + W$. Here g corresponds to

$$\begin{pmatrix} c_{k+1,k+1} & \cdots & c_{k+1,n} \\ c_{k+2,k+1} & \cdots & c_{k+2,n} \\ \hdotsfor{3} \\ c_{n,k+1} & \cdots & c_{n,n} \end{pmatrix}$$

Exercise

5 Let G be represented in $GL(V)$, and let W be a k-dimensional G-stable subspace. By picking the basis correctly we can represent G in $GL(n,F)$ as follows:

$$\alpha(g) = \left(\begin{array}{c|c} A(g) & 0 \\ \hline B(g) & C(g) \end{array} \right)$$

where $A(g)$ is a $k \times k$ matrix, $C(g)$ is an $(n-k) \times (n-k)$ matrix, 0 is a $k \times (n-k)$ matrix of all zeros, and $B(g)$ is an $(n-k) \times k$ matrix. If α is a homomorphism of G into $GL(n,F)$, show that α' and α'' are homomorphisms into $GL(k,F)$ and $GL(n-k,F)$, where $\alpha'(g) = A(g)$, and $\alpha''(g) = C(g)$.

If V is decomposable, then $V = W \oplus W'$, where W and W' are both S-stable. If B_W is a basis for W, and $B_{W'}$ is a basis for W', then $B_W \cup B_{W'}$ is a basis for V.

Proceeding as before, we examine $\theta(g)(w_i')$ for some w_i' in $B_{W'}$. Since W' is S-stable,

$$\theta(g)(w_i') = c_{i,k+1}w_{k+1}' + c_{i,k+2}w_{k+2}' + \cdots + c_{in}w_n'$$

Thus g is represented by $\begin{pmatrix} A(g) & 0 \\ 0 & B(g) \end{pmatrix}$.

Maschke's theorem says we often can assume that reducibility will give us decomposability. For the remainder of the chapter, decomposable and reducible will mean G-decomposable and G-reducible for some group G.

Theorem 1 (Maschke)   Let G be a finite group, and $\theta: G \to \mathrm{GL}(V)$ be a representation of G, where V is an n-dimensional vector space over a field of characteristic 0. Then if V is reducible, it is decomposable.

PROOF   Since V is reducible, there is a nontrivial G-subspace W of V. Going over to the matrix representation as described above, we have

$$\alpha(g) = \begin{pmatrix} \overbrace{A(g)}^{k} & \overbrace{0}^{n-k} \\ B(g) & C(g) \end{pmatrix} \begin{matrix} {\scriptstyle\}k} \\ {\scriptstyle\}n-k} \end{matrix}$$

Let $$Q = \sum_{g \in G} B(g)A(g^{-1}) \quad \text{and} \quad Q' = \frac{1}{|G|}Q$$

Then let $$R = \begin{pmatrix} I_k & 0 \\ Q' & I_{n-k} \end{pmatrix}$$

where I_j is the $j \times j$ identity matrix.   Then

$$\alpha'(g) = R^{-1}\alpha(g)R = \begin{pmatrix} A(x) & 0 \\ 0 & C(x) \end{pmatrix}$$

Thus, if we had used $(b_1R, b_2R, \ldots, b_nR)$ as a basis, we would have $\langle b_1R, b_2R, \ldots, b_kR \rangle$ spanning W. Letting $\langle b_{k+1}R, b_{k+2}R, \ldots, b_nR \rangle = W'$, we would have that W' is also G-stable, and $V = W \oplus W'$. Thus V is decomposed.   ////

Exercises

6   Fill in and check the details of the proof of Maschke's theorem. Where was it necessary to restrict the characteristic of the field? Where did we use the fact that $|G|$ is finite?

7 Show that the following representation, though reducible (in fact reduced), is not indecomposable: $(\mathbb{Z}_p, +) \xrightarrow{\theta} GL(2, \mathbb{Z}_p)$ is given by $\theta(k) = \begin{pmatrix} 1 & k \\ 0 & 1 \end{pmatrix}$. This shows that some restriction on the characteristic field is necessary. [It actually suffices for char $(F) \nmid |G|$.]

8 Find an example of a reducible, but indecomposable representation of an infinite group over a field of characteristic 0.

Let us now work through a concrete example using Maschke's theorem. If G is a permutation group on S, and we identify each s in S with a basis vector b_s, then $w = b_{s_1} + b_{s_2} + \cdots + b_{s_k} = \sum_{s \in S} s b_s$ is a G-stable vector. We earlier looked at S_3 acting on $\{b_1, b_2, b_3\}$, and we now return to this example.

We want a new basis starting with $c_1 = b_1 + b_2 + b_3$, so that $\langle c_1 \rangle = W$ will be G-stable. We complete the basis in any convenient manner, say with $c_2 = b_2$ and $c_3 = b_3$. This yields a representation β with

$$\beta(I) = \begin{pmatrix} 1 & 0 & 0 \\ 0 & 1 & 0 \\ 0 & 0 & 1 \end{pmatrix} \qquad \beta\begin{pmatrix} 1 & 2 & 3 \\ 2 & 1 & 3 \end{pmatrix} = \begin{pmatrix} 1 & 0 & 0 \\ 1 & -1 & -1 \\ 0 & 0 & 1 \end{pmatrix}$$

$$\beta\begin{pmatrix} 1 & 2 & 3 \\ 1 & 3 & 2 \end{pmatrix} = \begin{pmatrix} 1 & 0 & 0 \\ 0 & 0 & 1 \\ 0 & 1 & 0 \end{pmatrix} \qquad \beta\begin{pmatrix} 1 & 2 & 3 \\ 3 & 2 & 1 \end{pmatrix} = \begin{pmatrix} 1 & 0 & 0 \\ 0 & 1 & 0 \\ 1 & -1 & -1 \end{pmatrix}$$

$$\beta\begin{pmatrix} 1 & 2 & 3 \\ 2 & 3 & 1 \end{pmatrix} = \begin{pmatrix} 1 & 0 & 0 \\ 0 & 0 & 1 \\ 1 & -1 & -1 \end{pmatrix} \qquad \beta\begin{pmatrix} 1 & 2 & 3 \\ 3 & 1 & 2 \end{pmatrix} = \begin{pmatrix} 1 & 0 & 0 \\ 1 & -1 & -1 \\ 0 & 1 & 0 \end{pmatrix}$$

Maschke's theorem says that with a different choice of basis we get the same representation, except simplified in that

$$\begin{pmatrix} A(g) & 0 \\ \hline B(g) & C(g) \end{pmatrix} \qquad \text{becomes} \qquad \begin{pmatrix} A(g) & 0 \\ \hline 0 & C(g) \end{pmatrix}$$

Here, this yields

$$\beta^*(I) = \begin{pmatrix} 1 & 0 & 0 \\ \hline 0 & 1 & 0 \\ 0 & 0 & 1 \end{pmatrix} \qquad \beta^*\begin{pmatrix} 1 & 2 & 3 \\ 2 & 1 & 3 \end{pmatrix} = \begin{pmatrix} 1 & 0 & 0 \\ \hline 0 & -1 & -1 \\ 0 & 0 & 1 \end{pmatrix}$$

$$\beta^*\begin{pmatrix} 1 & 2 & 3 \\ 2 & 3 & 1 \end{pmatrix} = \begin{pmatrix} 1 & 0 & 0 \\ \hline 0 & 0 & 1 \\ 0 & -1 & -1 \end{pmatrix}$$

and so forth. [In this particular case, β^* can be obtained from the basis $(b_1 + b_2 + b_3, b_1 - 2b_2 + b_3, b_1 + b_2 - 2b_3)$.]

Exercises

9 Represent $\begin{pmatrix} 1 & 2 & 3 & 4 \\ 2 & 3 & 4 & 1 \end{pmatrix}$ and $\begin{pmatrix} 1 & 2 & 3 & 4 \\ 2 & 1 & 3 & 4 \end{pmatrix}$ with respect to (b_1, b_2, b_3, b_4)
and to $(b_1 + b_2 + b_3 + b_4, b_2, b_3, b_4)$. Here, Sym(4) acts on $\{1, 2, 3, 4\}$, and the
representation with respect to the basis (b_1, b_2, b_3, b_4) is the representation by
permutation matrices.

10 Verify that the matrices $\begin{pmatrix} 1 & 0 \\ 0 & 1 \end{pmatrix}$, $\begin{pmatrix} 0 & 1 \\ 1 & 0 \end{pmatrix}$, $\begin{pmatrix} 0 & 1 \\ -1 & -1 \end{pmatrix}$, $\begin{pmatrix} -1 & -1 \\ 0 & 1 \end{pmatrix}$,
$\begin{pmatrix} 1 & 0 \\ -1 & -1 \end{pmatrix}$, and $\begin{pmatrix} -1 & -1 \\ 1 & 0 \end{pmatrix}$ form a group isomorphic to Sym(3).

Definition Two representations, α and β, of a group G in $GL(n,F)$ are **equivalent** if there exists an invertible $n \times n$ matrix T such that $\alpha(g) = T^{-1}\beta(g)T$ for all g in G.

Definition If a representation of G on V is decomposable to $V = W \oplus W'$ with W and W' G-stable, then the representations of G on W and W' are called **components** or **subrepresentations** of the representation. That is, if

$$\alpha(g) = \left(\begin{array}{c|c} \beta(g) & 0 \\ \hline 0 & \gamma(g) \end{array} \right)$$

for all g in G, then β and γ are **components** or **subrepresentations** of α.

Definition Two representations, α and β, of a group G are **equivalent** if there is a nonsingular matrix T such that $\alpha(g) = T^{-1}\beta(g)T$ for all g in G. A representation of G by $m \times m$ matrices is called a representation of **degree** m. If two representations are equivalent, they must have the same degree.

Definition A representation is **faithful** if it is a monomorphism. Of our examples, A, C, D, E, F, and G are faithful. Let us now look at some unfaithful examples.

EXAMPLE H Let G be a group, and N a normal subgroup of index 2. Then let $\alpha: G \to GL(1,F)$ be given by $\alpha(n) = (1)$ for n in N, $\alpha(g) = (-1)$ for $g \in G - N$. Since $\ker(\alpha) = N$, this representation is not faithful unless $|G| = 2$.

EXAMPLE I Let C_n be a cyclic group of order n with c as a generator. Then $\alpha(c^j) = (e^{(2\pi i/n)j})$ defines an isomorphism from C_n into the 1×1 matrices over $\mathbb{C}$. Thus α is faithful.

Exercises

11 Let $C_n = \langle c \rangle$ be a cyclic group. Let $\beta_k(c^j) = e^{(2\pi i k/n)j}$. For which k is β_k faithful?

12 $G = \mathbb{Z}_4 \times \mathbb{Z}_2$ has eight irreducible representations, all of degree 1. What are these eight representations? How many are faithful?

EXAMPLE I′ Let A be a noncyclic finite abelian group. Then any one-dimensional representation of A is not faithful. To show this, we assume $\alpha: A \to \mathrm{GL}(1,F) \simeq F^*$ is an isomorphism. But any finite multiplicative subgroup of F^* must be cyclic (Chap. 7 contains a proof). This contradicts the fact that A is not cyclic.

We need a result due to Schur, and then we can handle all representations of finite abelian groups. For reasons unknown to the author, this result is always called a lemma.

The impatient reader can skip directly to the degree equation if he is willing to accept without proof the following four facts. First, every irreducible representation of an abelian group A is of degree 1. Second, A has $|A|$ distinct irreducible representations. Third, if G is a finite group, G has exactly $|G/G'|$ distinct representations of degree 1. Finally, if α is a representation of the factor group G/N, then α^* is a representation of G, where $\alpha^*(g) = \alpha(g + N)$.

Schur's Lemma If $\alpha: G \to \mathrm{GL}(n,F)$ and $\beta: G \to \mathrm{GL}(m,F)$ are irreducible representations of G, and L is an $m \times n$ matrix such that

$$L\alpha(g) = \beta(g)L \qquad \text{for all } g \text{ in } G$$

then either $L = 0$ or $m = n$, L is nonsingular, and α and β are equivalent.

PROOF For a proof by matrices, see Newman [64]. We shall go briefly through a different proof. L is a homomorphism from V to $\overline{V}$. Thus $L(V)$ is a subspace of $\overline{V}$ and since $\beta(g)L(V) = L\alpha(g)(V) \subseteq L(V)$ for all g in G $L(V)$ is a G-subspace of $\overline{V}$. By the irreducibility of $\overline{V}$ we either have $L(V) = \overline{V}$ or $\{0\}$. The latter case yields that $L(V) = \overline{V}$.

ker(L) is a subspace of V. Since $L\alpha(g)(\ker(L)) = \beta(g)L(\ker(L)) = \beta(g)\{0\} = \{0\}$, we see that $\alpha(g)(\ker(L)) \subseteq \ker(L)$ and thus ker(L) is a G-subspace. Thus ker(L) = $\{0\}$ or V and either L is one-to-one or equal to 0. Thus if L is not 0 it is one-to-one, onto, and thus nonsingular in turn implying $m = n$. Also L nonsingular implies $\alpha(g) = L^{-1}\beta(g)L$ so that α and β are equivalent. ////

Note that the L satisfying $L\alpha(g) = \beta(g)L$ form a ring a. Schur's lemma could be rephrased as saying this ring, if nontrivial, is a division ring.

We henceforth assume $F = \mathbb{C}$, which means all polynomials over $\mathbb{C}$ have roots, and all matrices have eigenvalues. We also assume G is a finite group.

Corollary 1 Suppose L is an $n \times n$ matrix such that $L\alpha(g) = \alpha(g)L$ for each g in G, where $\alpha(g)$ is an irreducible representation. Then

$$L = \begin{pmatrix} \lambda & & & \bigcirc \\ & \lambda & & \\ & & \ddots & \\ \bigcirc & & & \lambda \end{pmatrix} = \lambda I$$

for some λ in $\mathbb{C}$.

PROOF If L has λ as an eigenvalue, then $\alpha(g)(L - \lambda I) = (L - \lambda I)\alpha(g)$ for all g in G. If $v \neq 0$ is an eigenvector corresponding to λ, then $vL = \lambda v$ or $v(L - \lambda I) = 0$. Thus $v \neq 0$ is in the kernel of $L - \lambda I$, and $L - \lambda I$ is singular. By Schur's lemma, $L - \lambda I = 0$. Thus,

$$L = \lambda I = \begin{pmatrix} \lambda & & & \bigcirc \\ & \lambda & & \\ & & \ddots & \\ \bigcirc & & & \lambda \end{pmatrix}$$

////

Corollary 2 If z is in the center of G, and α is an irreducible representation of G, then $\alpha(z) = \lambda I$.

PROOF Since $zg = gz$ and α is a homomorphism, we have $\alpha(z)\alpha(g) = \alpha(g)\alpha(z)$ for all g in G. Let $L = \alpha(z)$ in the notation of the last corollary.

////

In Example C and Exercise 10 we have irreducible representations of D_8 and Sym(3). Using Corollary 2, we can see that $Z(D_8) = \{e, a^2\}$, and $Z(\text{Sym}(3)) = \left\{ \begin{pmatrix} 1 & 2 & 3 \\ 1 & 2 & 3 \end{pmatrix} \right\} = \{I\}$.

Corollary 3 An irreducible representation of an abelian group must be one-dimensional.

PROOF If A is abelian and α is a representation, then, by Corollary 2,

$$\alpha(g) = \lambda_g I = \begin{pmatrix} \lambda_g & & & \bigcirc \\ & \lambda_g & & \\ & & \ddots & \\ \bigcirc & & & \lambda_g \end{pmatrix}$$

for every g in A. This set of matrices is obviously reducible unless α has degree 1.

////

The next proposition is extremely easy and extremely useful.

Proposition 1   If $\theta: G \to H$ is a homomorphism, and $\alpha: H \to \mathrm{GL}(n,F)$ is a representation of H, then $\alpha\theta$ is a representation of G. If α is reducible, so is $\alpha\theta$.

Exercises

13   Prove Proposition 1.

14   Let $C_n = \langle c \rangle$ be a cyclic group of order n. Each of the following $\theta_i: C_n \to \mathbb{C}^*$ is a one-dimensional representation of C_n. Take $\theta_k(c^j) = \varepsilon^{kj}$, where $\varepsilon = e^{2\pi i/n}$. Find n representations of $G = C_n \times K$, where K is an arbitrary group, by first constructing a homomorphism from G onto C_n.

Proposition 2   A representation α of G yields a faithful representation α' of $G/\ker(\alpha)$.

PROOF   Since α is a homomorphism, the fundamental theorem of homomorphisms yields an isomorphism α' from $G/\ker(\alpha)$ onto the image of α in $\mathrm{GL}(n,F)$.   ////

Proposition 3   The number of representations of G of degree 1 is equal to the order of G/G', where G' is the commutator subgroup of G.

PROOF   Any one-dimensional representation of G is a homomorphism from G into the abelian group of 1×1 nonzero matrices over $\mathbb{C}$. The kernel of this representation must contain G', and thus any one-dimensional representation of G is a representation of G/G'. By Proposition 4 with $H = G/G'$, we see that any representation of G/G' is a representation of G. By proving the following lemma we shall finish the proof of this proposition, since G/G' is abelian.

Lemma   If A is a finite abelian group, every irreducible representation of A is one-dimensional, and there are $|A|$ distinct irreducible representations.

PROOF   Since A is finite and abelian, A is isomorphic to $C_1 \times C_2 \times \cdots \times C_l$, where each C_i is cyclic with generator c_i. Any homomorphism of A is determined by its value on each c_i. If we map c_i to $\mathbb{C}^*$, then $\phi(c_i) = \varepsilon^{(2\pi i/|C_i|)j}$ for $j = 0, 1, 2, \ldots, |C_i| - 1$. Thus we have $|C_i|$ choices for each $\phi(c_i)$, and $|C_1||C_2|\cdots|C_l| = |A|$ choices for a homomorphism from A to $\mathbb{C}^*$. We have already shown any irreducible representation of A is one-dimensional.   ////

Exercises

15 Find all the irreducible representations of $C_p \times C_p$, $C_p \times C_p \times C_p$, $C_{p^2} \times C_p$, and C_{p^3}, where C_n is a cyclic group of order n, and p is prime.

16 If A is an abelian group, show that the homomorphisms of A into $\mathbb{C}^*$ form a group. Which operation should you choose? This group is called $\text{Hom}(A,\mathbb{C}^*)$.

17 *If A is a finite abelian group, show that $\text{Hom}(A,\mathbb{C}^*) \simeq A$.

Having discussed some examples of representations, we come to the degree equation. Proofs of the degree equation can be found in Curtis and Reiner [18], Dornhoff [21], and Huppert [40].

Theorem 2 (the degree equation) Let G be a finite group, and let the degrees of the inequivalent irreducible representations of G be n_1, n_2, n_3, ..., n_k. Then,

(a) $|G| = n_1{}^2 + n_2{}^2 + n_3{}^2 + \cdots + n_k$.
(b) $k =$ the number of conjugate classes in G.
(c) $|G/G'|$ is the number of $n_i = 1$.
(d) Each n_i divides $|G/Z(G)|$, where $Z(G)$ is the center of G. In fact, each n_i divides $|G/A|$, where A is any abelian normal subgroup of G.

If G has N_i representations of degree i, then we can rewrite the degree equation as

$$|G| = N_1 \cdot 1^2 + N_2 \cdot 2^2 + \cdots + N_l \cdot l^2$$

with $k = N_1 + N_2 + \cdots + N_l$ and $N_1 = |G/G'|$. If $N_d \neq 0$, then $d \mid |G/Z(G)|$.

Utilizing the degree equation, we can prove or reprove some results on finite groups. We shall illustrate several of these results and give some others as exercises.

APPLICATION A Every group of order p^2 is abelian (where, as usual, p is prime.)

PROOF If P is a nonabelian group of order p^2, then P must have an irreducible representation of degree greater than 1. Otherwise, $|P/P'| = p^2$, so $P' = \{e\}$ and P is abelian. Let n be the degree of this representation. By part d, n certainly divides $|P|$, so $n = p$ or p^2. The principal representation taking every element to 1 is also an irreducible representation. Thus,

$$p^2 = |P| = 1^2 + n^2 + \sum (n_i)^2 \geq 1 + p^2$$

This contradiction completes the proof. ////

APPLICATION B If P is a nonabelian p-group of order p^n, then $|P/P'|$ has order greater than or equal to p^2. That is, the index of P' is at least p^2.

PROOF Let N_i denote the number of inequivalent irreducible representations of degree i. Thus

$$|P| = \sum_{d \mid |P|} N_d \cdot d^2 \qquad (*)$$

For this problem we obtain

$$|P| = p^n = N_1 + p^2 N_p + p^4 N_{p^2} + \cdots$$

Thus $N_1 \equiv 0 \pmod{p^2}$, and $N_1 = |P/P'|$ by c). $N_1 \geq 1$, since the principal representation is one-dimensional, and this completes the proof.

////

Exercises

18 Review the definition of solvable group, and use B to show that every finite p-group is solvable.

19 Show, for every nonabelian group P of order p^3, that $P' = Z(P)$ is of order p, and that there are $p^2 + p - 1$ conjugate classes. Check this for D_8 and the quaternions.

20 Show that no group of order p^4 exists with $|Z(P)| = |P'| = p$. [*Hint*: We start with groups of order 2^4, and work toward a contradiction. Every conjugate class contains one or two elements, since each conjugate class must have at most $|P'|$ elements. (Why?) Thus, in this case, P has nine conjugate classes. Show that every representation has degree 1 or 2, and thus P has 10 inequivalent irreducible representations.] The proof of this fact without using representations seems to be difficult.

21 Let G be a nonabelian group of order pq, where p and q are primes, and $p < q$.
(*a*) Since G is nonabelian, it must have an irreducible representation, and all irreducible representations of G have degree 1 or p.
(*b*) Show that $|G'| = q$.
(*c*) Show that $N_p = (q - 1)/p$ and thus that $q \equiv 1 \pmod{p}$.
(*d*) Show that G' is the only proper normal subgroup of G.
(*e*) Show that $Z(G) = \{1\}$.

22 No representation of the dihedral group D_n can be both of degree greater than 2 and irreducible.
(*a*) First recall that for D_{2n}, $a^n = b^2 = 1$ and $b^{-1}ab = a^{-1}$. Compute the number of conjugate classes in D_n.
(*b*) Show that D'_{2n} has order n (or $n/2$) when n is odd (n is even).

23 Show that Sym(4) has, among its irreducible representations, two of degree 1, one of degree 2, and two of degree 3. [Recall that $(\mathrm{Sym}(4))' = \mathrm{Alt}(4)$, so that $N_1 = 2$.]

9.2   CHARACTERS

The next circle of ideas is that surrounding characters. Character theory is vaguely miraculous in that it yields more results than it ought to. We shall only scratch the surface here, stating enough basic results so as to be able to construct some character tables and to prove Burnside's $p^a q^b$ theorem. We continue the numbering of theorems and propositions from the last section.

Definition   If $\alpha: G \to GL(n,\mathbb{C})$ is a representation of G, then χ_α, the **character** of α, is defined by $\chi_\alpha(g) =$ the trace of the matrix $\alpha(g)$. [If (a_{ij}) is an $n \times n$ matrix, the trace of (a_{ij}) is $a_{11} + a_{22} + \cdots + a_{nn}$.] We say χ_α is **irreducible** if α is irreducible.

We shall denote the trace of a matrix M by $\text{tr}(M)$. The next proposition enables us to use characters more freely.

Proposition 4   (*a*)   If A and B are $n \times n$ matrices over F, then $\text{tr}(AB) = \text{tr}(BA)$. [*Warning*: $\text{tr}(AB) \neq \text{tr}(A) \cdot \text{tr}(B)$.]

(*b*)   If C and D are $n \times n$ matrices and C is nonsingular, then $\text{tr}(D) = \text{tr}(C^{-1}DC)$.

(*c*)   If α and β are equivalent representations of G, then $\chi_\alpha = \chi_\beta$. Thus, χ is defined for a representation in $GL(V)$ and is independent of basis.

(*d*) $\chi_\alpha(g) = \chi_\alpha(h^{-1}gh)$ for any g and h in G. Thus, characters are constant functions on conjugate classes.

Exercise

1   Prove part *a* of the last proposition.

> PROOF OF PARTS *b*, *c*, AND *d*   Part *b* follows directly from part *a*, with $A = C^{-1}D$ and $B = C$.
>
> If α and β are equivalent, then there exists a nonsingular $n \times n$ matrix T such that $T^{-1}\alpha(g)T = \beta(g)$ for all g in G. Part *b* now implies that $\chi_\alpha = \chi_\beta$.
>
> To show part *d*, note first that $\alpha(h^{-1}gh) = \alpha(h^{-1})\alpha(g)\alpha(h)$, since α is a homomorphism. Part *b* now says that $\chi_\alpha(h^{-1}gh) = \chi_\alpha(g)$.   ////

We have already discussed three examples of characters in this book. For any one-dimensional representation, the representation is the character. This means that if we know G/G' we can compute all the characters of the one-dimensional representations. The second example arose in Sec. 4.3 on Burnside's counting theorem. The number of fixed points $\chi(g)$, where g is a permutation of a set S, is the character of the permutation representation given in Example E. Exercise 3 gives a third example, the Legendre symbol from Chap. 7.

Exercises

2 Prove that $\chi(g)$ is a character in Example E. Also compute χ in Example E'.

3 If $G = \mathbb{Z}_p^*$, show that the Legendre symbol $\left(\dfrac{\cdot}{p}\right)$ is a character of G. (See H. Cohn [16] for many generalizations of this exercise in number theory.)

We gather many of the basic statements about characters into the following theorem.

Theorem 3 (character relations) Let G be a finite group with conjugate classes $C_1, C_2, \ldots, C_k$. Let $|C_i| = c_i$, and let $\chi_1, \chi_2, \ldots, \chi_k$ be the characters of a complete set of inequivalent, irreducible representations.

(a) $\sum_{g \in G} \chi_\alpha(g)\chi_\beta(g^{-1}) = 0$ if β and α are inequivalent, irreducible representations.

(b) $\sum_{g \in G} \chi_\alpha(g)\chi_\alpha(g^{-1}) = |G|$ if α is irreducible.

(c) $\chi_\alpha(g^{-1}) = \overline{\chi_\alpha(g)}$, the complex conjugate of $\chi_\alpha(g)$.

(d) $\chi_\alpha(e) =$ degree of the representation α.

(e) $\sum_{i=1}^{k} \chi_i(g)\chi_i(h^{-1}) = 0$ if g is not conjugate to h.

(f) $\sum_{i=1}^{k} \chi_i(g)\chi_i(g^{-1}) = |G|/c_i$ if g is in C_i.

(g) $\sum_{g \in G} \chi(g)\chi(g^{-1}) = |G|$ if and only if the character is irreducible.

REMARKS Part d is easy to prove. Just note that

$$\alpha(e) = \begin{pmatrix} 1 & & & \bigcirc \\ & 1 & & \\ & & \ddots & \\ \bigcirc & & & 1 \end{pmatrix}$$

The other parts are proved in Herstein [39], Newman [64], Curtis and Reiner [18], Serre [79], Dornhoff [21], and Huppert [40]. Note that part g is a convenient test for irreducibility.

As a first example, we can compute the characters of all the irreducible representations of D_8. The conjugate classes of D_8 are $C_1 = \{e\}$, $C_2 = \{a^2\}$, $C_3 = \{a, a^3\}$, $C_4 = \{b, ba^2\}$, and $C_5 = \{ba, ba^3\}$. By Theorem 2b, there are five irreducible representations of D_8 (up to equivalence). We construct a 5×5 character table as follows: The (i,j)th entry is $\chi_i(g_j)$, where g_j is any element of C_j. A table is given as follows:

	C_1	C_2	C_3	C_4	C_5
χ_1	1	1	1	1	1
χ_2					
χ_3					
χ_4					
χ_5					

It is convenient to always let $C_1 = \{e\}$, and let χ_1 be the character of the principal representation.

Since $G' = \{e,a^2\}$ and $|G/G'| = 4$, we know by Theorem 2c that G' has four one-dimensional representations. Now, using Theorem 2a, we obtain $8 = |G| = 1^2 + 1^2 + 1^2 + 1^2 + n^2$, so the remaining irreducible representation has degree 2. Since $G/G' \simeq (\mathbb{Z}_2, +) \times (\mathbb{Z}_2, +)$, we can easily compute the one-dimensional characters. Using Theorem 3d and the one-dimensional characters, we have

	C_1	C_2	C_3	C_4	C_5
χ_1	1	1	1	1	1
χ_2	1	1	-1	-1	1
χ_3	1	1	-1	1	-1
χ_4	1	1	1	-1	-1
χ_5	2				

Note that $\{e,a^2\} = C_1 \cup C_2 = G'$, so that the first two columns coincide for these four representations derived from G/G'.

To fill in the next row, we use part e of the last theorem. For instance, to find $\chi\,(C_3)$ we have

$$\chi_1(C_1)\chi_1(C_3) + \chi_2(C_1)\chi_2(C_3) + \chi_3(C_1)\chi_3(C_3) + \chi_4(C_1)\chi_4(C_3) + \chi_5(C_1)\chi_5(C_3) = 0$$
$$= 1 \cdot 1 + 1 \cdot (-1) + 1 \cdot (-1) + 1 \cdot 1 + 2 \cdot \chi_5(C_3)$$

so $\chi_5(C_3) = 0$. Similarly, $\chi_5(C_2) = -2$, and $\chi_5(C_4) = \chi_5(C_5) = 0$, completing our table.

Character tables are defined similarly for all finite groups. Note that solving the degree equation will, by part d of the character-relations theorem, give us the first column of the character table.

Exercise

4 If $\mathbb{Z}_3$ denotes the cyclic group of order 3, show that its character table is:

	$C_1 = \{e\}$	$C_2 = \{c\}$	$C_3 = \{c^2\}$
χ_1	1	1	1
χ_2	1	ω	ω^2
χ_3	1	ω^2	ω

where $\omega = e^{2\pi i/3}$.

Utilizing this exercise, we can write a character table for $A = \text{Alt}(4)$ (which is also the group of the tetrahedron). We start, as usual, by determining the conjugate classes:

$$C_1 = \{I\}$$
$$C_2 = \{(12)(34),\ (13)(24),\ (14)(23)\}$$
$$C_3 = \{(123),\ (124),\ (134),\ (234)\}$$
$$C_4 = \{(132),\ (142),\ (143),\ (243)\}$$

$A' = C_1 \cup C_2$, so there are $|A/A'| = 3$ one-dimensional representations. Using (III) now yields

$$12 = 1^2 + 1^2 + 1^2 + n^2$$

Thus, we have

	C_1	C_2	C_3	C_4
χ_1	1	1	1	1
χ_2	1			
χ_3	1			
χ_4	3			

Using the last exercise, the fact that $A/A' \simeq C_3$, and Proposition 1, we can incorporate the character table of Exercise 4, obtaining

	C_1	C_2	C_3	C_4
χ_1	1	1	1	1
χ_2	1	1	ω	ω^2
χ_3	1	1	ω^2	ω
χ_4	3			

Using Theorem 3e thrice, we find that the complete bottom row is

χ_4	3	-1	0	0

Exercises

5 Write a character table for C_2, and then one for $\text{Sym}(3)$.

6 Let $G = \text{Sym}(4)$.
   (a) Show that G has five conjugate classes.
   (b) Show that $G' = \text{Alt}(4)$, so that G has one-dimensional characters.

(c) Show that the remaining representations of G have degrees 2, 3, and 3.

(d) If $C_1 = \{I\}$ and $C_2 = \{(12)(34), (13)(24), (14)(23)\}$, then $N = C_1 \cup C_2$ $\triangleleft\ G$ and $G/N \simeq \text{Sym}(3)$. Use this to compute three rows of the character table of Sym(4).

(e) The row

	I	$(ab)(cd)$	(abc)	$(abcd)$	(ab)
χ_4	3	-1	0	-1	1

will be given by some discussion of doubly transitive permutation groups. Use this row plus part e of the theorem to finish the character table.

Theorem 4 (auxiliary theorem)

(a) $\chi_i(g)$ is always an algebraic integer.

(b) $|\chi_i(g)| \le \chi_i(e)$.

(c) If g has order 2, then $\chi_i(g) = \chi_i(e) - 2k$, where k is a nonnegative integer.

(d) If G is a doubly transitive permutation group on S (or group acting on S), and $\chi(g)$ is the number of fixed points of g, then $\chi^\circ(g) = \chi(g) - 1$ is an irreducible character of G.

(e) Let $\gcd(|C_j|, \text{degree } \chi_\alpha) = 1$, for χ_α a character of an irreducible representation α. Then either $\chi_\alpha(C_j) = 0$ or $\alpha(C_j) = \lambda I$, so that $\alpha(C_j)$ is in the center of $\alpha(G)$.

Definition a is an **algebraic integer** if it is a root of some polynomial $x^n + z_1 x^{n-1} + \cdots + z_n$ with z_i in $\mathbb{Z}$. The algebraic integers can be shown to be a subring of $\mathbb{C}$.

This auxiliary theorem lists a few devices used for computing character tables. A more systematic method is first to compute the character table for a subgroup H, and then use the Frobenius reciprocity theorem. We won't discuss this important theorem, but, again, see any of the references.

Exercises

7 Check part c of the auxiliary theorem on the character tables for D_8, Alt(4), and Sym(3).

8 Do likewise for parts b and e.

As a last application, we prove the famous Burnside $p^a q^b$ theorem. This depends on part e of the auxiliary theorem.

Lemma Let G be a finite group containing a conjugate class of order p^a, where p is a prime, and $a \geq 1$. Then G contains a proper normal subgroup (that is, G is not simple).

PROOF   Let $\chi_1, \chi_2, \ldots, \chi_k$ be a complete set of inequivalent, irreducible representations of G. Then since $|G/G'| = 1$ we have all $n_i > 1$ and

$$|G| = 1^2 + n_2{}^2 + \cdots + n_k{}^2$$

where χ_1 is principal, and n_i is the degree of χ_i. $p \mid |G|$, so p must not divide some n_i where χ_i is not principal.

$$\sum_{i=1}^{k} \chi_i(1)\chi_i(c_j) = 0$$

where $c_j \in C_j$ and $|C_j| = p^a$, by part e of the character-relations theorem. Rephrased, this says

$$1 + \sum_{i=2}^{k} n_i\chi_i(c_j) = 0$$

so it is not possible that $\chi_i(c_j)$ always equals 0 when $\gcd(\deg(\chi_i), p) = 1$.

Thus, for at least one irreducible representation α, we must have $p \nmid \deg(\chi_\alpha)$ and $\chi_\alpha(c_j) \neq 0$. Now we use part e of the auxiliary theorem and note that the only possibility left is

$$\alpha(c_j) = \lambda I$$

Suppose H consists of those elements of G such that $\alpha(h)$ is scalar. Then, since $\alpha(H)$ is normal in $\alpha(G)$, we have $H \triangleleft G$, and we have just shown $H \neq \{e\}$. However, $H \neq G$, since α is irreducible.   ////

Theorem 5 (Burnside)   If G is a group of order $p^a q^b$, where p and q are primes, then G is solvable.

PROOF   If $p = q$, we have shown both in Chap. 5 and earlier in this chapter that G is solvable. If $p \neq q$, then let $|G| = 1 + |C_2| + |C_3| + \cdots + |C_k|$ be the class equation for G. Obviously, pq cannot divide each of $C_2, C_3, \ldots, C_k$. Thus, either some $|C_i| = 1$, whence C_i is contained in the center of G, or some $|C_i|$ is a power of p or q and the lemma applies. Either way, we produce a normal subgroup N. Induction applied to N and G/N shows they are solvable, and thus G itself is solvable (cf. Chap. 5).   ////

Exercise

9 Show that the smallest simple group not of prime order has order 60. [*Hint*: In a group of order 30, there is either only one (thus normal) subgroup of order 5 (by Sylow) or exactly six elements not having order 5.]

In the first edition of Burnside's book [14, 1897], he spent many pages on special cases of the $p^a q^b$ theorem and indicated that the theory of group representations was at that time not directly useful. By the second edition (1911) the special cases were gone, the $p^a q^b$ theorem was included, and his opinion of group representations was higher.

Burnside's $p^a q^b$ theorem is a major steppingstone in the finer theory of solvable groups. There are several other accessible and important theorems which could be done at this point. We define a *Frobenius group* to be a transitive permutation group with no nonidentity element fixing more than one element. Then those elements with no fixed points, together with the identity, can be proved to form a normal subgroup called the *Frobenius kernel*. Frobenius, using character theory, proved this result in 1901.

Exercises

10 Show that Sym(3) and D_{10} are Frobenius groups. What are the Frobenius kernels in each case?

Another theorem, again due to Burnside, is that any transitive permutation group on p symbols (p prime) is either doubly transitive or solvable.

11 Sym(5), Alt(5), D_{10}, and $\mathbb{Z}_5$ can all be regarded as permutation groups on five symbols. Check Burnside's theorem for these groups.

I hope this presentation, largely by exercises and examples, will have whetted the reader's desire to learn more about the subject. I am aware that I skipped most of the proofs and have only scratched the surface of a vast and challenging subject. To the interested reader I suggest Newman's pamphlet [64] which is available from the Superintendent of Documents, U.S. Government Printing Office, Washington, D.C. 20402, for 60 cents. I also recommend learning more algebra, particularly a little module theory, and then reading Curtis and Reiner [18], Herstein [39], Dornhoff [21], Serre [79], or Passman [69]. Burnside's book [14] is a classic, well written, and available in paperback. Lang [53] has a very concise account of this theory. Littlewood [57] has an entirely different approach and contains an extensive set of character tables.

10

A SURVEY OF GALOIS THEORY

10.1 SOME FIELD THEORY

This chapter is the most important one for connecting modern and classical algebra. The basic goal of the chapter is to examine the solution of polynomials by the methods of modern algebra. In doing this, we use groups, rings, fields, solvable groups, permutation groups, maximal ideals, quotient rings, and vector spaces. It is a beautiful theory.

In this section we set up some definitions and some basic theorems to handle field extensions. In the next section we survey how this field theory can be used to prove some famous impossibility theorems. These are the duplication of the cube, squaring the circle, and trisecting an arbitrary angle. In the last sections we outline the procedure for demonstrating the insolvability of a quintic equation by radicals. These are fascinating questions, owing to their simplicity, their long history, and especially to the intriguing idea of proving insolvability. Some historical remarks are to be found at the end of Sec. 10.2 and the beginning of Sec. 10.3.

Before proceeding, the reader might reread Chap. 3 and those parts of Chaps. 2 and 6 that concern fields.

Definition If F is a subfield of K, then K is called an **extension field** of F. In such a case, K is a vector space over K, and the dimension of K as a vector space over F is the **degree of K over F**. This degree is denoted $[K:F]$.

Exercise

1 Find $[\mathbb{C}:\mathbb{R}]$, $[\mathbb{Q}(\sqrt{3}):\mathbb{Q}]$, $[\mathbb{R}:\mathbb{Q}]$, and $[F:F]$, where F is any field.

Our first result strongly resembles Lagrange's theorem.

Theorem 1 If F, K, and L are fields, with $F \subseteq K \subseteq L$, then we have $[L:K][K:F] = [L:F]$. This is of interest primarily when all three degrees are finite. It is also true, however, that the left-hand side is infinite if and only if the right-hand side is infinite.

PROOF We assume $[L:K] = n$ and $[K:F] = m$. Let $\{l_1,l_2,...,l_n\}$ be a basis for L over K, and $\{k_1,k_2,...,k_m\}$ be a basis for K over F. We want a basis for L over F, and $\{l_1k_1,l_1k_2,...,l_2k_1,l_2k_2,...,l_nk_m\}$ seems a logical choice. In fact, it works in a straightforward way and is left as the next two exercises.

If $[L:K] = \infty$, then let $\{l_1,l_2,...\}$ be a basis for L over K. Then $\{l_1,l_2,...\}$ must still be linearly independent over F, since $\sum_i c_i l_i \neq 0$ for all nontrivial choices of c_i in K yields $\sum_i f_i l_i \neq 0$ for all nontrivial choices of f_i in F. (Here, $\sum_i$ denotes finite sums, and "nontrivial" means not selecting all the c_i or f_i equal to 0.)

But $\{l_1,l_2,...\}$ spans a subspace S and is a basis for S over F. By a theorem quoted in Chap. 3, $\{l_1,l_2,...\}$ can be expanded to a basis for L over F. Thus, L has an infinite basis over F, and $[L:F] = \infty$.

Similarly (Exercise 4), if $[K:F] = \infty$, then $[L:F] = \infty$, and this completes the proof. ////

Exercises

2 If $\{l_1,l_2,...,l_n\}$ is a basis for L over K, and $\{k_1,k_2,...,k_m\}$ is a basis of K over F, show that $B = \{l_1k_1,l_1k_2,...,l_1k_m,l_2k_1,l_2k_2,...,l_nk_m\}$ is a linearly independent set over F.

3 Show that the set B spans L over F (notation from Exercise 2).

4 Show that, if $F \subseteq K \subseteq L$ are fields with $[K:F] = \infty$, then $[L:F] = \infty$.

Corollary If $F \subseteq K \subseteq L$ are fields, then $[K:F] \mid [L:F]$.

Definition If K is an extension field of F, and a is in K, then $F(a)$ is the smallest subfield of K containing both F and a.

Exercise

5   If $\mathscr{C}$ is any set of subfields of K, show that the intersection of all the subfields in $\mathscr{C}$ is again a subfield of K.

Using Exercise 5, we obtain one of two characterizations of $F(a)$, the "outside" characterization.

Proposition   If $F \subseteq K$ are fields, and a is in K, then:
(a) $F(a)$ is the intersection of all subfields of K containing both F and a.
(b) $F(a)$ consists of all elements of the form $p(a)/q(a)$, where p and q are polynomials with coefficients in F, and $q(a) \neq 0$. This is the "inside" characterization.

PROOF   (a)   Let I be the intersection of all subfields of K that contain both F and a. By the last exercise, I is a field. Since I must contain F and a, I contains $F(a)$, the smallest subfield containing F and a.

Conversely, since $F(a)$ contains F and a,

$$I = F(a) \cap (\text{other subfields})$$

so $F(a)$ contains I. Thus $I = F(a)$.

(b) $F(a)$ contains a and is a field, so $F(a)$ contains a, a^2, a^3, a^4, ... $F(a)$ also contains F, so all polynomials $f_0 + f_1 a + f_2 a^2 + \cdots + f_k a^k = p(a)$ must also be in $F(a)$. Similarly, multiplicative inverses must be in $F(a)$, which gives us elements of the form $1/q(a)$, where $q(a) \neq 0$ is a polynomial. Finally, we must have all elements of the form $p(a)/q(a)$. It is now easy to check additive and multiplicative closure and inverses.   ////

Exercises

6   Complete the verifications in the proof of part b.
7   In the field $\mathbb{R}$, find $\mathbb{Q}(2)$, $\mathbb{Q}(\sqrt{2})$, and $\mathbb{Q}(\pi)$. (*Fact:* Lindemann proved there is no nontrivial polynomial with coefficients in $\mathbb{Q}$ having π as a root; that is, π is not algebraic. For a proof, see Hardy and Wright [35].)

Definition   Let $F \subseteq K$ be fields. If $F(a)$ is an extension of finite degree over F, we say a is **algebraic** over F. If every element of K is algebraic, we say K is **algebraic** over F. If a is not algebraic, it is **transcendental**.

In $\mathbb{R}$, let us look at $\mathbb{Q}(\sqrt[3]{2})$. We want to see if $\sqrt[3]{2}$ is algebraic over $\mathbb{Q}$ and, if so, of what degree. Since $\sqrt[3]{2}$ is in $\mathbb{Q}(\sqrt[3]{2})$, so is $(\sqrt[3]{2})^2 = \sqrt[3]{4}$. However, $(\sqrt[3]{2})^3 = 2$ is already in $\mathbb{Q}$, so we need go no further in this direction. If $F = \{q_1 + q_2 \sqrt[3]{2} + q_3 \sqrt[3]{4} \mid q_i \text{ in } \mathbb{Q}\}$ is a field, then it will be the smallest containing $\mathbb{Q}$ and $\sqrt[3]{2}$. Thus we will have $F = \mathbb{Q}(\sqrt[3]{2})$; $\{1, \sqrt[3]{2}, \sqrt[3]{4}\}$ will be a basis for $\mathbb{Q}(\sqrt[3]{2})$ over $\mathbb{Q}$, and $\sqrt[3]{2}$ will have degree 3 over $\mathbb{Q}$.

Additive and multiplicative closure are easily checked, as is the existence of additive inverses. This leaves the problem of expressing $1/(a + b\sqrt[3]{2} + c\sqrt[3]{4})$ in the form $q_1 + q_2\sqrt[3]{2} + q_3\sqrt[3]{4}$. However,

$$(a + b\sqrt[3]{2} + c\sqrt[3]{4})^{-1} = (1/N)[(a^2 - 2bc) + (2c^2 - ab)\sqrt[3]{2} + (b^2 - ac)\sqrt[3]{4}]$$

where $N = a^3 + 2b^3 + 4c^3 - 6abc$. N cannot be zero unless $a = b = c = 0$ (Exercise 8), and this completes the example. Of course, the method for finding the inverse is completely unexplained, and we would like an easier method for finding the degree of a given algebraic number.

Exercises

8 Show that for no three rational numbers a, b, and c does $a^3 + 2b^3 + 4c^3 - 6abc = 0$ except if $a = b = c = 0$. [Show first that it is sufficient to take a, b, and c in $\mathbb{Z}$ and gcd $(a,b,c) = 1$.]

9 What is the degree of $\sqrt{7}$ over $\mathbb{Q}$? Of i over $\mathbb{R}$?

10 What is the degree of $\sqrt{2} + \sqrt{3}$ over $\mathbb{Q}$? [What is $(a + b\sqrt{2} + c\sqrt{3} + d\sqrt{6})(a - b\sqrt{2} + c\sqrt{3} - d\sqrt{6})$?]

Our next theorem gives us both an easier way of determining if an element a is algebraic and a simpler description of $F(a)$ when this happens.

Theorem 2 Let K be an extension field of F, and let a be an element of K. Then a is algebraic over F if and only if there is a polynomial $p(x)$ with coefficients in F such that $p(a) = 0$. [Put in other words, such that a is a root of $p(x)$.] If a is algebraic, then $F(a)$ is composed of all $r(a)$, where $r(x)$ is a polynomial over F of degree less than the degree of $p(x)$.

PROOF The elements $\{1, a, a^2, a^3, \ldots\}$ are either linearly independent or dependent. If they are independent, $[F(a):F] = \infty$, a is transcendental, and there is no $p(x)$ such that $p(a) = 0$. If they are dependent, then some set $\{1, a, a^2, \ldots, a^n\}$ is linearly dependent, and there exist f_i in F, not all zero, such that $f_0 1 + f_1 a + f_2 a^2 + \cdots + f_n a^n = 0 = p(a)$. Dividing through, if necessary, we may safely assume $f_0 \neq 0 \neq f_n$ and that no polynomial of smaller degree has a for a root.

This tells us that $p(x)$ is an irreducible polynomial because if $p(x)$ factors properly as $p(x) = q(x)r(x)$, then $0 = p(a) = q(a)r(a)$. Since $q(x)$ and $r(x)$ have lower degree than $p(x)$, we have produced zero divisors in a field, which is impossible.

We now recall that $F[x]$ is a euclidean domain and thus a PID. Those polynomials having a as a root form an ideal, and $p(x)$ is the generator of this ideal.

What is left to prove? We want to show $\{1, a, a^2, \ldots, a^{n-1}\} = B$ is a basis for $F(a)$ over F. Since $F(a) = \{r(a)/q(a) \mid q(a) \neq 0\}$, we can proceed first to take care of all $r(a)$, and then all $1/q(a)$. To show that all polynomials are spanned by B, we note first that if

$$p(a) = 0 = f_0 1 + f_1 a + f_2 a^2 + \cdots + f_n a^n$$

then
$$a^n = \frac{-1}{f_n}(f_0 1 + f_1 a + \cdots + f_{n-1} a^{n-1})$$

is in the span of B.

$$a^{n+1} = \frac{-1}{f_n}(f_0 a + f_1 a^2 + \cdots + f_{n-1} a^n)$$

is then also in the span of B. As soon as a^m is in the span of B, we can multiply this relation by another power of a and show that a^{m+1} is in the span of B. [We are using $p(a) = 0 = \sum_{i=0}^{n} f_i a^i$ as a recursion formula.] Since all the a^i are in the span of B, so are all polynomials $r(a)$.

Assume $q(a) \neq 0$. Then $p(x)$ does not divide $q(x)$, and, since $p(x)$ is irreducible, this gives us gcd $(p(x), q(x)) = 1$. Since $F[x]$ is euclidean, we have the existence of $c(x)$ and $b(x)$ such that

$$c(x)p(x) + b(x)q(x) = 1$$

Thus
$$c(a)p(a) + b(a)q(a) = 1 = c(a) \cdot 0 + b(a)q(a) = b(a)q(a)$$

and $b(a) = (q(a))^{-1}$.

Using the recursion relation as often as necessary, we can make $b(a)$ into a polynomial of degree less than n. ////

We can return now to our previous example and see if the work is less. If $a = \sqrt[3]{2}$, then $a^3 = 2$, $a^3 - 2 = 0$, and $p(x) = x^3 - 2$ is our irreducible polynomial over $\mathbb{Q}$ by Eisenstein's criterion. Thus $[\mathbb{Q}(\sqrt[3]{2}):\mathbb{Q}] = 3$, and $\mathbb{Q}(\sqrt[3]{2}) = \{q_0 1 + q_1 \sqrt[3]{2} + q_2 (\sqrt[3]{2})^2 \mid q_i \in \mathbb{Q}\}$.

Exercises

11 Find, using this last theorem, the degrees of $\sqrt[5]{7}$ over $\mathbb{Q}$, i over $\mathbb{R}$, and $\sqrt{2} + \sqrt{3}$ over $\mathbb{Q}$.

12 Find the degree of $e^{2\pi i/5}$ over both $\mathbb{Q}$ and $\mathbb{R}$.

13 In the last proof, show that the polynomials having a as a root form an ideal, that $p(x)$ is a generator of this ideal, and thus that $p(x)$ divides every polynomial in this ideal.

Recapping Theorem 2 briefly, it says a is algebraic of degree n over $F \Leftrightarrow [F(a):F] = n \Leftrightarrow$ an irreducible polynomial $p(x)$ exists in $F[x]$ of degree n with $p(a) = 0 \Leftrightarrow 1, a, a^2, \ldots, a^{n-1}$ is a basis for $F(a)$ over F. Using Theorem 2, we can quickly prove some other theorems.

Theorem 3   If K is a finite extension of F, then K is algebraic over F.

PROOF   Let $[K:F] = n$. We show K is algebraic by showing each element b in K is algebraic. $[F(b):F] \mid n$ by the corollary to Theorem 1, so $[F(b):F] \leq n$ is finite and, by Theorem 2, b is algebraic over F.   ////

At this point all should be warned that the converse statement, though tempting, is false. For instance, $\mathbb{Q}(\sqrt{2},\sqrt{3},\sqrt{5},\sqrt{7},\sqrt{11},\ldots)$ is algebraic over $\mathbb{Q}$ but is not a finite extension of $\mathbb{Q}$.

As motivation for the next theorem, the reader is asked to prove the following just using irreducible polynomials. If a and b are algebraic over F, then $a + b$ is algebraic over F. Similarly, ab is algebraic. This can be done using some linear algebra but is not completely trivial.

Theorem 4   If K is an extension field of F, and a, b in K are algebraic over F, then $a + b$ and ab are algebraic over K. If, in addition, a is non-zero, then $1/a$ is algebraic over F.

PROOF $F(a, b)$ is the smallest subfield of K containing F, a, and b. Thus $F(a,b) = F(a)(b) = F(b)(a)$. Certainly both ab and $a + b$ are in $F(a,b)$. $F(a)$ is a subfield of $F(a,b)$, so

$$[F(a,b): F] = [F(a,b): F(a)][F(a):F]$$

$[F(a):F]$ is finite so we must examine $[F(a,b):F(a)] = [F(a)(b):F(a)]$.

There is some irreducible polynomial $p(x)$ with coefficients in F so that $p(b) = 0$. These coefficients are also in $F(a)$, so that $p(x)$ is a polynomial over $F(a)$ that has b as a root. [Polynomials of smaller degree than $p(x)$ might also work, but this does not matter.] Thus $[F(a)(b):F(a)]$ is finite, and so is $[F(a,b):F] = [F(a)(b):F(a)][F(a):F]$. Thus $a + b$, ab, and $1/a$ all lie in a finite extension of F and therefore are algebraic over F by Theorem 3.   ////

Corollary   If K is an extension field of F, and A_F is the set of elements in K algebraic over F, then A_F is a field.

PROOF   Immediate from Theorem 4.   ////

The primary example is those numbers in $\mathbb{C}$ that are algebraic over $\mathbb{Q}$. This is called the field of *algebraic numbers* and is denoted by $\mathbb{A}$. $\mathbb{A}$ is a second example of an algebraic extension of $\mathbb{Q}$ that is not a finite extension of $\mathbb{Q}$.

Exercises

14 By examining $\sqrt[n]{2}$ as n goes to infinity, show that $[\mathbb{A}:\mathbb{Q}]$ cannot be finite.
15 If K is an algebraic extension of F, and L is an algebraic extension of K, show that L is an algebraic extension of K. [If $b \in L$, and $p(x) = k_0 + k_1 x + k_2 x^2 + \cdots + k_n x^n$ satisfies $p(b) = 0$, what can be said about $F(k_0, k_1, ..., k_n, b)$?]

If no particular fields are referred to, then a number in $\mathbb{C}$ is transcendental if it is not algebraic over $\mathbb{Q}$. e and π were proved to be transcendental in the late nineteenth century. In 1851, Liouville produced the first transcendental numbers, including $10^{-1} + 10^{-2} + 10^{-3!} + 10^{-4!} + \cdots$. Hermite proved e was transcendental in 1873, Lindemann in 1882 proved π^2, and thus also π, is transcendental.

More recently, it has been shown that a^b is transcendental, where a and b are algebraic and b is also irrational. In general, these questions are both fascinating and very difficult to answer. It is unknown, for instance, if $e + \pi$ is transcendental.

Exercises

16 Use the power-series expansion for e to show that e is irrational.
17 Using DeMoivre's theorem, show that $\sin 1°$ and $\cos 1°$ are algebraic numbers.
18 If K is an extension field of F, and a is in K, show that $[F(a):F(a^2)] = 1$ or 2. Give examples of both.
19 Let a be an element of odd degree over F, where a is in K, and K is an extension field of F. Show that $F(a^2) = F(a)$.
20 Which of the following elements in $\mathbb{C}$ are algebraic over $\mathbb{Q}$: $\sqrt{17}$, $\sqrt{1} + \sqrt{2}$, $(i + 5)/\sqrt{2}$, π^3?
21 $\varepsilon = (i + 1)/\sqrt{2}$ is a root of $z^8 - 1$, and this is algebraic over $\mathbb{Q}$. What is the degree of ε over $\mathbb{Q}$, $\mathbb{Q}(\sqrt{3})$, $\mathbb{Q}(\sqrt{2})$, $\mathbb{R}$, and $\mathbb{C}$, respectively?

Earlier, we defined the characteristic of a ring R as the smallest positive integer n such that $nr = r + r + r + \cdots + r = 0$ for all r in R. If no such integer exists, we say char (R), the characteristic of R, is zero. Any integral domain has characteristic 0 or p, where p is prime, and this includes all fields.

If K is a field, then the *prime field* of K is the intersection of all the subfields of K. Alternatively, we could say the prime field is the smallest subfield contained in K. Prime fields are easily classified.

Proposition   If K is a field, the prime field of K is isomorphic to $\mathbb{Q}$ if char $(K) = 0$, and to $\mathbb{Z}_p$ if char $(\mathbb{Z}_p) = p$.

SKETCH OF PROOF   Every subfield of K must contain 0 and 1 and thus must contain $1 + 1 = 2$, $1 + 1 + 1 = 3$, $1 + 1 + 1 + 1 = 4$, ... If char $(K) = p$, we obtain $0, 1, 2, 3, \ldots, p - 1$ as distinct elements this way, and they do form a field isomorphic to $\mathbb{Z}_p$. Otherwise, our infinite list has distinct entries $0, 1, 2, 3, \ldots$, and since the prime field is itself a field, it must contain $-1, -2, -3, \ldots$ and then a/b, where $a,\ b \in \mathbb{Z}$, and $b \neq 0$. This gives us a subfield isomorphic to $\mathbb{Q}$, and we are done.   ////

This proposition is helpful when we want to count various things connected with finite fields.

Proposition   If F is a finite field, then $|F| = p^n$ for some prime p and some positive integer n.

PROOF F contains some prime field $\mathbb{Z}_p$ by the last proposition. Thus F is a vector space of dimension n over $\mathbb{Z}_p$. If $b_1, b_2, \ldots, b_n$ is a basis of F over $\mathbb{Z}_p$, then every element in F can be uniquely expressed as $z_1 b_1 + z_2 b_2 + \cdots + z_n b_n$ with z_i in $\mathbb{Z}_p$.

This gives p choices for each of n locations, so $|F| = p^n$.   ////

Exercises

22 Show that there is no field of order 6.

23 What is the prime field of $\mathbb{Q}(\sqrt{3})$? Of $\mathbb{R}$? Of a field of order 243? Of the quotient field of the integral domain $\mathbb{Z}_5[x]$?

24 Let K be an algebraic extension of F, and let T be an integral domain, so that $F \subseteq T \subseteq K$. Show that T is a field.

25 In this exercise we use some linear algebra as a systematic method for determining minimal polynomials.

    (a) If K is an extension field of F, and a is an element of K, show that $T_a \colon K \to K$ is a linear transformation with respect to F if $T_a(k) = ak$.

    (b) If K is a finite extension of F, and $B = (b_1, b_2, \ldots, b_n)$ is an ordered basis, then T_a is represented by the matrix whose ith row is $(\lambda_{i1}, \lambda_{i2}, \ldots, \lambda_{in})$, given by $ab_i = \lambda_{i1} b_1 + \lambda_{i2} b_2 + \cdots + \lambda_{in} b_n$. Call this matrix M_a.

    (c) The characteristic polynomial $|xI - M_a|$ is a polynomial with coefficients in F, and a is a root of this polynomial. Thus this characteristic polynomial is divided by the minimal polynomial.

(*d*)   Find the minimal polynomials for $2 + 3i$ and $a + bi$ over $\mathbb{R}$, using this method with $(1, i)$ as the basis for $\mathbb{C}$ over $\mathbb{R}$.

(*e*)   Find the minimal polynomial for $a + b\sqrt[3]{2} + c\sqrt[3]{4}$ over $\mathbb{Q}$.

(*f*)   Use the minimal polynomials in parts *d* and *e* to find multiplicative inverses for the elements mentioned.

10.2   GEOMETRIC CONSTRUCTIONS AND IMPOSSIBILITY THEOREMS

Assume we have a straightedge and compass, and, starting with a line L, we mark off two points and call them 0 and 1. Using only the straightedge and compass, which other points on L can we determine? We shall call these points $\mathbb{G}$ and prove $\mathbb{G}$ is a field. Actually we say a is in $\mathbb{G}$ if we can mark $(a, 0)$ on the plane.

Certainly we can determine $2, 3, 4, 5, \ldots$ and $-1, -2$, using only the compass. Thus $Z \subseteq \mathbb{G}$. Also, we can erect perpendiculars, so we can first find $\sqrt{2}$ in the plane, and then use the compass to mark of $\sqrt{2}$ on L. Thus $\sqrt{2}$ is in $\mathbb{G}$.

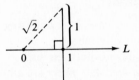

We shall first show that $\mathbb{G}$ is a subfield of $\mathbb{R}$, and then look at $[\mathbb{Q}(g):\mathbb{Q}]$, where $g \in \mathbb{G}$.

To show $\mathbb{G}$ is a subfield, we proceed briefly as follows:

1   If g_1 and g_2 are in $\mathbb{G}$, then so are $g_1 + g_2$ and $g_1 - g_2$. Fix the compass with its ends at 0 and g_2, and then move it so one point is on g_1.

2   If g_1 and g_2 are in $\mathbb{G}$, we construct a line segment of length $g_1 g_2$ as follows: Draw another line M through 0. Mark off lengths $0B = 1$

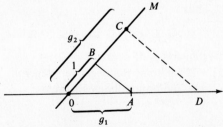

and $0C = |g_2|$ along M. Connect A and B. Draw a line through C parallel to the line AB. By similar triangles, $0D$ has the desired length $g_1 g_2$.

3 g_1/g_2 is given by construction *2*, except that AC is the line segment to construct first.   Then draw a line parallel to AC through B.

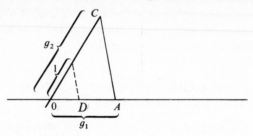

Putting together constructions *1*, *2*, and *3*, we see that $\mathbb{G}$ is indeed a subfield of $\mathbb{R}$.

Since we can draw perpendiculars, we can construct any point in the plane with coordinates (g_1,g_2), where the g_i are in $\mathbb{G}$.

There are three ways to obtain new constructible points from old:

1 Intersecting two lines
2 Intersecting a circle with a line
3 Intersecting two circles

and case *3* can be replaced by case *2* by choosing the line determined by the common chord between the two circles.

Does case *1* product new points for us? No, since if $ax + by = c$ and $dx + ey = f$ are lines, and a, b, c, d, e, and f are all in F, then the coordinates of their point of intersection is also in F.

Case *2* does produce some new points.   Let $ax + by = c$ and $(x - d)^2 + (y - e)^2 = f^2$, where a, b, c, d, e, and f are in F.   Then $(x - d)^2 + [(c - ax)/b - e]^2 = f^2$ is quadratic, so its solutions are in $F(\sqrt{\alpha})$, where α is in F.   So at each stage we either stay in F or move to $F(\sqrt{\alpha},)$ where α is in F.

Thus, starting with $\mathbb{Q}$, we next obtain $\mathbb{Q}(\sqrt{q}) = F_1$, then $F_1(\sqrt{f_1}) = F_2$, and so on.   If a point g is in $\mathbb{G}$ then it has to be constructible in a finite number of steps.   Each of these steps either does not change our field or gives us a quadratic extension. Thus

$$\mathbb{Q} \subseteq \mathbb{Q}(a_1) \subseteq \mathbb{Q}(a_2) \subseteq \cdots \subseteq \mathbb{Q}(a_n) \subseteq \mathbb{Q}(g)$$

and each individual extension is of degree 1 or 2.

Thus $[\mathbb{Q}(g):\mathbb{Q}] = 2^n$ for some nonnegative integer n.   This little numerical result enables us to prove the impossibility theorems.

Exercise

1 If g is constructible, show that $\sqrt{|g|}$ is also constructible.

Immediately after the first geometers learned to bisect an arbitrary angle, they started to ask if an arbitrary angle could also be trisected. They succeeded in finding good approximations with only a straightedge and compass. They succeeded if they could use a ruler and compass, or a fixed hyperbola, straightedge, and compass, or various other combinations. As the next theorem necessitates, however, they did not succeed in finding an exact trisection method with only a straightedge and compass.

Theorem   It is impossible to trisect every angle using straightedge and compass alone.

PROOF   We shall show that it is impossible to trisect 60°. If we could trisect 60°, then cos 20° would be an element of $\mathbb{G}$, as the following diagram indicates.

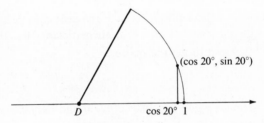

$(\cos 20°, \sin 20°)$

D $\cos 20°$ 1

Recall that DeMoivre's formula gives the triple-angle formula $\cos 3\theta = 4 \cos^3 \theta - 3 \cos \theta$, and that $\cos 60° = \frac{1}{2}$. We have $\frac{1}{2} = 4x^3 - 3x$, where $x = \cos 20°$. However $8x^3 - 6x - 1$ is irreducible over $\mathbb{Q}$, so $[\mathbb{Q}(x):\mathbb{Q}] = 3$, and x is not constructible.   ////

The rumored origin of the second of these problems was the plague that struck Athens in 429 BC, claiming one quarter of the population. A delegation was sent to the oracle of Apollo at Delos and was told that, for the plague to be averted, the cubical altar to Apollo must be doubled. The townspeople doubled the size of each edge, which of course increased the volume eight times. This did not avert the plague, and it was thought that perhaps the oracle intended the *volume* of the altar to be doubled. To do this meant increasing each side to $\sqrt[3]{2}$ times its original size. The second problem, often called the *Delian* problem, is to construct $\sqrt[3]{2}$ by straightedge and compass.

Theorem   It is impossible to construct $\sqrt[3]{2}$ exactly using only a straightedge and compass.

PROOF   If $x = \sqrt[3]{2}$, then $x^3 = 2$, and $x^3 - 2$ is irreducible over $\mathbb{Q}$. Thus $[\mathbb{Q}(\sqrt[3]{2}):\mathbb{Q}] = 3$, which again is not a power of 2, and $\sqrt[3]{2}$ is not constructible.   ////

It would be interesting to know if Apollo would have settled for an approximation or would have allowed a ruler instead of a straightedge.

The third problem was reportedly conceived by the philosopher Anaxagoras while he was in prison. This problem was "squaring the circle." This meant, given a circle, to find a square having the same area, and came to mean doing this with straightedge and compass.

Theorem   Given a circle, it is impossible to construct a square having the same area as the given circle, using only a straightedge and compass.

PROOF   If we assume the circle has radius 1, then its area is π. Thus we must construct a side of the square of length $\sqrt{\pi}$. However, π and $\sqrt{\pi}$ are transcendental. So $\mathbb{Q}[(\sqrt{\pi}):\mathbb{Q}] = \infty$ and certainly not a power of 2, so $\sqrt{\pi}$ is not constructible.   ////

The popularity of these problems over the years has been enormous, and it is a tribute to the curiosity of Greek mathematicians that they did not say that $\sqrt[3]{2}$ equals 1.259 and forget the matter.

Exercises

2   Show that a regular n-gon is constructible if and only if either $\cos(2\pi/n)$ or $\sin(2\pi/n)$ is in $\mathbb{G}$.

3   Show that a regular 9-gon is not constructible.

4   *Show that a regular pentagon is constructible. (It was shown that every n-gon is constructible if $n = 2^k + 1$ is a prime. These primes are called *Fermat primes* and are discussed briefly in the number-theory chapter.)

5   Assuming the last exercise as well as the rest of the section, which of the following angles are constructible: 15°, 36°, 20°, 21°, 10°?

The construction by straightedge and compass of regular polygons is another famous source of problems. Construction of a regular n-gon depends on constructing the points $(\cos(2\pi/n), \sin(2\pi/n))$. Considering the plane as the complex plane, this is the construction of $\gamma = e^{2\pi i/n}$. The smallest irreducible polynomial over $\mathbb{Q}$ having γ as a root is $\Phi_n(Z) = (Z - \gamma) \cdots (Z - \gamma^k) \cdots (Z - \gamma^{n-1})$, where gcd $(k,n) = 1$. In Warner [87] and Adamson [1] it is shown that $\Phi_n(Z)$ is irreducible over $\mathbb{Q}$, and obviously the degree of $\Phi_n(Z)$ is $\phi(n)$ (Euler's phi function). See Exercise 7 for the key case.

Gauss [28] indicated in 1826 that an n-gon is constructible if n is a prime of the form $2^k + 1$. He also was the first to construct a 17-gon. The construction of n-gons where $n = 2^k$, $3 \cdot 2^k$, or $5 \cdot 2^k$ was known to the Greeks, and this was the first step beyond. Pierre L. Wantzel, in 1837 [86], proved that an n gon is constructible if and only if $n = 2^k p_1 p_2 \ldots p_l$, where the p_i's are distinct primes of the form $2^n + 1$. He also proved the impossibility theorems.

Exercises

6   Compute $\phi(2^k)$, $\phi(3)$, $\phi(15)$, $\phi(257)$, $\phi(p^2)$.

7   (Half the result of Gauss and Wantzel)

   (a)   Show, for a prime p, that a regular p-gon is constructible only if $p = 2^n + 1$.

   (b)   Show that if an mk-gon is constructible, then both m-gons and k-gons are constructible.

10.3   GALOIS THEORY (PRELIMINARIES)

The ancient Babylonians knew what is today called the quadratic formula. In fact, they even had tables for $n^3 + n^2$ which enabled them to find approximate solutions for cubic equations of the form $ax^3 + bx^2 + c = 0$. Not much further progress was made for 2,000 years, until in 1545 Cardan published a method for solving the general cubic. Cardan had learned the method from Tartaglia and had sworn not to publish it. (In the middle ages money was made in mathematics problem-solving contests, not by the doctrine of publish or perish.) Scipione del Ferra seems to have known the method before either Tartaglia or Cardan. Cardan's secretary, Ferrari, came up with the general solution of the quartic shortly thereafter, and it was also published in Cardan's "Ars Magna" in 1545. For a further discussion of this history, see Boyer [12] or Ore's "Cardan, The Gambling Scholar" [66]. For the method of solving cubics and quartics, see Boyer or Birkhoff and MacLane [9].

     After this, attention focused on quintics, and no one could solve the general case. In 1799 and 1824 Ruffini and Abel, respectively, published proofs of the impossibility of solving the general quintic equation by radicals, thus settling the matter. At about the same time, Galois was finding results in group theory and in applications of these groups to field extensions that are the basis for the current proof of the impossibility theorem. (Incidentally, both Abel and Galois died while very young.) Another famous problem was settled positively around this time, and Galois theory gives an easy, though nonconstructive, proof. Note that $e^{2\pi i/3} = (-1 + \sqrt{-3})/2$ is expressible by radicals, as are $e^{2\pi i/4} = \sqrt{-1}$, and $e^{2\pi i/6} = (1 + \sqrt{-3})/2$. The problem is to show that all $e^{2\pi i/n}$ are expressible in terms of radicals.

Exercise

1   If $\gamma = e^{2\pi i/5}$, show that γ is a root of $x^4 + x^3 + x^2 + x + 1 = f(x)$. Let $y = x + 1/x$, and reduce this $f(x)$ to a quadratic. Solve the quadratic, and solve for x, thus producing $\gamma = e^{2\pi i/5}$ in terms of radicals.

In this section we first discuss the number of solutions of a polynomial. We then prove Kronecker's theorem saying that, for any polynomial $f(x)$ in $F[x]$, we can find a finite extension field of F in which $f(x)$ has a root. Then, putting together several such extensions, we find a finite extension field of F where $f(x)$ factors (splits) into linear factors. We indicate that minimal fields with this "splitting" property are unique up to isomorphism. Finally, if $f(x)$ is irreducible, we find that permuting the roots of $f(x)$ leads to field automorphisms of the splitting field which fix all elements in F. These automorphisms form the Galois group of $f(x)$ over F, which is discussed in the next section. The diligent reader might read this section once at home to get his bearings, and a second time at the library to fill in some missing proofs.

Proposition (remainder theorem) Let F be a field, $F[x]$ the ring of polynomials over F, and $p(x)$ a nonconstant element in $F[x]$. Then for any element a in F we have

$$p(x) = (x - a)q(x) + p(a)$$

and the degree of $q(x)$ is one less than the degree of $p(x)$.

PROOF The division algorithm for $F[x]$ gives $p(x) = (x - a)q(x) + r(x)$, where either $r(x) = 0$ or degree $(r(x)) <$ degree $(q(x)) = 1$. In either case, $r(x) = r$ is a constant, and we have $p(x) = (x - a)q(x) + r$. Substituting a for x, we see that $p(a) = (a - a)q(a) + r = r$, and we have the remainder theorem. Checking degrees on both sides of the equation tells us the degree of $q(x)$ is one less than the degree of $p(x)$. ////

Corollary If a is a root of $p(x)$, then $(x - a)|p(x)$.

PROOF $p(x) = (x - a)q(a) + p(a)$, but $p(a) = 0$ since a is a root of $p(x)$. ////

Definition If a is a root of $p(x)$ and $(x - a)^m|p(x)$ but $(x - a)^{m+1} \nmid p(x)$, then we say a is a **root of multiplicity** m.

When counting the number of roots of a polynomial, we always count each root of multiplicity m as m roots. Even so, we have the following proposition.

Proposition A polynomial of degree n over a field has at most n roots.

PROOF Let $p(x) \in F[x]$ be a polynomial of degree n. If $p(x)$ has no roots we are done, so we may assume a is a root of $p(x)$ of multiplicity m. Then $p(x) = q(x)(x - a)^m$, and $q(x)$ is a polynomial of degree $n - m$. By induction we can assume $q(x)$ has at most $n - m$ roots, and none of these is equal to a. Every root of $q(x)$ is a root of $p(x)$, and a root of $p(x)$ is a root of $q(x)$ or is equal to a. For instance, if b is a root of $p(x)$, then $0 = p(b) = (b - a)^m q(b)$, so $b = a$ or $q(b) = 0$. Alternatively, $F[x]$ is a

UFD, so if b were a root of $p(x)$ not equal to a and not a root of $q(x)$, we would have

$$(x - b)r(x) = p(x) = (x - a)^m q(x)$$

contradicting unique factorization. Thus we have m roots at a and at most $n - m$ roots for $q(x)$, and $p(x)$ has at most n roots.   ////

Exercises

2 Complete the proof by induction of this last proposition by showing all polynomials of degree 1 over a field have exactly one root.

3 Show this last proposition holds if we change "field" to "integral domain."

4 If R is a commutative ring with 1 but not an integral domain, show that there is a $p(x) = ax$ with two roots.

5 If $R = \mathcal{Q}$, the quaternions, find a quadratic polynomial with at least six roots.

Our next goal is to prove a famous theorem due to Kronecker. If F is a field, and $p(x)$ a polynomial in $F[x]$, then we can find a finite extension E of F so that $p(x)$ will factor completely in E. Our entering step is the following proposition.

Proposition Let $p(x)$ be an irreducible polynomial of degree n, where $p(x)$ is in $F[x]$, and F is a field. Then there exists a field E such that $[E:F] = n$, and there is an element r in E such that $p(r) = 0$.

PROOF We work in the ring $F[x]$, and try to construct an ideal I. If I is maximal, $F[x]/I$ will be a field. To make the image of $p(x)$ equal to 0 in $F[x]/I$, we only have to make sure $p(x)$ is in I. It seems reasonable to try $I = \langle p(x) \rangle$. Since $p(x)$ is irreducible, $\langle p(x) \rangle$ is a prime ideal; since $F[x]$ is a PID, $\langle p(x) \rangle$ is maximal. (See Sec. 6.3, Exercise 16.)

The mapping $\theta: F \to E = [F[x]/\langle p(x) \rangle$ given by $\theta(f) = f + \langle p(x) \rangle$ is a monomorphism, so F can be considered a subfield of E.

Finally, 1, $\bar{x}$, $\bar{x}^2$, ..., $\bar{x}^{n-1}$ is a basis for E, where $\bar{x} = x + \langle p(x) \rangle$. If the $\bar{x}^i$ were linearly dependent, then $\bar{0} = q(\bar{x}) = a_0 + a_1\bar{x} + \cdots + a_{n-1}\bar{x}^{n-1}$, and $q(x)$ would be in $\langle p(x) \rangle$ so that $p(x) | q(x)$, But $q(x)$ has lower degree than $p(x)$, so this cannot happen.

If $f(x)$ is in $F[x]$, then

$$f(x) = c(x)p(x) + r(x)$$

where $r(x)$ has degree less than $p(x)$. Thus,

$$f(\bar{x}) = c(\bar{x}) \cdot \bar{0} + r(\bar{x}) = r(\bar{x})$$

and $r(\bar{x})$ is in the span of $\{1, \bar{x}, \bar{x}^2, ..., \bar{x}^{n-1}\}$. Hence 1, $\bar{x}$, $\bar{x}^2$, ..., $\bar{x}^{n-1}$ is a basis for E over F, and $[E:F] = n$, completing the proof.   ////

We can now go to work on an arbitrary polynomial $f(x)$ of degree n in $F[x]$. If $f(x)$ is not irreducible, let $p(x)$ be an irreducible factor of $f(x)$. We can construct an extension E_1 of F so that $[E_1:F] \leq n$ and E_1 contains a root r_1 of $p(x)$, and thus of $f(x)$. We then look at $f_1(x) = f(x)/(x - r_1)$, which is a polynomial in $E_1(x)$ of degree $n - 1$. This $f_1(x)$ has a root r_2 in E_2, where $1 \leq [E_2:E_1] \leq n - 1$. $f_2(x) = f_1(x)/(x - r_2)$ has a root in E_3 with $1 \leq [E_3:E_2] \leq n - 2$. We continue until we have all n roots of $f(x)$ in a field E_n. In addition, we have

$$1 \leq [E_n:F] = [E_n:E_{n-1}][E_{n-1}:E_{n-2}]\cdots[E_2:E_1][E_1:F]$$

$$\leq 1 \cdot 2 \cdot 3 \cdots (n - 1) \cdot n = n!$$

This proves Kronecker's theorem, which we now state formally.

Kronecker's theorem   If $f(x)$ is a polynomial of degree n over a field F, then there exists a field E such that $f(x)$ has all n roots in E, and $[E:F] \leq n!$.

Definition   Let $f(x)$ be a polynomial of degree n over a field F. Suppose E is an extension field of F such that $f(x)$ has n roots in E. Suppose also that E is the smallest field that has this property, so that if $F \subseteq L \subsetneq E$ then $f(x)$ has fewer than n roots in L. Then E is a **splitting field for** $f(x)$ **over** F.

A restatement of Kronecker's theorem is as follows: For every $f(x)$ over F there exists a splitting field E of $f(x)$, and $[E:F] \leq n!$.

Several examples may be illustrative here.

EXAMPLE A   Let $f(x) = ax^2 + bx + c$ be a quadratic equation over $\mathbb{Q}$. Then $\alpha = (-b + \sqrt{b^2 - 4ac})/2a$ and $\beta = (-b - \sqrt{b^2 - 4ac})/2a$ are the roots of $f(x)$. If $b^2 - 4ac = m$ is not a square of $\mathbb{Q}$, then $\mathbb{Q}(\sqrt{m})$ is a splitting field for $f(x)$ over $\mathbb{Q}$. If m is a square, then $\mathbb{Q}$ itself is a splitting field. Since $[\mathbb{Q}(\sqrt{m}):\mathbb{Q}] \leq 2 = 2!$, we see that, indeed, Kronecker's theorem holds.

EXAMPLE B   Having covered $n = 2$, let us look at a case where $n = 3$. The polynomial $g(x) = x^3 - 2$ has one root $\sqrt[3]{2}$, where $\sqrt[3]{2}$ is a real number; its other two roots are $\varepsilon\sqrt[3]{2}$ and $\varepsilon^2\sqrt[3]{2}$, where $\varepsilon = e^{2\pi i/3}$. By Kronecker's theorem, a splitting field for $x^3 - 2$ over $\mathbb{Q}$ has degree $\leq 6 = 3!$ and, in fact, must divide 6. The field $\mathbb{Q}(\sqrt[3]{2})$ contains only one of the three roots of $x^3 - 2$, and $[\mathbb{Q}(\sqrt[3]{2}):\mathbb{Q}] = 3$. The splitting field S must thus contain $\mathbb{Q}(\sqrt[3]{2})$ properly, and this leaves no choice. We must have $[S:\mathbb{Q}] = 6$. Obviously, $\mathbb{Q}(\sqrt[3]{2}, \varepsilon)$ will contain all the roots of $x^3 - 2$; since $\varepsilon = (-1 + \sqrt{-3})/2$ has degree 2 over $\mathbb{Q}$, $S = \mathbb{Q}(\sqrt[3]{2}, \varepsilon)$. We can check that $[S:\mathbb{Q}] = [S:\mathbb{Q}(\sqrt[3]{2})][\mathbb{Q}(\sqrt[3]{2}):\mathbb{Q}] = 2 \cdot 3 = 6$.

Exercises

6 *If a, b, and c are picked at random in $\mathbb{Q}$, $x^3 + ax^2 + bx + c$ will almost always have a splitting field of degree 6. Can you find a, b, and c such that the splitting field of $x^3 + ax^2 + bx + c$ has degree 3?

7 Find the splitting fields for $x^3 - 8$ and $x^3 + x^2 + x + 1$ over $\mathbb{Q}$.

EXAMPLE C   The polynomial $h(x) = x^4 + x^2 + 1$ is an excellent one to remember in order to distinguish between "having roots" and "being irreducible." There is no way $x^4 + x^2 + 1$ can have a real root, but nonetheless $x^4 + x^2 + 1 = (x^2 + x + 1)(x^2 - x + 1)$. Let us use this to find a splitting field for $h(x)$ over $\mathbb{Q}$. Factoring completely gives $h(x) = (x - \omega)(x - \omega^2)(x - \omega^4)(x - \omega^5)$ with $\omega = e^{2\pi i/6} = (1 + \sqrt{-3})/2$. Thus, $\mathbb{Q}(\omega) = \mathbb{Q}(\sqrt{-3})$ is a splitting field for $h(x)$, but $[\mathbb{Q}(\omega):\mathbb{Q}] = 2 \leq 4! = 24$.

Exercises

8 Show that if $p(x)$ is irreducible of degree n over F, and if S is a splitting field for $p(x)$, then $n \leq [S:F] \leq n!$.

9 What is the degree of a splitting field of $x^4 - 2$ over $\mathbb{Q}$.

EXAMPLE D   Over $\mathbb{C}$ every polynomial has $\mathbb{C}$ itself as its splitting field. This fact is called the *fundamental theorem of algebra* and is proved using analysis. There is an easy proof using Liouville's theorem, and both are done in any first course in complex variables. See, for instance, Ahlfors [2] or Pennisi [70]. A more topological proof is given in Birkhoff and MacLane [9]. A charming induction proof using very little analysis is given in van der Waerden [85] and Lang [53]. We will not prove the result here. An alternative formulation is that every polynomial with coefficients in $\mathbb{C}$ factors completely in $\mathbb{C}$.

Exercise

10 Show that if $p(x)$ is in $\mathbb{R}[X]$, then the splitting field of $p(x)$ is either $\mathbb{R}$ or $\mathbb{C}$. Use this to show that every cubic polynomial in $\mathbb{R}(X]$ must have a root. Why do a rough graph and the intermediate-value theorem indicate the same result? Can this be generalized?

The next theorem says that splitting fields are isomorphic. This can even be generalized a bit as follows:

Let F and $\bar{F}$ be isomorphic fields, where $\theta: F \to \bar{F}$ is the isomorphism. Denote $\theta(f)$ by $\bar{f}$ for all f in F. We can extend θ to an isomorphism from $F[x]$ to $\bar{F}[y]$ by defining $\theta(\sum_{i=0}^{n} f_i x^i) = \sum_{i=0}^{n} \bar{f}_i y^i$. We shorten this by saying $\theta(p(x)) = \bar{p}(y)$.

Theorem   Let $p(x)$ be a polynomial in $F[x]$, and $\bar{p}(y)$ the corresponding polynomial in $\bar{F}[y]$. If E is a splitting field for $p(x)$, and if $\bar{E}$ is a splitting field for $\bar{p}(y)$ over $\bar{F}$, then E and $\bar{E}$ are isomorphic.

BRIEF OUTLINE OF PROOF   (See Herstein [38] or Paley and Weichsel [68] for a complete proof.) Basically, the idea is to start with the first isomorphism listed below, and use it to show each of the following maps is also an isomorphism. Also assume $q(x)$ is an irreducible factor of $p(x)$.

$$\theta: F \to \bar{F} \qquad \text{where } \theta(f) = \bar{f}$$

$$\theta: F[x] \to \bar{F}[y] \qquad \text{where } \theta(\sum f_i x^i) = \sum \bar{f}_i y^i$$

$$\theta_1: \frac{F[x]}{\langle q(x) \rangle} \simeq E_1 \to \frac{\bar{F}[y]}{\langle \bar{q}(y) \rangle} \simeq \bar{E}_1$$

If α is a root of $q(x)$, and β a root of $\bar{q}(y)$, then we have $\theta_1(\alpha) = \beta$ as well as $\theta_1(f) = \bar{f}$ for all f in F.

E_1 is an extension of F, and $\bar{E}_1$ of $\bar{F}$, and at least one root of $p(x)$ lies in E_1. We repeat our procedure until all roots of $p(x)$ lie in some $E_n = E$.   ////

An examination of this proof allows us to introduce automorphisms. Let $p(x)$ be an irreducible polynomial with α and β as distinct roots. Letting $F = \bar{F}$, $p(y) = p(x)$, and $E = \bar{E}$, we see that we can find an isomorphism $\phi: E \to E$ such that $\phi(f) = f$ for all f in F, but with $\phi(\alpha) = \beta$. An automorphism of E satisfying $\phi(f) = f$ for all f in F is called an *F-automorphism*.

Corollary   If $p(x)$ is an irreducible polynomial in $F[x]$ with distinct roots α and β in the splitting field E, then there is a F-automorphism ϕ of E with $\phi(\alpha) = \beta$.

Informally, this might be stated as " ϕ fixes every element in F but moves α to β." This also says that the automorphisms of E are transitive on the roots of $p(x)$.

Ignoring the polynomials for a minute, we have an interesting question. If E is an extension of F, what automorphisms of E fix all elements in F?

Exercises

11   Let E be a field, and S any set of automorphisms of E. Also let E_S be those elements of E fixed by all elements of S. Is E_S a subfield of E?

12   Let E be an extension field of F, and let $G(E/F)$ be the set of all automorphisms of E that fix every element in F. Show that $G(E/F)$ is a group with composition of functions as its operation.

Using these last two exercises, we see that if E is an extension field of F, then $G(E/F) = G$ is a group. Then E_G is a subfield of E and, obviously, $F \subseteq E_G$.

The key question is, when do we have $F = E_G$? When this equality holds, we call E a *normal extension* of F. We would like to check some examples, and we present some useful background information as a series of propositions.

Proposition 1   The only automorphism of $\mathbb{Q}$ is the identity. Similarly, if E is an extension of $\mathbb{Q}$, any automorphism of E is the identity when restricted to $\mathbb{Q}$.

    PROOF   If α is an automorphism of $\mathbb{Q}$, then $\alpha(1) = 1$, $\alpha(1 + 1) = \alpha(2) = 2$, and $\alpha(n) = n$ for all integers n. Then $\alpha(nm^{-1}) = \alpha(n)\alpha(m)^{-1} = nm^{-1}$ for all $m \neq 0$, for n, m in $\mathbb{Z}$.   ////

Proposition 2   The only automorphism of $\mathbb{Q}(\sqrt[3]{2})$ is the identity.

    PROOF   If α is an automorphism of $\mathbb{Q}(\sqrt[3]{2})$, then certainly $\alpha(q) = q$ for all q in $\mathbb{Q}$. Also, α takes cube roots of, say, 2, to themselves. So $\alpha(\sqrt[3]{2}) = \sqrt[3]{2}$, $\varepsilon\sqrt[3]{2}$, or $\varepsilon^2\sqrt[3]{2}$, where $\varepsilon = e^{2\pi i/3}$. However, $\mathbb{Q}(\sqrt[3]{2}) \subseteq \mathbb{R}$, so we must have $\alpha(\sqrt[3]{2}) = \sqrt[3]{2}$, from which it follows that α is the identity.   ////

Proposition 3 $\mathbb{Q}(\sqrt[3]{2},\varepsilon)$ is an extension of $\mathbb{Q}$, and $G(\mathbb{Q}(\sqrt[3]{2},\varepsilon)/\mathbb{Q}) \simeq$ Sym (3).

    PROOF   Every automorphism of $\mathbb{Q}(\sqrt[3]{2},\varepsilon)$ automatically is a $\mathbb{Q}$-automorphism. Also, any automorphism must preserve cube roots and thus takes $\sqrt[3]{2}$ to $\sqrt[3]{2}$, $\sqrt[3]{2}\,\varepsilon$, or $\sqrt[3]{2}\,\varepsilon^2$. Similarly, ε goes to ε or ε^2, since they are the two roots of $x^2 + x + 1$. Let

$$\alpha_j{}^i(\sqrt[3]{2}) = \sqrt[3]{2}\,\varepsilon^i \qquad i = 0, 1, 2$$

while

$$\alpha_j{}^i(\varepsilon) = \varepsilon^j \qquad j = 1, 2$$

This gives six automorphisms $\alpha_j{}^i$. It is straightforward to show that each $\alpha_j{}^i$ is an automorphism and that they form a nonabelian group. Since Sym (3) is, up to isomorphism, the only nonabelian group of order 6, we are done.   ////

Proposition 4   The field $\mathbb{R}$ admits only the identity automorphism.

    PROOF   Let α be any automorphism of $\mathbb{R}$. Since $\alpha(x^2) = \alpha(x)\alpha(x)$, we see that α takes positive numbers to positive numbers. If $y > x$, then $y - x > 0$, so $\alpha(y - x) = \alpha(y) - \alpha(x) > 0$, and α respects order. If $\alpha(x) > x$, then let q_0 be a rational so that $\alpha(x) > q_0 > x$. By Proposition 1 we have

$\alpha(q_0) = q_0$; this gives our contradiction, since $q_0 > x$ but $\alpha(x) > \alpha(q_0) = q_0$. A similar contradiction comes from $\alpha(x) < x$. ////

When we rephrase these propositions in terms of normal extensions, we have:

Proposition 2' $\mathbb{Q}(\sqrt[3]{2})$ is a nonnormal extension of $\mathbb{Q}$. $G(\mathbb{Q}(\sqrt[3]{2})/\mathbb{Q})$ contains only the identity, and the fixed field of the identity is all of $\mathbb{Q}(\sqrt[3]{2})$.

Proposition 3' $\mathbb{Q}(\sqrt[3]{2},\varepsilon)$ is a normal extension of $\mathbb{Q}$. Every element of $(\sqrt[3]{\mathbb{Q}2},\varepsilon)$ is expressible as

$$q_1 + q_2\sqrt[3]{2} + q_3\sqrt[3]{4} + q_4 \varepsilon + q_5 \varepsilon\sqrt[3]{2} + q_6 \varepsilon\sqrt[3]{4}$$

(Since $\varepsilon^2 = -1 - \varepsilon$, we do not need ε^2 terms.) Of these, the only elements invariant under all the $a_j{}^i$ are those with $q_2 = q_3 = q_4 = q_5 = q_6 = 0$. Since the $\alpha_j{}^i$ comprise $G(\mathbb{Q}(\sqrt[3]{2},\varepsilon)/\mathbb{Q})$, the fixed field of this G is $\mathbb{Q}$.

Proposition 4' $\mathbb{R}$ is a nonnormal extension of $\mathbb{Q}$ (or of any other subfield of $\mathbb{R}$). The proof is similar to that of Proposition 2'.

The following theorem, which we again only briefly outline, yields the connection between polynomials and normal extensions.

Theorem Let F be any field of characteristic 0 or any finite field. Then a finite extension E of F is normal if and only if E is the splitting field for some polynomial in $F[x]$.

OUTLINE OF PROOF (See Herstein [38], Fraleigh [23], Paley and Weischel [68], or Dean [20] for a complete proof.)

First we find an element α in E such that $E = F(\alpha)$. This part requires some restriction of F, but either characteristic 0 or finiteness suffices. If E is a normal extension, let $p(x) = [x - \beta_1(\alpha)][x - \beta_2(\alpha)] \cdots [x - \beta_n(\alpha)]$, where $G(E/F) = \{\beta_1, \beta_2, ..., \beta_n\}$. E turns out to be precisely the splitting field of this polynomial.

In the other direction, let E be a splitting field of a polynomial $q(x)$. If $p(x)$ is an irreducible factor of $q(x)$, then let $r_1, r_2, ..., r_k$ be the roots of $p(x)$. We look at E as an extension of $F(r_1)$, and E is still a splitting field of $q(x)$. Applying induction, we can first show any element fixed by all elements of $G(E/F)$ is in $F(r_1)$. We can find enough automorphisms of E to take r_1 to any of $r_1, r_2, ..., r_k$. and this suffices to locate the fixed points of $G(E/F)$ in F itself. Thus, E is normal over F. ////

When we have this happy situation of a normal extension E over F that is also a splitting field, we say $G(E/F)$ is the *Galois group* of E over F. Equivalently, if we start with a polynomial $q(x)$ in $F[x]$, we let E be the splitting field of $q(x)$ and now call $G(E/F)$ the Galois group *associated with* $f(x)$. In the next section we shall examine the information that can be exchanged between the field extension E over F and the Galois group $G(E/F)$.

Exercise

13 Let G be any group of automorphisms of a field E, and let $E_G = F = \{f \,|\, f \in E,$ $\sigma(f) = f \,\forall \sigma$ in $G\}$. Note that E is normal over F.

    (*a*)   Let $G = \{I, \sigma\}$ and $E = \mathbb{C}$, where $\sigma(x + iy) = x - iy$. Find a subfield such that $\mathbb{C}$ is a normal extension of this subfield.

    (*b*)   If $E = \mathbb{Q}(\sqrt{2}, \sqrt{3})$, then let $G_1 = \{I, \alpha\}$. Define α by $\alpha(q_0 + q_1\sqrt{2} + q_3\sqrt{3} + q_4\sqrt{6}) = q_0 - q_1\sqrt{2} + q_3\sqrt{3} - q_4\sqrt{6}$. Find a subfield such that E is a normal extension. Changing $\sqrt{2}$ to $\sqrt{3}$, find another such subfield. Letting $G = \{I, \alpha, \beta, \alpha\beta\}$, show that E is normal over $\mathbb{Q}$. (An appropriate β must be defined.)

10.4 GALOIS THEORY (FUNDAMENTAL THEOREM)

The main theorem of Galois theory says that the lattice of subfields between E and F, where E is a finite normal extension of F, is the same as the lattice of subgroups of $G(E/F)$ turned upside down. Since $Q(\sqrt[3]{2}, \varepsilon)$ is a normal extension of Q, we have the following lattices:

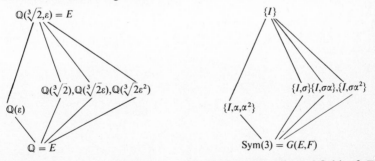

If H is subgroup of $G(E,F)$, it corresponds to E_H, the subfield of E consisting of those elements fixed by each element of H. Order is also preserved, and normal subgroups correspond to normal extensions of the base field. Writing all this out gives our theorem. Going the other way, an intermediate field K corresponds to the subgroup G_K of all automorphisms of E fixing all elements of K.

Fundamental theorem of Galois theory Let E be a finite normal extension of a field F, where F is either of characteristic 0 or finite. In this case E is called a *Galois extension*, and $G(E/F)$ is called the *Galois group* of E over F. There is a one-to-one correspondence between E and F and the subgroups of $G(E/F)$. The lattice of subfields is isomorphic to the lattice of corresponding subgroups if we order the subfields by $\subseteq$ and the subgroups by $\supseteq$. If $K_1 \supseteq K_2$, then $G_{K_2} \supseteq G_{K_1}$ and $[K_1:K_2] = [G_{K_2}:G_{K_1}]$. Also, K is a normal extension if and only if G_K is normal in G. Finally, if K is a normal extension of F, then $G(K/F) \simeq G(E/F)/G_K$.

The proof can be found in many texts, including Herstein [38], Dean [20], Fraleigh [23], Paley and Weichsel [68], and Warner [87]. A more extensive treatment, using a little topology to cover the infinite case, is given in Kaplansky [46], Lang [53], and Goldhaber, and Ehrlich [30]. The proof is lengthy, rather than hard, as is often the case with multipart theorems. Some parts, like $K_1 \supseteq K_2 \Leftrightarrow G_{K_2} \supseteq G_{K_1}$, are immediate from the definitions.

Exercise

1 Assuming $[K_1:K_2] = [G_{K_1}:G_{K_2}]$, show that if H is a subgroup of $G(E/F)$, then $G_{(E_H)} = H$. Similarly, $K = E_{(G_K)}$.

We shall do another example. The polynomial we want to examine is $x^4 - 7$. We can factor this over $\mathbb{C}$ as

$$\left(x^2 - \sqrt{7}\right)\left(x^2 + \sqrt{7}\right) = \left(x - \sqrt[4]{7}\right)\left(x + \sqrt[4]{7}\right)\left(x - i\sqrt[4]{7}\right)\left(x + i\sqrt[4]{7}\right).$$

We let $\alpha = \sqrt[4]{7}$, the positive real fourth root of 7. The splitting field for $x^4 - 7$ over $\mathbb{Q}$ is $(\mathbb{Q}\alpha, i)$. (See Exercise 2.) The vaious subfields of $\mathbb{Q}(\alpha, i)$ are

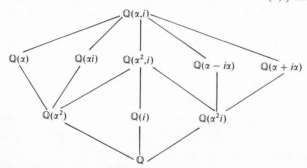

We imitate closely the procedure we followed in computing the Galois group of $\mathbb{Q}(\sqrt[3]{2}, \omega)$ over $\mathbb{Q}$. Any automorphism of $\mathbb{Q}(\alpha, i)$ must take α to α,

$-\alpha$, $i\alpha$, or $-i\alpha$ and must take i to i or $-i$. Define $g_j{}^k$ to be the automorphism such that

$$g_j{}^k(\alpha) = \alpha(i)^k \qquad k = 0, 1, 2, 3$$
$$g_j{}^k(i) = (i)^j \qquad j = 1 \text{ or } 3$$

It can be shown that the $\{g_j{}^k\}$ form a group isomorphic to D_8, and this is the content of Exercise 3. Letting $g_1{}^1 = a$ and $g_3{}^0 = b$, we have $a^4 = I = b^2$ and $b^{-1}ab = a^3$.

Writing the lattice of subgroups of D_8 upside down, we have

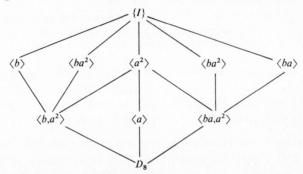

By just writing down these lattices we have concealed many verifications. To do one in detail, let us check that $\langle ba \rangle = \{I, ba\}$ corresponds to the subfield $\mathbb{Q}(\alpha + i\alpha)$. All the elements in $\mathbb{Q}(\alpha, i)$ are fixed by I, so we merely have to look at ba and its fixed points. $ba = g_3{}^0 g_1{}^1$ so $ba(\alpha) = b(a(\alpha)) = b(\alpha i) = b(\alpha)b(i) = \alpha(-i) = -i\alpha$. Similarly, $ba(i) = b(i) = -i$. Now a typical element of $\mathbb{Q}(\alpha, i)$ has the form

$$q^* = q_1 + q_2\alpha + q_3\alpha^2 + q_4\alpha^3 + q_5 i + q_6\alpha i + q_7\alpha^2 i + q_8\alpha^3 i$$
$$ba(g^*) = q_1 - q_2\alpha i - q_3\alpha^2 + q_4 i\alpha^3 - q_5 i + q_6(-\alpha) + q_7\alpha^2 i + q_8\alpha^3$$

If $q^* = baq^*$, we have $q_3 = 0 = q_5$, $q_4 = q_8$, and $q_2 = -q_6$. In this case, $q^* = q_1 + q_2(\alpha - \alpha i) + q_4(\alpha^3 + \alpha^3 i) + q_7\alpha^2 i$. But $(\alpha - \alpha i)^2 = \alpha^2(-2i) = -2\alpha^2 i$, and $(\alpha - \alpha i)^3 = -2\alpha^2 i(\alpha - \alpha i) = -2(\alpha^3 + \alpha^3 i)$. Thus, q^* is in $\mathbb{Q}(\alpha - i\alpha)$.

Exercises

2   (a)   Show that the splitting field for $x^4 - 7$ over $\mathbb{Q}$ is $\mathbb{Q}(\alpha, i)$, where $\alpha = \sqrt[4]{7}$ is the real positive fourth root of 7.

     (b)   Show that $[\mathbb{Q}(\alpha, i): \mathbb{Q}] = 8$.

     (c)   Note that $\mathbb{Q}(\alpha, i)$ is a normal extension of $\mathbb{Q}$.

     (d)   Note that $G(\mathbb{Q}(\alpha, i)/\mathbb{Q})$ is a group of order 8.

3   (a)   Show that $G(\mathbb{Q}(\alpha, i)/\mathbb{Q})$ is D_8, the dihedral group of order 8.

     (b) D_8 has four nonnormal subgroups, each of order 2. What does this tell us about the subfields of $\mathbb{Q}(\alpha, i)$?

(c) In any group, all subgroups of index 2 are normal. What does this tell us about subfields in a finite normal field extension?

4 If K is the splitting field of $x^5 - 7$ over $\mathbb{Q}$, show that $[K:\mathbb{Q}] = 20$. [*Hint*: $\mathbb{Q}(\varepsilon)$ is an intermediate field, where $\varepsilon = e^{2\pi i/5}$.]

5 *If K is the splitting field of $x^n - 7$ over $\mathbb{Q}$, show that $[K:\mathbb{Q}] = n\phi(n)$.

6 We have shown earlier that any polynomial $\Phi(x) = x^{p-1} + x^{p-2} + \cdots + x + 1$ is irreducible over $\mathbb{Q}$, where p is a prime. If $\varepsilon = {}^{2\pi i/p}$ show that $\mathbb{Q}(\varepsilon)$ is a splitting field for $\Phi_p(x)$.

Many terms have been transferred from group theory to Galois extensions and conversely. For instance, if E is a normal extension of F, then E is called a cyclic, abelian, or solvable *extension* if $G(E/F)$ is, respectively, a cyclic, abelian, or solvable group.

7 If E is a splitting field over F, and $G = G(E/F)$, then what subgroup of G corresponds to the largest abelian extension field of F in E? Show that $G(\mathbb{Q}(\varepsilon)/\mathbb{Q})$ is abelian by showing that $\alpha(\varepsilon) = \varepsilon^i$ for some i and that this completely determines the automorphism α. Here $\varepsilon = e^{2\pi i/n}$.

At this point let me mention one problem I have been trying to give away as a master's-thesis topic for many years. Find a polynomial $f(x)$ such that, if E is the splitting field for $f(x)$ over $\mathbb{Q}$, then $G(E/\mathbb{Q})$ is a dihedral group of given order $2n$. Some recent work of Shafarevich [80] shows this is definitely possible.

8 *Dirichlet's theorem (a hard theorem to prove; see Landau [52] or LeVeque [55]) implies that the sequence $(1, 1 + n, 1 + 2n, 1 + 3n, \ldots)$ contains an infinite number of primes. If $q = 1 + kn$, where q is prime, look at $G(\mathbb{Q}(\varepsilon)/\mathbb{Q})$, where $\varepsilon = e^{2\pi i/q}$. Show that $G(\mathbb{Q}(\varepsilon)/\mathbb{Q})$ is cyclic and that there exists a subfield H such that $G(H/\mathbb{Q})$ is cyclic of order n. (This seems a lot of work for a mere cyclic group.)

UNSOLVED PROBLEM If G is an arbitrary finite group, can a Galois extension E of $\mathbb{Q}$ be found so that $G(E/\mathbb{Q})$ is isomorphic to G? This is probably the best-known unsolved problem in Galois theory.

If F is a finite field, we know it must have order p^n. F contains a prime subfield which we can call $\mathbb{Z}_p$. What can we say about $G(F/\mathbb{Z}_p)$? First of all, is F normal over Z_p? Every element of $(F - (0), \cdot)$ satisfies $x^{p^n-1} - \bar{1} = 0$ by Lagrange's theorem. This yields $p^n - 1$ distinct solutions, so F is a splitting field for this polynomial. Since $[F:\mathbb{Z}_p] = n$, we know that $|G(F/\mathbb{Z}_p)| = n$. Our old acquaintance, the Frobenius homomorphism, now reappears. Let $\theta(x) = x^p$, so that θ is the Frobenius homomorphism. Since F is a commutative ring of characteristic p, θ is a ring homomorphism. Since E is a field, $\ker(\theta) = 0$, and θ is an automorphism. It turns out that $G(F/\mathbb{Z}_p) = \{I, \theta, \theta^2, \ldots, \theta^{n-1}\}$. To show this, it suffices to show that the order of θ is n, since n is the order of $G(F/\mathbb{Z}_p)$.

Let a be a generator of the cyclic group F^*.

$$\theta(a) = a^p \qquad \theta^2(a) = a^{p^2} \qquad \cdots \qquad \theta^j(a) = a^{p^j}$$

For $\theta^k(a) = a$, we must have $a^{p^k} = a$, which is equivalent to $a^{p^k-1} = 1$. This will happen first for $k = n$, since the generator a of F^* has order $p^n - 1$.

Exercises

9   Consider $x^{p^m-1} - 1$ as a polynomial over $\mathbb{Z}_p$. Let E be a splitting field for this polynomial. If E has p^m elements, show that the $p^m - 1$ nonzero elements of E all satisfy $x^{p^m-1} - \bar{1}$. Why can't E have fewer than p^m elements? Why is $[E:\mathbb{Z}_p] = m$ rather than $p^m - 1$, the degree of the polynomial?

10   Let F and K both be fields of order p^n. We want to show that they are isomorphic. Do this by showing they both are splitting fields of the same polynomial over $\mathbb{Z}_p$.

11   (a)   Show that $\mathbb{Q}(\varepsilon)$ is cyclic over $\mathbb{Q}$, where $\varepsilon = e^{2\pi i/3}$. (Use the first lattice in this section, if you wish to.)

      (b)   Show that $\mathbb{Q}(i)$ and $\mathbb{Q}(\alpha^2 i)$ are both cyclic extensions of $\mathbb{Q}$, where $\alpha = \sqrt[4]{7}$.

      (c)   Find an abelian extension that is not cyclic.

"Solvable" is a concept that started with polynomials and eventually became a group-theory concept. A polynomial $p(x)$ in $F[x]$ is *solvable by radicals* or, allowing some imprecision, *solvable*, if all its roots can be found by field operations in F plus extraction of nth roots. For instance, $ax^2 + bx + c$ is solvable by radicals over $\mathbb{Q}$, since its roots are given by $(-b \pm \sqrt[2]{b^2 - 4ac})/2a$. Similarly, $x^3 + ax + b$ has, as a root,

$$\sqrt[3]{-\frac{b}{2} + \sqrt{\frac{a^3}{27} + \frac{b^2}{4}}} + \sqrt[3]{-\frac{b}{2} - \sqrt{\frac{a^3}{27} + \frac{b^2}{4}}}$$

For this example to work, we must have the characteristic of F not equal to 2, or 3, but, given this, the solution is obtainable by field operations plus the extraction of square and cube roots.

Definition   A polynomial $f(x)$ in $F[x]$ is **solvable by radicals** if all its roots are in $F(\alpha_1, \alpha_2, ..., \alpha_n)$, where, for each i, $\alpha_i^{k_i}$ is in $F(\alpha_1, \alpha_2, ..., \alpha_{i-1})$ for some positive integer k_i.

We examined solvable groups back in Chap. 5, and now is an appropriate time to recall some of the results we obtained there:

1   If G is solvable, so are any subgroup of G and any factor group G/K.

2   If $K \lhd G$ and both K and G/K are solvable, then so is G.

3   If G is solvable and finite, then there exist subgroups K_i such that $\{e\} = K_0 \subset K_1 \subset K_2 \subset \cdots \subset K_n = G$, where each $K_i \triangleleft K_{i+1}$, and K_{i+1}/K_i is cyclic of prime order.

4   G is solvable if there exist subgroups $\{e\} = H_0 \subset H_1 \subset H_2 \subset \cdots \subset H_k = G$, and each $H_i \triangleleft H_{i+1}$, and H_{i+1}/H_i is abelian.

5   Sym(5) is not solvable.

The theorem that relates solvable groups to solvability by radicals is as follows:

Theorem   Let F be a field of characteristic 0, and let $f(x)$ be a polynomial over F. Let E be the splitting field for $f(x)$. Then $f(x)$ is solvable by radicals over F if and only if $G(E/F)$ is a solvable group.

OUTLINE OF PROOF   As a first step, we extend F to $U = F(\varepsilon_3, \varepsilon_4, \ldots, \varepsilon_n)$, where n is the degree of $f(x)$, and ε_i is an ith root of unity. This has the effect of giving us all the nth roots we need, and U is a normal and abelian extension of F (compare Exercise 6).

Now let us construct a radical extension over U. If $x^k - a$ is to be solved (that is, kth roots are to be used), then a splitting field for $x^k - a$ comes about in one step. For if r is a root of $x^k - a$, then $r, r\varepsilon_k, r(\varepsilon_k)^2, \ldots, r(\varepsilon_k)^{k-1}$ is the complete set of roots, and ε_k is already in U. This kind of extension is normal and even cyclic, with $\alpha(r) = r\varepsilon_k$ defining the generator α. We thus have the following field lattice:

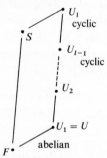

S is a splitting field for $f(x)$. Since $f(x)$ splits in U_l, we have $S \subset U_l$. We go over to the group lattice and have

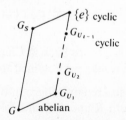

By result 4, G is solvable. However, the group we are interested in is $G(S/F)$, which is isomorphic to G/G_S and is thus also solvable by result 1.

Starting with a solvable Galois group, we now assume F contains all necessary roots of unity. (Basically, if $\{e\} \subset G \subset G^*$, where G is solvable, $G \lhd G^*$, and G^*/G is abelian, then G^* is solvable, so we may assume roots of unity are included without losing solvability.) Using result 3, we can look at subgroups K_i and K_{i+1}. We know $K_i \lhd K_{i+1}$, and K_{i+1}/K_i is cyclic of order p. This gives us fields F_i and F_{i+1} such that F_i is a normal extension of F_{i+1}, and $[F_i{:}F_{i+1}] = p$. If α generates $G(F_i/F_{i+1})$, then there is a nonzero element c in F_i such that $\alpha(c) = \varepsilon_p c$. Then $\alpha^k(c^p) = c^p$ for all k, so that $x^p - c^p$ is an irreducible polynomial over F_{i+1}, and F_i is the splitting field. Thus we have adjoined a series of pth roots, and S is contained in a radical extension. Thus $f(x)$ is solvable by radicals. ////

Exercises

12 If F_i is a normal extension of F_{i+1}, and $[F_i{:}F_{i+1}] = n$, then let b be an element of F_i. Set

$$c = b + \varepsilon\alpha^{-1}(b) + \varepsilon^2\alpha^{-2}(b) + \cdots + \varepsilon^{n-1}\alpha^{n-1}(b)$$

where α is the generator of $G(F_i/F_{i+1})$, and ε is an nth root of unity contained in F_{i+1}. Show that $\alpha(c) = \varepsilon c$.

13 We have not included finite fields in the theorem on solvability by radicals. The results are true there, but our proof will not work, because of the absence of nth roots. Show that if F is a field of order p^n, then it has no pth root of 1 other than 1 itself.

Let us first proceed to the famous problem of expressing $e^{2\pi i/n}$ by radicals over $\mathbb{Q}$. Since $e^{2\pi i/n} = \varepsilon$ is a root of $x^n - 1$, we have to see if $x^n - 1$ is solvable by radicals. Since the roots of $x^n - 1$ are ε, ε^2, ..., ε^{n-1}, $\varepsilon^n = 1$, we see that the splitting field for $x^n - 1$ is $\mathbb{Q}(\varepsilon)$; we merely have to show $G(\mathbb{Q}(\varepsilon)/\mathbb{Q})$ is solvable.

Any automorphism α of $\mathbb{Q}(\varepsilon)$ is determined by the value of $\alpha(\varepsilon)$. However, α must take ε to ε^k, where $\gcd(k,n) = 1$. Otherwise, $\alpha(\varepsilon)$ would not be a primitive nth root of unity as ε is.

Thus, let α and β be any two automorphisms in $G(\mathbb{Q}(\varepsilon)/\mathbb{Q})$ with, say, $\alpha(\varepsilon) = \varepsilon^i$ and $\beta(\varepsilon) = \varepsilon^j$. Then $(\alpha \circ \beta)(\varepsilon) = \alpha(\varepsilon^j) = \varepsilon^{ij} = \beta(\varepsilon^i) = (\beta \circ \alpha)(\varepsilon)$, and $\alpha \circ \beta = \beta \circ \alpha$. Thus $G(\mathbb{Q}(\varepsilon)/\mathbb{Q})$ is abelian and certainly solvable. Therefore, $\varepsilon = e^{2\pi i/n}$ is solvable by radicals over $\mathbb{Q}$.

Exercises

14 Show that this $G(\mathbb{Q}(\varepsilon)/\mathbb{Q})$ is isomorphic to the group $\Phi(n)$ discussed in Chap. 6. Thus $|G(\mathbb{Q}(\varepsilon)/\mathbb{Q}| = \phi(n)$.

15 Show that all polynomials over $\mathbb{R}$ or $\mathbb{C}$ are solvable by radicals. (The fundamental theorem of algebra, discussed in Sec. 10.3, might be helpful.)

Before we can go on to a proof of the insolvability of quintics, we need a rather technical proposition about symmetric groups.

Proposition If H is a subgroup of $\mathrm{Sym}(p)$ containing both a transposition and an p-cycle, then $H = \mathrm{Sym}(p)$. Here p denotes a prime.

PROOF   Relabeling as necessary, we can assume that the transposition is (12). The notation of the p-cycle can be cycled so that it reads $(1ab\cdots)$; taking an appropriate power, we can assume the p-cycle is $(1,2,c,d,\ldots)$, and we may as well relabel the last elements in order.

First $(12)(123\cdots p) = (23\cdots p)$ is in H. Then $(23\cdots p)(12)(23\cdots p)^{-1} = (13)$ is in H, as are $(23\cdots p)^k(12)(23\cdots p)^{-k} = (1, k+1)$. Hence (12), (13), $\ldots$, $(1k)$, $\ldots$, $(1p)$ are all in H. Then $(1k)(1j)(1k) = (kj)$ is in H for all k and j.

Any cycle $(abc\cdots l) = (al)\cdots(ac)(ab)$ is now also in H, since all the transpositions are in H. Since any permutation is a product of disjoint cycles, and all the cycles are in H, we have $\mathrm{Sym}(p) = H$. [We use the fact that p is prime to be able to relabel the p-cycle as $(1,2,c,d,\ldots)$.]   ////

Theorem (Abel-Ruffini)   The general equation of degree 5 or higher is not solvable by radicals. This is true, in particular, over the field $\mathbb{Q}$.

PROOF   It will suffice to produce a single polynomial that is not solvable by radicals. We choose $f(x) = x^5 - 10x + 2$, partly because it is irreducible by Eisenstein's criterion. By differentiating $f(x)$, we find that $f(x)$ has a minimum at $\sqrt[4]{2}$ and a maximum at $-\sqrt[4]{2}$. Hence, we can quickly ascertain that $f(x)$ has three real roots and thus one pair of complex roots.

Let E be a splitting field of $f(x)$ over $\mathbb{Q}$. Since $f(x)$ is irreducible, $G(E/F)$ is transitive on the five roots of $f(x)$, and $G(E/F)$ must contain a 5-cycle. Complex conjugation is also an element of $G(E/F)$ and is represented by the transposition of the two complex roots. By the last proposition $G(E/F) = \mathrm{Sym}(5)$, and, by result 5, $\mathrm{Sym}(5)$ is not solvable.   ////

Exercises

16 Show that $2x^5 - 5x^4 + 5$ is not solvable by radicals over $\mathbb{Q}$.

17 Show that $ax^4 + bx^2 + c$ is always solvable by radicals over any field of characteristic not equal to 2.

18 The following fact about permutation groups on p elements (p again prime) is true and was first proved by Galois: A transitive permutation group on p elements is solvable if and only if the only subgroup fixing any two elements is the identity. Use this fact to prove *Galois's theorem on insolvability.* Let $f(x)$ be an irreducible polynomial of prime degree p over $\mathbb{Q}$. Then $f(x)$ is solvable by radicals if and only if the splitting field E of $f(x)$ equals $\mathbb{Q}(r_i, r_j)$, where r_i and r_j are any two distinct roots of $f(x)$.

19 We proved in Chap. 5 that Sym(3) and Sym(4) are solvable. Use this to show that every cubic and quartic polynomial over a field F is solvable by radicals.

20 Show that $ax^8 + bx^6 + cx^4 + dx^2 + f$ is solvable by radicals over any field F.

BIBLIOGRAPHY

1 Adamson, I. T.: "Introduction to Field Theory," Wiley, New York, 1964.
2 Ahlfors, L.: "Complex Analysis," 2nd ed., McGraw-Hill, New York, 1966.
3 Archibald, R. G.: "An Introduction to the Theory of Numbers," Merrill, Columbus, Ohio, 1970.
4 Artin, E.: The Influence of J. H. M. Wedderburn on the Development of Modern Algebra, *Am. Math. Soc. Bull.* 56, 65–72 (1950).
5 Artin, E.: Uber einen Satz von Herrn J. H. M. Wedderburn, *Abh. Hamburg Math. Sem.*, 5: 245–250 (1928).
6 Artin, E.: "Galois Theory," Univ. Notre Dame Press, Notre Dame, Ind., 1964.
7 Benson, C. T., and Grove, L. C.: "Finite Reflection Groups," Bogden and Quigley, Tarrytown-on-Hudson, N.Y., 1971.
8 Berlekamp, E. R.: "Algebraic Coding Theory," McGraw-Hill, New York, 1968.
9 Birkhoff, G., and MacLane, S.: "A Survey of Modern Algebra," 3d ed. Macmillan, New York, 1965.
10 Bolker, E.: "Elementary Number Theory," Benjamin, New York, 1970.
11 Borevich, Z., and Shafarevich, I. R.: "Number Theory," Academic, New York, 1966.

12 Boyer, C. B.: "A History of Mathematics," Wiley, New York, 1968.

13 Brinkmann, H. W., and Klotz, E. G.: "Linear Algebra and Analytic Geometry, Addison-Wesley, Reading, Mass., 1971.

14 Burnside, W.: "Theory of Groups of Finite Order," 2d ed. (reprint), Dover, New York, 1955.

15 Clay J. R., *J. Number Theory*, The Punctured Plane is Isomorphic to the Unit Circle. **1**:500–501 (1964).

16 Cohn, H.: "A Second Course in Number Theory," Wiley, New York, 1962.

17 Cohn, P.: Unique Factorization Domains, *Am. Math. Mon.*, **80**:1–17 (1973).

18 Curtis, C. W., and Reiner, I.: "Representation Theory of Finite Groups and Associative Algebras," Wiley, New York, 1962.

19 Davenport, M.: "The Higher Arithmetic," Harper Torchbooks, New York, 1960.

20 Dean, R. A.: "Elements of Abstract Algebra," Wiley, New York, 1966.

21 Dornhoff, L.: "Group Representation Theory," vol. I, Dekker, New York, 1971.

22 Feit, W., and Thompson, J.: Solvability of Groups of Odd Order, *Pac. J. Math.*, **13**:775–1029 (1963).

23 Fraleigh, J.: "A First Course in Abstract Algebra," Addison-Wesley, Reading, Mass., 1967.

24 Freyd, P.: "Abelian Categories," Harper Row, New York, 1964.

25 Frobenius, F. G.: Neuer Beweis der Sylowschen Satzes, *Gesam. Abh.*, **2**:301–303 (1968).

26 Frobenius, F. G., and Stickelberger, L.: Uber Gruppen von vertauschbaren Elementen, *Gesam. Abh.*, **1**:545–591 (1968).

27 Fuchs, L.: "Infinite Abelian Groups," Academic, New York, vol. 1, 1970, vol. 2, 1973 (first edition, Pergamon, New York, 1960).

28 Gauss, C. F.: "Disquistiones Arithmeticae" (English translation), Yale, New Haven, Conn., 1966.

29 Goldhaber, J. K., and Ehrlich, G.: "Algebra," Macmillan, New York, 1970.

30 Golod, E. S., and Shafarevich, I. R.: On Towers of Class Fields, *Izv. Akad. Nauk. S.S.R.*, Ser. Mat. 28:261–272 (1964).

31 Griffith, P. H.: "Infinite Abelian Group Theory," Univ. Chicago, Chicago Press, 1970.

32 Hall, M., Jr.: "The Theory of Groups," Macmillan, New York, 1959.

33 Halmos, P.: "Finite-dimensional Vector Spaces," Springer, New York, 1958.

34 Harary, F., and Palmer, E. M.: "Graphical Enumeration," Academic, New York, 1973.

35 Hardy, G. H., and Wright, E. M.: "An Introduction to the Theory of Numbers," 4th ed. Clarendon, Oxford, 1960.

36 Hartshorne, R.: "Foundations of Projective Geometry," Benjamin, New York, 1967.

37 Henderson, D.: A Short Proof of Wedderburn's Theorem, *Am. Math. Mon.*, **72**:385–386 (1965).

38 Herstein, I. N.: "Topics in Algebra," Xerox, New York, 1964.

39 Herstein, I. N.: "Noncommutative Rings," Math. Assoc. Am. (Wiley, distributor), New York, 1968.

40 Huppert, B.: "Endliche Gruppen," vol. 1, Springer, New York, 1967.

41 Ireland, K., and Rosen, M.: "Elements of Number Theory," Bogden and Quigley, Tarrytown-on-Hudson, N.Y., 1972.

42 Jacobson, N.: "Lectures in Abstract Algebra," Van Nostrand, Princeton, N.J., vol. 1, 1951, vol. II, 1953, vol. 3, 1964.

43 Jans, J. P.: "Rings and Homology," Holt, New York, 1964.

44 Kamke, E.: "Theory of Sets" (English translation of 2d ed.), Dover, New York, 1950.

45 Kaplansky, I. J., "Infinite Abelian Groups," 2d ed. Univ. Michigan Press, Ann Arbor, 1969.

46 Kaplansky, I. J.: "Fields and Rings," 2d ed. Univ. Chicago Press, Chicago, 1972.

47 Kline, M.: "Mathematical Thought from Ancient to Modern Times," Oxford Univ. Press, New York, 1972.

48 Knopp, M.: "Modular Functions in Analytic Number Theory," Markham, Chicago, 1970.

49 Kronecker, L.: *J. für Math.*, Ein Fundamentalsate der allgemein Arithmetik, **100**:490–510 (1887).

50 Kurosh, A. G.: "The Theory of Groups" (English translation), Chelsea, New York, vol. I, 1955, vol. II, 1956.

51 Lagrange, J-L.: "Reflexions sur la resolution algebrique des equations," Oeuvres, vol. 3, pp. 205–421

52 Landau, E. G. H.: "Elementary Number Theory" (English translation of 2d ed.), Chelsea, New York, 1966.

53 Lang, S.: "Algebra," Addison-Wesley, Reading, Mass., 1965.

54 Lehner, J.: "A Short Course in Automorphic Functions," Holt, New York, 1966.

55 LeVeque, W. J : "Topics in Number Theory," 2 vols., Addison-Wesley, Reading, Mass., 1956.

56 Levinson, N.: Coding Theory, *Am. Math. Mon.*, **77**:249–258 (1970).

57 Littlewood, O. E.: "The Theory of Group Characters," 2d ed. Oxford, New York, 1950.

58 Liu, C. L.: "Introduction to Combinatorial Mathematics," McGraw-Hill, New York, 1968.

59 MacDonald, I. O.: "The Theory of Groups," Oxford Univ. Press, New York, 1968.

60 McCoy, N. H.: "The Theory of Rings," Macmillan, New York, 1964.

61 Miller, G. A.: "Collected Works," Univ. Illinois Press, Urbana, vol. I, pp. 1–9, 1935.

62 Mitchell, B.: "Theory of Categories," Academic, New York, 1965.

63 Mordell, L: "Diophantine Equations," Academic, New York, 1962.

64 Newman, M.: "Matrix Representations of Groups," U.S. Gov. Printing Office, Washington, D.C., 1968.

65 Niven, I., and Zuckerman, H. S.: "An Introduction to the Theory of Numbers," Wiley, New York, 1972.

66 Ore, O.: "Cardano, the Gambling Scholar," Dover, New York, 1953.

67 Ore, O.: "Number Theory and Its History," McGraw-Hill, New York, 1948.

68 Paley, H., and Weichsel, P. M.: "A First Course in Abstract Algebra," Holt, New York, 1966.

69 Passman, D.: "Permutation Groups," Benjamin, New York, 1968.

70 Pennisi, L.: "Elements of Complex Variables," Holt, New York, 1966.

71 Polya, G.: Kombinatorische Anzahlbestimmungen fur Gruppen, Graphen, und Chemische Verbindungen, *Acta Math.*, **68**:145–254 (1937).

72 Redei, L.: "Algebra," vol. I, Pergamon, New York, 1967.

73 Rotman, J. J.: "The Theory of Groups, An Introduction," 2d ed. Allyn and Bacon, Boston, 1973.

74 Sah, G. H.: "Abstract Algebra," Academic, New York, 1967.

75 Samuel, P.: Unique Factorization, *Am. Math. Mon.*, **75**:945–952 (1968).

76 Samuel, P.: "Lectures on Unique Factorization Domains," Tata Institute, Bombay, 1964.

77 Schreier, O., and Sperner, E.: "Introduction to Modern Algebra and Matrix Theory" (English translation), 2d ed. Chelsea, New York, 1959.

78 Scott, W. R.: "Group Theory," Prentice-Hall, Englewood Cliffs, N.J., 1964.

79 Serre, J-P.: "Representation lineares des groupes finis," Hermann, Paris, 1967.

80 Shafarevich, I. R.: Construction of Fields of Algebraic Numbers with a Given Solvable Galois Group, A.M.S. Translations, series 2, vol. 4, pp. 185–237, 1956.

81 Shanks, D.: "Solved and Unsolved Problems in Number Theory," Spartan, Washington, 1962.

82 Stark, H: *Proc. Natl. Acad. Sci.*, vol. 57, pp. 216–221,

83 Stoll, R. R., and Wong, E. T.: "Linear Algebra," Academic, New York, 1968.

84 Thompson, J. G.: Nonsolvable Finite Groups All of Whose Local Subgroups are Solvable, (I) *Bull. Am. Math. Soc.*, **74**:383–437 (1968); (II) *Pac. J. Math.*, **33**: 451–536 (1970).

85 van der Waerden, B. L.: "Modern Algebra" (English translation), Ungar, New York, vol. I, 1949, vol. II, 1950.

86 Wantzel, P.: *J. de Math.*, **2**:366–72 (1837).

87 Warner, S.: "Classical Modern Algebra," Prentice-Hall, Englewood Cliffs, N.J., 1971.

88 Wedderburn, J. H. M.: A Theorem on Finite Algebras, *Am. Math. Soc. Trans.*, **6**:349–52 (1905).

89 Wielandt, H.: Ein Beweis fur die Existenz der Sylow Gruppen, *Arch. Math.*, **10**: 401–402 (1959).

90 Wilder, R. L.: "Introduction to the Foundations of Mathematics," Wiley, New York, 1952.

91 Yale, P. B.: "Geometry and Symmetry," Holden-Day, San Francisco, 1968.

92 Kiernan, B. M.: The Development of Galois Theory from Lagrange to Artin, *Arch. Hist. Exact Sci.*, **8**:40–154 (1971).

93 Motzkin, T.: The Euclidean Algorithm, *Bull. Amer. Math. Soc.* **55**:1142–1146 (1949).

94 Wilson, J. C.: A Principal Ideal Ring That Is Not a Euclidean Ring, *Math. Mag.* **46**:34–38 (1973).

SELECTED ANSWERS AND
OCCASIONAL HINTS AND COMMENTS

CHAPTER 1

Section 1.1

1 As with all if-and-only-if statements, there are two proofs to do. We shall prove that if $A \subseteq B$ then $A \cap B = A$: First, if $x \in A \cap B$, then x is in A, so $A \cap B \subseteq A$. If $y \in A$, then y is also in B, since $A \subseteq B$. Thus, $y \in A \cap B$ and, as y was an arbitrary element, we have $A \subseteq A \cap B$. $A \cap B \subseteq A$ and $A \subseteq A \cap B$ together show that $A = A \cap B$.

3 If $S = \{a,b,c\}$, then S has eight subsets. If $S = \{a_1,a_2,a_3,a_4,a_5\}$, then S has 32 subsets.

6 (*a*) is not closed; (*c*) and (*d*) are closed.

7,8 (*a*) is neither commutative nor associative.

     (*c*) is both commutative and associative.

Section 1.2

1 C Every even integer can be written in the form $2n$, and this is often useful. $0 = 2 \cdot 0$ is the identity, since $0 + 2n = 2n + 0 = 2n$ for all even numbers $2n$. The associative law for addition holds for any subset of the real numbers, including the even integers. Since $2n + 2(-n) = 0$, we have that $2(-n) = -2n$ is the inverse for $2n$. Adding, $2n + 2m = 2(n + m)$, so closure holds.

D Two arbitrary odd numbers can be written $2n + 1$ and $2m + 1$. Adding them yields $2n + 1 + 2m + 1 = 2(n + m + 1)$ which is not odd. Therefore, closure doesn't hold. To particularize this, just note that, say, $3 + 5 = 8$ so closure does not hold.

3 For associativity, see page 4. If a and b are nonzero reals, so is $3ab = a * b$. To find an identity we solve $a = a * e = 3ae$ and find that $e = \frac{1}{3}$. (Check that this works.) Solving $\frac{1}{3} = a * a' = 3aa'$ will yield a candidate for an inverse. Checking these yields that $(G, *)$ is a group.

4 (a) and (c) are groups.

6 (b) ϕ is the identity and $A = A'$.

(e) 2^n.

Section 1.3

2 Note that $(ab)^2 = abab$.

4 First note that $x^2 = e$ if and only if $x = x^{-1}$. Thus we want the number of subsets $\{x, x^{-1}\}$ that consist of one element. Write $G = \{e\} \cup \{x, x^{-1}\} \cup \{y, y^{-1}\} \cup \cdots$, and count both sides.

6 Use the fact that $(xy)^2 = e$.

7 We have only to establish the existence of an identity and inverses. If $G = \{a_1, a_2, \ldots, x, \ldots, a_n\}$, then also $G = \{x * a_1, x * a_2, \ldots, x * a_n\}$. Thus $x = x * a_i$ for some a_i. This a_i is our candidate for an identity. Any g in G can be written as $a_j * x$, so $g * a_i = a_j * x * a_i = a_j * x = g$. The a_i is a right identity. Similarly, we can produce a left identity a_k so that $a_k * g = g$ for all g in G. Since $a_i = a_k * a_i = a_k$, we have a true identity. The existence of inverses is similar but easier.

13 $(\mathbb{Z}_5, \odot)$ could be modified by eliminating the first row and column. $(\mathbb{Z}_6, \odot)$ needs to have the $\bar{0}$, $\bar{2}$, $\bar{3}$, and $\bar{4}$ rows and columns deleted.

Section 1.4

2 The infinite sequence $\{x, x^2, x^3, \ldots, x^n, \ldots\}$ can contain only a finite number of different elements, so there must be duplication. If $x^i = x^j$, where $j > i$, then $e = x^i * x^{-i} = x^j * x^{-i} = x^{j-i}$; letting $n = j - i$ finishes the problem.

4

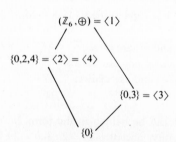

6   In $(\mathbb{R} - (0), \cdot)$, $H = \{x \mid |x| < 1\}$ and $J = \{5, \frac{1}{5}\}$ will suffice. Even simpler examples
    exist in $(\mathbb{Z}, +)$.

8   $H = \{a/2^n 3^m \mid a,n,m \in \mathbb{Z}\}$ is a noncyclic subgroup of $(\mathbb{Q}^+, \cdot)$. Why?

9   Let C be cyclic with generator g, and let x and y be two elements of C. Then
    $x = g^n$ and $y = g^m$, where $m,n \in \mathbb{Z}$. Thus, $x * y = g^n * g^m = g^{n+m} = g^{m+n} = g^m * g^n$
    $= y * x$. Thus C is abelian.

11 If $G = \langle g \rangle = \{e, g, g^{-1}, g^2, g^{-2}, \ldots\}$ is cyclic, then $\langle g^{-1} \rangle = \{e, g^{-1}, (g^{-1})^{-1} = g, g^{-2},$
    $g^2, \ldots\}$. Thus g^{-1} is a second generator for G unless $g = g^{-1}$. This yields $g^2 = e$
    and $G = \{e\}$ or $G = \{e, g\}$.

14 Assume G_p is cyclic with a/p^b as its generator. Then $1/p^{b+1}$ is also in G_p but, since
    no integer n exists with $n(a/p^b) = 1/p^{b+1}$, we see that a/p^b cannot be a generator.
    [Since $+$ is the operation, we use $n(a/p^b)$ instead of $(a/p^b)^n$.]

17 Exercise 2 provides a positive n such that $a^n = e$. Then $a^{n-1} \cdot a = e$ so $a^{-1} = a^{n-1}$.
    In a similar manner $a^{-k} = a^{n-k}$ so all the negative integers can be provided for.

Section 1.5

1   (a) and (e) are not functions; (d) is a function.

2   (a) and (e) are functions; (d) is not but can be made into one by requiring $|x| \le 3$
    and $f(x) \ge 0$.

3   A function $f : S \to T$ is one-to-one if (s_1, t) and (s_2, t) both in f requires that $s_1 = s_2$.
    Equivalently, if $f(s_1) = f(s_2)$ implies $s_1 = s_2$, then f is one-to-one.

4   1(d) and 2(e) are not 1-1; 1(a) and 1(e) are 1-1 but are not functions; 2(a) is 1-1;
    2(d) is not a function and is not 1-1, even if we fix it as above to be a function.

7   1(a) is onto but not a function; 1(d), 1(e), 2(d), and 2(e) are not onto; 2(a) is an
    onto function. A function from $\mathbb{R}$ to $\mathbb{R}$ is onto if any horizontal line intersects its
    graph at least once.

9   $f(n) = 2n$ is a 1-1 function from $\mathbb{Z}$ to $\mathbb{Z}$ that is not onto.

$$g(n) = \begin{cases} n - 5 & n \ge 0 \\ n & n < 0 \end{cases}$$

is an onto function that is not 1-1.

10 If $f = \{(s, f(s))\}$, then $f^{-1} = \{(f(s), s)\}$, and $(f^{-1})^{-1} = \{s, f(s)\} = f$.

12 Either $m = -1$, b arbitrary, or $m = 1$, $b = 0$, is necessary.

14 (a) is not a function; neither is its inverse; it does equal its inverse. (c) is a function;
    its inverse isn't. (e) isn't a function; neither is its inverse, which it doesn't equal.

16 If $f_x(g_1) = f_x(g_2)$, then $g_1 * x = g_2 * x$, and by the right cancellation property
    $g_1 = g_2$. Thus f_x is 1-1. (This really amounts to a rephrasing of the right cancella-
    tion law in terms of the right multiplication function being 1-1.) If g is any
    element of g, we have $f_x(g * x^{-1}) = g$ so f_x is onto.

Section 1.6

1   (a) $\begin{pmatrix} 3 & 5 \\ 0 & 0 \end{pmatrix}$; (c) $\begin{pmatrix} ad & ae + bf \\ 0 & cf \end{pmatrix}$; (e) $\begin{pmatrix} ab & ac \\ ad & af \end{pmatrix}$.

3 Since there are many subgroup verifications in this book, we do this rather typical verification in some detail: Let $H = \left\{ \begin{pmatrix} a & b \\ 0 & d \end{pmatrix} \middle| ad \neq 0 \right\}$ be our subset of $GL(2, \mathbb{R})$. We start with $I = \begin{pmatrix} 1 & 0 \\ 0 & 1 \end{pmatrix}$. The key fact is not the computation of I but checking that it is in H. Since $\begin{pmatrix} 1 & 0 \\ 0 & 1 \end{pmatrix}$ has a zero in the bottom left-hand corner, and since $ad = 1 \cdot 1 = 1 \neq 0$, I is indeed in H.

$$\begin{pmatrix} a & b \\ 0 & d \end{pmatrix}^{-1} = \begin{pmatrix} d/ad & -b/ad \\ 0/ad & a/ad \end{pmatrix} = \begin{pmatrix} 1/a & -b/ad \\ 0 & 1/d \end{pmatrix}$$

Again we have a 0 in the bottom left-hand corner. Also, $1/a \cdot 1/d = 1/ad \neq 0$. Thus the inverse property holds for H.

$$\begin{pmatrix} a & b \\ 0 & d \end{pmatrix}\begin{pmatrix} e & f \\ 0 & g \end{pmatrix} = \begin{pmatrix} ae & af + bg \\ 0 & dg \end{pmatrix}$$

Thus 0 shows up in the correct corner again. If $\begin{pmatrix} a & b \\ 0 & d \end{pmatrix}$ and $\begin{pmatrix} e & f \\ 0 & g \end{pmatrix}$ are in H, then we have $ad \neq 0 \neq eg$, and this yields $ae \cdot dg \neq 0$. This is sufficient for closure and thus for H to be a subgroup.

7 Case 1: $a + d = 0 \Rightarrow d = -a$, $\quad c = (1-a)^2/b$
Case 2: $b = c = 0 \Rightarrow a = \pm 1$, $\quad d = \pm 1$

8 Consider $\begin{pmatrix} 2 & 3 \\ 0 & 5 \end{pmatrix}$.

10 Referring back to Exercise 7, we find case 1 gives $M = \begin{pmatrix} a & b \\ c & d \end{pmatrix} = \begin{pmatrix} a & b \\ (1 - a^2)/b & -a \end{pmatrix}$, which is not in $SL(2, \mathbb{R})$. Case 2 yields $\begin{pmatrix} 1 & 0 \\ 0 & 1 \end{pmatrix}, \begin{pmatrix} -1 & 0 \\ 0 & 1 \end{pmatrix},$ $\begin{pmatrix} 1 & 0 \\ 0 & -1 \end{pmatrix}, \begin{pmatrix} -1 & 0 \\ 0 & -1 \end{pmatrix},$ and $\begin{pmatrix} -1 & 0 \\ 0 & -1 \end{pmatrix}$ is the only one of order 2 and determinant 1.

11 (b) $f: S \to T$ and $g: T \to U$ are both onto. Picking any u in U, we can find t in T such that $g(t) = u$. Similarly, since f is onto, we can find s in S such that $f(s) = t$. Then, however, $(g \circ f)(s) = g(f(s)) = g(t) = u$, and $g \circ f$ is onto.

Section 1.7

2 (b) If $f(x) = 3^x$ and $g(x) = x^2$, then $(g \circ f)(x) = 3^{2x}$, which is a 1-1 function from $\mathbb{Z}^+$ to $\mathbb{Z}^+$, where $f: \mathbb{Z}^+ \to \mathbb{Z}^+$ and $g: \mathbb{Z} \to \mathbb{Z}^+$; g is not 1-1 on $\mathbb{Z}$, but $g \circ f$ is 1-1.

3 $\begin{pmatrix} 1 & 2 & 3 & 4 & 5 \\ 2 & 3 & 1 & 4 & 5 \end{pmatrix}$

5 Let $H = \{e, a^2, ba, ba^3\}$ denote the subset of D_8 under consideration. Using the

table on page 37, we have

	e	a^2	ba	ba^3
e	e	a^2	ba	ba^3
a^2	a^2	e	ba^3	ba
ba	ba	ba^3	e	a^2
ba^3	ba^3	ba	a^2	e

Thus H is closed. We now observe that $e \in H$ and that $e^{-1} = e$, $(a^2)^{-1} = a^2$, $(ba)^{-1} = ba$, and $(ba^3)^{-1} = ba^3$ are all in H; thus H is indeed a subgroup. This last sentence could be replaced by quoting Exercise 4 of Sec. 1.4, since H is both finite and closed under an associative operation.

7   There is one such nonidentity element in D_8, none in Sym (3).

8   Generalize Sym (3) as on page 34, D_8, and D_{10} (Exercise 6), which yield examples of order 6, 8, and 10.

11   If $a \in A$, $H = \{f \mid f$ is a permutation of $A, f(a) = a\}$.

Closure: Let f and g be in H. Then $f \circ g$ is also one-to-one and onto, as f and g are both 1-1 and onto (cf., Sec. 1.6). Thus $f \circ g$ is a permutation. Since $(f \circ g)(a) = f(g(a)) = f(a) = a$, $f \circ g$ is in H.

Identity: I is a permutation, and $I(a) = a$, so $I \in H$, where I is the identity function on A.

Inverse: Since $f \in H$ is 1-1 and onto, f^{-1} exists and is also 1-1 and onto. Since $f(a) = a$, we apply f^{-1} to both sides and obtain $a = I(a) = f^{-1}(f(a)) = f^{-1}(a)$. Thus f^{-1} is a permutation of A, and $f^{-1}(a) = a$, so $f^{-1} \in H$.

13   (c)   If e is a fixed edge of the cube, then $H_e = \{$all rigid motions taking the edge to itself$\}$. Obviously H_e consists of I and the motion turning the edge around $180°$. Thus $|H_e| = 2$.

15   Exchange the two rows. Thus, if $\beta = \begin{pmatrix} 1 & 2 & 3 & 4 \\ 3 & 1 & 2 & 4 \end{pmatrix}$, $\beta^{-1} = \begin{pmatrix} 3 & 1 & 2 & 4 \\ 1 & 2 & 3 & 4 \end{pmatrix}$.

Straightening the top row out yields $\beta^{-1} = \begin{pmatrix} 1 & 2 & 3 & 4 \\ 2 & 3 & 1 & 4 \end{pmatrix}$.

Section 1.8

1   (a) and (c) are equivalence relations.

3   (a) $\bar{0} = \{0, 5, -5, 10, -10, 15, \ldots\}$
$\bar{1} = \{1, 6, -4, 11, -9, 16, \ldots\}$
$\bar{2} = \{2, 7, -3, 12, -8, 17, \ldots\}$
$\bar{3} = \{3, 8, -2, 13, -7, 18, \ldots\}$
$\bar{4} = \{4, 9, -1, 14, -6, 19, \ldots\}$

Again we run into the phenomenon of many names for the same equivalence class. For instance, $-3 \in \bar{2}$ so $\overline{-3} = \bar{2}$, as we can also see directly by comparing $\overline{-3} = \{-3, 2, -8, 7, -13, 12, \ldots\}$ with $\bar{2}$.

(c) $\overline{f(x)} = \{f(x) + c \mid c \in \mathbb{R}\}$

Section 1.9

1 Let $H = \{e,ab\}$ and $K = \{e,b\}$. Then $HK = \{ee,abe,eb,ab^2\} = \{e,ab,b,a\}$, which is not a subgroup since it isn't closed.

2 If $J = \{e,b\} = Jb$ then $Ja = \{a,ba\} = \{a,a^3b\} = Ja^3b$; $Ja^2 = \{a^2,ba^2\} = \{a^2,a^2b\} = Ja^2b$; $Ja^3 = \{a^3,ba^3\} = \{a^3,ba\} = Jba$. We have a total of four right cosets.

3 D_8 contains suitable candidates for H and g. So does Sym (3).

8 If x has order n, then n is the smallest positive integer such that $x^n = e$. Since d divides n, n/d is a positive integer, and $x^{n/d}$ is an element of G. Now we first note that $(x^{n/d})^d = e$. If k were a positive integer smaller than d, then $(x^{n/d})^k = e$ would contradict the minimality of n, since $(n/d)k < n$. Thus d is the order of $x^{n/d}$.

REMARK   The results of Exercises 10 to 18 will be covered in Sec. 4.3.

10 $Z(D_8) = \{e,a^2\}$ and $Z(A) = A$ for any abelian group.

13 The conjugate classes in Sym (3) are $\{e\}$, $\{a,a^2\}$, and $\{b,ba,ba^2\}$. D_8 has five conjugate classes. In D_8 we have $C_{D_8}(a^2) = D_8$ and $C_{D_8}(b) = \{e,a^2,b,ba^2\}$.

$$C_{GL(2,\mathbb{R})}\begin{pmatrix}1 & 1 \\ 0 & 1\end{pmatrix} = \left\{ \begin{pmatrix} a & b \\ 0 & a \end{pmatrix} \,\middle|\, a \neq 0 \right\}.$$

23 The trouble here is that the first four axioms of Euclid certainly don't guarantee that we can label any point in the plane (p,q). In fact, in Chap. 10 we will show that, say, $(\sqrt[3]{2}, \pi)$ cannot be constructed in the plane using straightedge and compass.

Section 1.10

2 We list the elements of $\mathbb{Q}^+$ as follows:

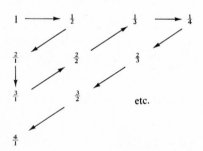

We then list the elements of $\mathbb{Q}^+$ by following the arrows, except that we delete duplicates. Thus we obtain

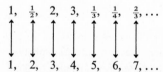

For details see Kamke [44] or Wilder [90] or a variety of other books.

3 We shall demonstrate the symmetric property, leaving the reflexive law and transitive law to the reader: If $S \sim T$, then there exists a 1-1, onto function

$f: S \to T$. Since f is 1-1 and onto, f^{-1} is a function from T to S. We recall from Sec. 1.5 that f^{-1} is automatically 1-1 and onto and thus $T \sim S$.

4 Applying part e, we see that D_8 and GL(2, $\mathbb{R}$) cannot be isomorphic to any of the others. Using part c in addition, we see that $\mathbb{Z}_{20}$, $\mathbb{Z}_8$, D_8, and GL(2,$\mathbb{R}$) are each isomorphic only to themselves. The isomorphism $\phi: \mathbb{Z} \to 2\mathbb{Z}$ given by $\phi(n) = 2n$ shows $\mathbb{Z}$ and $2\mathbb{Z}$ to be isomorphic. Counting solutions of $x^2 = 2$ in $(\mathbb{Q}^+, \cdot)$ and $2x = \phi(2)$ shows that $(\mathbb{Q}^+, \cdot)$ and $(\mathbb{Q}, +)$ are not isomorphic to each other. Since neither is cyclic, neither is isomorphic to $(\mathbb{Z}, +)$.

5 Yes, H is isomorphic to G.

6 $x^2 = 4$ has two solutions in $(\mathbb{R} - (0), \cdot)$. How many solutions does $2x = \phi(4)$ have in $(\mathbb{R}, +)$?

9 $\phi(r) = \begin{pmatrix} 1 & r \\ 0 & 1 \end{pmatrix}$ yields the isomorphism between $(\mathbb{R}, +)$ and a subgroup of GL(2, $\mathbb{R}$).

10 Is $(x_1, y_1)M + (x_2, y_2)M = [(x_1, y_1) + (x_2, y_2)]M$? For M to be 1-1 and onto, recall from Sec. 1.6 that M^{-1} must exist or, alternatively, that $ad - bc \neq 0$.

Section 1.11

1 gcd $(5{,}291, 4{,}514) = 37 = 34 \cdot 4{,}514 - 29 \cdot 5{,}291$.

3 $(a + k) - a = k$, so any number dividing $a + k$ and a must divide k. If $g = $ gcd $(a + k, a)$, it divides both a and $a + k$, and thus it divides k.

5 Using the linear property (or guessing), we find $1 = \gcd(7,15) = 1 \cdot 15 - 2 \cdot 7$. Divide both sides by 105 for (a). Then multiply through by 23 for (b).

Section 1.12

1 $\mathbb{Z}_{101}$
   |
   $\langle \bar{0} \rangle$

3 Sym (3) can have proper subgroups only of order 2 or 3 by Lagrange's theorem. Any subgroup of order 2 or 3 is cyclic by a corollary of Lagrange's theorem. Sym (3) itself is nonabelian.

8 (a) $Z(V) = V$.

   (b) $Z((\mathbb{Z}_2, \oplus) \times $ Sym (3)$) = \mathbb{Z}_2 \times \{e\} = \{(\bar{0}, e), (\bar{1}, e)\}$.

   (c) $Z($Sym (4)$) = \{e\}$. You might start by eliminating from the center all those elements that do not commute with $\begin{pmatrix} 1 & 2 & 3 & 4 \\ 2 & 3 & 4 & 1 \end{pmatrix}$.

9 We have that

$$(a,b) \in Z(A \times B)$$

iff $(a,b)(a',b') = (a',b')(a,b)$ $\forall a' \in A, b' \in B$

or $(aa', bb') = (a'a, b'b)$

iff $aa' = a'a$ and $bb' = b'b$ $\forall a' \in A, b' \in B$

iff $a \in Z(A)$ and $b \in Z(B)$

iff $(a,b) \in Z(A) \times Z(B)$

This exercise generalizes Exercise 7, since, if A is abelian, then $A = Z(A)$.

14 Since $ab = ba$, any element in $\langle a,b \rangle$ can be written as $a^i b^j$, with $0 \leq i < n$, $0 \leq j < m$. Thus $\langle a,b \rangle$ has at most nm elements. It will suffice then to show that ab has order nm. Since $ab = ba$, we have $(ab)^k = a^k b^k$ for any integer k. Also, if $a^k b^k = e$ then we have, in this particular case, that $a^k = e = b^k$. Otherwise, $a^k = b^{-k}$ and a^k will have order dividing m as well as n. Since $\gcd(n,m) = 1$, this says a^k has order 1, and thus $a^k = e$. Similarly, $b^k = e$. Thus $(ab)^k = e = a^k b^k$ implies that both m and n divide k and the smallest positive k that will work is mn.

Section 1.13

1 We show that f_x is a homomorphism and onto. Since $f_x(gh) = x^{-1}(gh)x = x^{-1}gxx^{-1}hx = f_x(g)f_x(h)$, we have that f_x is a homomorphism. Since for any y we have $f_x(xyx^{-1}) = x^{-1}(xyx^{-1})x = y$, f_x is indeed onto.

2 All subgroups of D_8 are normal except for $\{e,b\}$, $\{e,ba\}$, $\{e,ba^2\}$, and $\{e,ba^3\}$.

3 No. See Exercise 2.

4 Let $N = \{e,n\}$. Then $xN = x\{e,n\} = \{x,xn\} = \{x,nx\} = \{e,n\}x = Nx$ for all x in G. Thus xn equals x or nx. Since $xn \neq x$, we must have $xn = nx$ for all x in G, and thus $n \in Z(G)$. Therefore, $\{e,n\} \subseteq Z(G)$.

6 We look at any element $m^{-1}n^{-1}mn$, where $m \in M$ and $n \in N$. Since $(m^{-1}n^{-1}m)n = n'n \in N$ and $m^{-1}(n^{-1}mn) = m^{-1}m' \in M$, we have that $m^{-1}n^{-1}mn \in M \cap N = \{e\}$. Thus $m^{-1}n^{-1}mn = e$, which is equivalent to $mn = nm$.

9 (c) and (e) represent homomorphisms; (a) doesn't. Neither (c) nor (e) is an isomorphism.

10 In (c), $\ker(\phi) = \left\{ \begin{pmatrix} a & b \\ c & d \end{pmatrix} \Big| ad - bc = 1 \right\} = SL(2, \mathbb{R})$ as in Exercise 3 of Sec. 1.6.

In (e), $\ker(\phi) = \left\{ \begin{pmatrix} 1 & b \\ 0 & 1 \end{pmatrix} \right\}$.

12 (a) ϕ is a homomorphism.

(b) $\phi(5x^2 + 2) = d(5x^2 + 2)/dx = d(5x^2)/dx = \phi(5x^2)$, so ϕ is not 1-1. Since $\phi(\int p(x)\,dx) = p(x)$, we see that ϕ is onto.

14 (a)

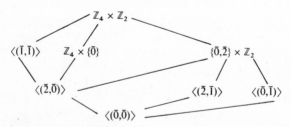

(b) If $N = \langle (\bar{0},\bar{1}) \rangle = \{(\bar{0},\bar{1}),(\bar{0},\bar{0})\}$, then $\mathbb{Z}_4 \times \mathbb{Z}_2/N \simeq \mathbb{Z}_4$.

15 (a) Let $n_1 = c_1 m + r_1$ and $n_2 = c_2 m + r_2$. Then

$$\phi(n_1 + n_2) = \phi((c_1 + c_2)m + r_1 + r_2)$$
$$= \begin{cases} \bar{r}_1 + \bar{r}_2 & \text{if } r_1 + r_2 < m \\ \bar{r}_1 + \bar{r}_2 - \bar{m} & \text{if } r_1 + r_2 \geq m \end{cases}$$

$$= \begin{cases} \phi(n_1) + \phi(n_2) & \text{if } r_1 + r_2 < m \\ \phi(n_1) + \phi(n_2) - \bar{0} & \text{if } r_1 + r_2 \geq m \end{cases}$$

$$= \phi(n_1) + \phi(n_2)$$

(b) $\ker(\phi) = m\mathbb{Z} = \{0, m, -m, 2m, -2m, \ldots\}$.

(c) $\mathbb{Z}/m\mathbb{Z} \simeq \mathbb{Z}_m$, which could be called the fundamental theorem of elementary number theory.

17 (a) We show that HK is a subgroup. Is $h_1 k_1 h_2 k_2$ in HK? Since $K \lhd G$ we have $h_1 k_1 h_2 k_2 = h_1 h_2 (h_2^{-1} k_1 h_2) k_2 = h_1 h_2 k_3 k_1 \in HK$. Also, $e = ee \in HK$. Finally, $(hk)^{-1} = k^{-1} h^{-1} = h^{-1}(hk^{-1}h^{-1}) = h^{-1}k'$, since $K \lhd G$. Thus $(hk)^{-1} = h^{-1}k' \in HK$.

(e) $H \cap K \lhd H$. Thus $h_1(H \cap K) = (H \cap K)h_1 = (H \cap K)h_2 = h_2(H \cap K) \Leftrightarrow h_2^{-1} h_1 (H \cap K) \Leftrightarrow h_2^{-1} h_1 \in H \cap K$. However, $h_1 k_1 = h_2 k_2$ implies $h_2^{-1} h_1 = k_2 k_1^{-1} \in H \cap K$.

18

$$\phi(x + y) = \begin{pmatrix} \cos(x+y) & \sin(x+y) \\ -\sin(x+y) & \cos(x+y) \end{pmatrix}$$

$$= \begin{pmatrix} \cos x \cos y - \sin x \sin y & \sin x \cos y + \sin y \cos x \\ -\sin x \cos y - \sin y \cos x & \cos x \cos y - \sin x \sin y \end{pmatrix}$$

$$= \begin{pmatrix} \cos x & \sin x \\ -\sin x & \cos x \end{pmatrix} \begin{pmatrix} \cos y & \sin y \\ -\sin y & \cos y \end{pmatrix} = \theta(x)\theta(y)$$

Also, $\ker(\theta) = \{2\pi n \mid n \in \mathbb{Z}\}$.

21 (a) $9 \nmid 73{,}127{,}894$, since $9 \nmid 7 + 3 + 1 + 2 + 7 + 8 + 9 + 4$.

Section 1.14

2 $\begin{pmatrix} 1 & a \\ 0 & b \end{pmatrix}^{-1} \begin{pmatrix} 1 & c \\ 0 & d \end{pmatrix}^{-1} \begin{pmatrix} 1 & a \\ 0 & b \end{pmatrix} \begin{pmatrix} 1 & c \\ 0 & d \end{pmatrix} = \begin{pmatrix} 1 & c(1-b) - a(1-d) \\ 0 & 1 \end{pmatrix}$ and by varying

a, b, c, d as necessary we arrive at $\left\{ \begin{pmatrix} 1 & f \\ 0 & 1 \end{pmatrix} \middle| f \in \mathbb{R} \right\} = G'$.

3 We know that G/G' is abelian, and we have the canonical homomorphism $\phi: G \to G/G'$. If $H \supseteq G'$ then $\phi(H)$ is a subgroup of G/G', and thus $\phi(H)$ is normal, since G/G' is abelian. But inverse images of normal subgroups are normal, so $\phi^{-1}(\phi(H))$ is normal. Since $H \supseteq G'$, we have $\phi^{-1}(\phi(H)) = H$. (The lattice of subgroups over G' in G equals that of G/G'.)

7 Let $G = \mathcal{Q}_{4n}$. Since $a^{-1}b^{-1}ab = a^{-2}$, we see that $\langle a^{-2} \rangle = \langle a^2 \rangle \subseteq G'$. However, $|\langle a^2 \rangle| = n$, so $G/\langle a^2 \rangle$ has order 4 and is thus abelian. Therefore, $G' \subseteq \langle a^2 \rangle$ and $G' = \langle a^2 \rangle$. What is the gap here? We haven't shown that $\langle a^2 \rangle \lhd G$. However, $b^{-1}a^2 b = b^{-1}abb^{-1}ab = (b^{-1}ab)^2 = a^{-2} \in \langle a^2 \rangle$, so $b^{-1}\langle a^2 \rangle b \subseteq \langle a^2 \rangle$. This implies $b^{-k}\langle a^2 \rangle b^k \subseteq \langle a^2 \rangle$ and $(a^j b^k)^{-1} \langle a^2 \rangle a^j b^k \subseteq \langle a^2 \rangle$. Thus $\langle a^2 \rangle \lhd G$, and the demonstration is complete.

8 The identity for G is $(0,0)$, so $(a,b)^{-1} = (a(-1)^{b+1}, -b)$. Thus

$$(1,1)^{-1}(2,2)^{-1}(1,1)(2,2) = (2,0).$$

We can show that $\langle(2,0)\rangle \lhd G$ and that $G/\langle(2,0)\rangle \simeq \mathbb{Z}_2 \times \mathbb{Z}$. The latter being abelian yields the upper bound for G', and $G' = \langle(2,0)\rangle$.

10 Sym $(3)' = $ Alt (3). Sym $(4)' = $ Alt (4), the subgroup of order 12 in Sym (4). See Sec. 1.15 for further details on this.

12 $$G = G' \xrightarrow{\ \alpha\ } G/K = \overline{G} \xrightarrow{\ \beta\ } \overline{G}/\overline{G}'$$
$$\big|$$
$$K$$

Let $\overline{G} = G/K$. By Proposition 1, $\overline{G}/\overline{G}'$ is abelian, and using two successive canonical homomorphisms, we have a homomorphism $\beta \circ \alpha \colon G \to \overline{G}/\overline{G}'$. If $\overline{G}/\overline{G}'$ is not the identity group, then $G \neq G'$. Thus $\overline{G}/\overline{G}'$ is just one element, so $\overline{G} = \overline{G}'$ or $G/K = (G/K)'$.

13 This exercise has an interesting application. In the case of Sym (n) (see next section), we have $Z(\text{Sym}(n)) = \{e\}$ and Sym $(n)' = $ Alt (n). Can we (using Exercise 13) have other proper normal subgroups?

Section 1.15

4 To write $(\mathbb{Q},+)$ as $H_1 \times H_2$ we would need nontrivial nonintersecting subgroups. Since $(\mathbb{Q},+)$ is abelian, normality is automatic. However, assume a_1/b_1 is in H_1 and a_2/b_2 is in H_2. Assume b_1 and b_2 are positive integers. Then $a_1 = b_1(a_1/b_1)$ is in H_1, $a_2 = b_2(a_2/b_2)$ is in H_2, and $a_1 a_2$ is in $H_1 \cap H_2$. Thus we can never have $H_1 \cap H_2 = \{0\}$ if H_1 and H_2 are nontrivial.

6 (a) If f is not 1-1, then $f(a_1) = f(a_2)$, where $a_1 \neq a_2$. Thus $(g \circ f)(a_1) = (g \circ f)(a_2)$, and $g \circ f$ is not an isomorphism.

   (c)* We know that both $f(A)$ and ker (g) are normal in B. Let $b \in f(A) \cap \text{ker}(g)$. Then $b = f(a)$, and $g(b) = e_C$. Thus $(g \circ f)a = g(f(a)) = g(b) = e_C$, and, $g \circ f$ being an isomorphism, this forces $a = e_A$. But $b = f(e_A) = e_B$, and we have shown $f(A) \cap \text{ker}(g) = \{e_B\}$. If we can show that $B = f(A) \cdot \text{ker}(g)$, we will be finished. We first note that if $b \in B$ then $g(b) = c = (g \circ f)(a)$ for some a in A, since $g \circ f$ is an isomorphism onto C. Now g will take both b and $f(a)$ to c, so $g((f(a)^{-1})b) = e_C$ or $f(a)^{-1}b = k \in \text{ker}(g)$. Thus $b = f(a)k \in f(A)\, \text{ker}(g)$, and we're done.

7 *Lemma*: If we are given $h_1 \colon J \to A$ and $h_2 \colon J \to B$, we can define $h \colon J \to A \times B$ by $h(j) = (h_1(j), h_2(j))$. This is quickly verified to be a homomorphism. Also, $(\pi_i \circ h)(j) = \pi_i(h_1(j), h_2(j)) = h_i(j)$ for $i = 1$ or 2. Thus, $\pi_i \circ h = h_i$.

   Assume now that $\pi_1 \circ h$ and $\pi_2 \circ h$ are homomorphisms. By our lemma, there exists a homomorphism $h^* \colon J \to A \times B$ such that $\pi_i \circ h^* = \pi_i \circ h$ for $i = 1,2$. But $h^*(j) = ((\pi_1 \circ h)(j), (\pi_2 \circ h)(j)) = h(j)$ for all j in J, and thus $h = h^*$ and is a homomorphism.

8 (a) (12345); (c) $(123)(67) = (123)(4)(5)(67)$.

9 (a) $\begin{pmatrix} 1 & 2 & 3 & 4 & 5 & 6 & 7 & 8 \\ 3 & 4 & 5 & 6 & 7 & 8 & 2 & 1 \end{pmatrix}$; (c) $\begin{pmatrix} 1 & 2 & 3 & 4 & 5 & 6 & 7 & 8 \\ 8 & 7 & 6 & 5 & 4 & 3 & 2 & 1 \end{pmatrix}$.

10 Let $\alpha = (a_1 a_2 \cdots a_n)$ and $\beta = (b_1 b_2 \cdots b_n)$, where $\{a_1, a_2, \ldots, a_n\} = A$ and $\{b_1, b_2, \ldots, b_n\} = B$ are disjoint sets. If α and β are permutations of S, then three

situations can occur: If $s \in S$, then s can be an element of A, of B, or neither. In all three cases, show that $(\alpha \circ \beta)(s) = (\beta \circ \alpha)(s)$.

11 (12345). $(a_1 a_2 \cdots a_n) = ?$

13 (a) and (c) are even.

15   If α and β are even, then $\alpha = (a_1 a_2)(a_3 a_4) \cdots (a_{4m-1} a_{4m})$ and $\beta = (b_1 b_2)(b_3 b_4) \cdots (b_{4n-1} b_{4n})$.   Thus $\alpha \circ \beta = (a_1 a_2)(a_3 a_4) \cdots (a_{4m-1} a_{4m})(b_1 b_2)(b_3 b_4) \cdots (b_{4n-1} b_{4n})$, which comprises $2n + 2m$ transpositions and is thus even.

19   The key fact is that Alt (n), for $n \neq 4$, has no proper subgroups. If ϕ: Alt $(n) \to G$ is a homomorphism, then ker $(\phi) \lhd$ Alt (n). Thus either ker $(\phi) = \{e\}$ or Alt (n). If ker $(\phi) =$ Alt (n), then ϕ (Alt $(n)) = \{e_G\}$. If ker $(\phi) = e$, then ϕ is 1-1 by Exercise 23 of Sec. 1.13.

20

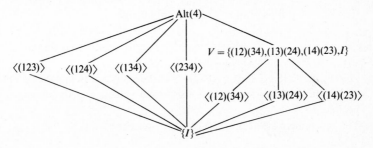

$V = \{(12)(34), (13)(24), (14)(23), I\}$

V is normal. Assume H is a subgroup of order 6. Then H is normal in Alt (4). (Why?) $H \cap V$ is normal in Alt (4). (Why?) But $H \cap V$ has order 2 and thus is one of the three subgroups of order 2, none of which is normal.

21   The key fact is that Alt (5) has no proper normal subgroups.

22 (a) $S \simeq$ Alt(3) $\times$ Alt(3)

  (b) $S =$ Sym(6)

23   Label the vertices of T as follows:

What are the 12 elements of T, written out as permutations of the vertices?

24   Let E be the set of even elements of G, and let x be an odd element. Show that $|E| = |xE|$ and that, if $\mathcal{O}$ is the set of odd elements, then $|x\mathcal{O}| = |\mathcal{O}|$. Also $xE \subseteq \mathcal{O}$ and $x\mathcal{O} \subseteq E$.

25 (b) If your answer doesn't involve least common multiples, it's incorrect.

Section 1.16

2   This f can take on four formal values.

3   We show closure. If α and β are in H so that $f(c_1, c_2, \ldots, c_n) = f(\alpha(c_1), \alpha(c_2), \ldots, \alpha(c_n)) = f(\beta(c_1), \beta(c_2), \ldots, \beta(c_n))$, then

$$f((\alpha \circ \beta)(c_1), (\alpha \circ \beta)(c_2), \ldots, (\alpha \circ \beta)(c_n)) = f(\alpha(\beta(c_1)), \alpha(\beta(c_2)), \ldots, \alpha(\beta(c_n))).$$

Before proceeding, note that $(\beta(c_1), \beta(c_2), \ldots, \beta(c_n))$ is also a generic point, and that elements of H must hold constant the value at any generic point. Thus,

$$f((\alpha \circ \beta)(c_1),(\alpha \circ \beta)(c_2),\ldots,(\alpha \circ \beta)(c_n)) = f(\beta(c_1),\beta(c_2),\ldots,\beta(c_n)) = f(c_1,c_2,\ldots,c_n),$$

and $\alpha \circ \beta$ is in H.

4 (a) If $G/Z(G)$ is cyclic, then it has a generator $aZ(G)$. Then a commutes with all elements $a^n z$, where $z \in Z(G)$. Thus a commutes with all of G, and a is in $Z(G)$, which implies $G = Z(G)$, which contradicts G being nonabelian.

(b) and (c) follow from (a).

(d) If $|G| = 15$ and G is nonabelian, then $Z(G) = e$. Thus $15 = 1 + \sum_{a \in G} |\mathrm{Cl}(a)|$, where $|\mathrm{Cl}(a)| = 3$ or 5. The only possible break-up is $15 = 1 + 3 + 3 + 3 + 5$. Show this can't happen. You may assume from Chap. 4 (Sylow theorems) that the subgroups of order 3 and 5 are normal.

5 If $\alpha = (a_1,a_2,\ldots,a_n)$ and $\beta = (b_1,b_2,\ldots,b_m)$ are disjoint cycles, then $\alpha\beta$ has order $\mathrm{lcm}(n,m)$. Thus, if γ is an element of order 15 and is written as disjoint cycles, it will require a disjoint 5-cycle and 3-cycle.

8 (a) $H = \{I, (12)(34), (13)(24), (14)(23)\} = $ is the subgroup not changing the value of f. Thus f takes on at most six formal values, and these can be easily produced.

10 (a) The remainders are $\{1,3,4,9,10,12\}$; (c) yes (why?).

CHAPTER 2

Section 2.1

2 It is true that $\binom{n}{i} + \binom{n}{i+1} = \binom{n}{i+1}$, which can be shown directly or by interpreting $\binom{n}{k}$ as the number of ways to select k objects from n objects. Now a straightforward induction on n completes the proof.

3 $\mathbb{Z}_{10}$ has no nilpotent elements other than 0. In $\mathbb{Z}_{20}$, 0 and 10 are the nilpotent elements.

4 If $X \cdot Y = X \cap Y = S$, then $X = Y = S$, so $\mathscr{U}(R) = \{S\}$.

5 $\{2,4,5,6,8\} = $ the set of zero divisors in $\mathbb{Z}_{10}$. In β we have $X \cap Y = X \cdot Y = \varnothing$ whenever X and Y are disjoint sets.

6 Assume R is an integral domain and $a \cdot b = 0$, with $a \neq 0$, $b \neq 0$. Then $a \cdot b = 0 = a \cdot 0$. Canceling a, we find that $b = 0$, which is a contradiction. We leave the converse proof to the reader.

7 $x^2 = x$ for $x \neq 0$ implies $x = 1$ by Exercise 1 of Sec. 1.2. Thus $x^2 = x$ implies $x = 0$ or $x = 1$, and these are the only two idempotents in a division ring D. The same result holds in an integral domain, since $x^2 = x$ implies $x(x - 1) = 0$.

8 The idempotents are $\{(1,1), (1,0), (0,1), (0,0)\}$.

10 For this definition of $+$ we still have $0 = \varnothing$. However, we no longer have additive inverses, for if X is a nonempty subset, then $X \cup Y = \varnothing$ can't be solved.

Section 2.2

1 The function

$$f(x) = \begin{cases} 0 & 0 \le x \le \tfrac{1}{2} \\ x - \tfrac{1}{2} & \tfrac{1}{2} \le x \le 1 \end{cases}$$

is in $C[0,1]$.

2 The function e^x has a multiplicative inverse but x^2 doesn't, since $1/x^2$ is not continuous at 0.

5 We note that $f(x)$ in $C[0,1]$ has an inverse if $1/f(x)$ is continuous. For this we need $f(x) \ne 0$ for all $x \in [0,1]$. By the intermediate-value theorem we can say that $f(x)$ is either always positive or always negative on $[0,1]$.

8 Any element of the form

is in $I(-\infty,\infty)$. The multiplicative identity has to be $f(x) = 1 \; \forall x \in \mathbb{R}$. However, $\int_{-\infty}^{\infty} f(x)\,dx = \int_{-\infty}^{\infty} 1\,dx = \infty$.

9 We demonstrate two of the axioms here and leave the rest to the reader: If $f\colon S \to R$ and $g\colon S \to R$ are two elements of R^S, then $(f+g)(s) = f(s) + g(s)$, while $(f \cdot g)(s) = f(s) \cdot g(s)$. Since $f(s)$ and $g(s)$ are elements of R, both $f(s) + g(s)$ and $f(s)g(s)$ are well-defined. Letting s vary through S will give us the new functions $f + g$ and $f \cdot g$, and we have verified both additive and multiplicative closure.

We also verify a distributive law: Let f, g, and h all be elements of R^S. We want to show that $(f + g) \cdot h = f \cdot h + g \cdot h$. For any s in S we have

$$[(f + g) \cdot h](s) = (f+g)(s) \cdot h(s) = [f(s) + g(s)] \cdot h(s)$$
$$= f(s) \cdot h(s) + g(s) \cdot h(s) \qquad \text{since the distributive}$$
$$\text{law holds in } R$$
$$= (f \cdot h)(s) + (g \cdot h)(s) = (f \cdot h + g \cdot h)(s)$$

As this holds for all s in S, we have $f \cdot h + g \cdot h = (f + g) \cdot h$.

11 Let $m = 2 + 2i$. Then $m^2 = 8i$, $m^4 = -64$, and the powers of m^4 never reach 1. Thus $\langle m \rangle$ is infinite, and $\langle m \rangle \subseteq M$.

13
$$\theta((a + bi)(c + di)) = \theta(ac - bd + (bc + ad)i)$$
$$= \sqrt{(ac - bd)^2 + (bc + ad)^2}$$
$$= \sqrt{a^2c^2 + b^2d^2 + b^2c^2 + a^2d^2}$$
$$= \sqrt{(a^2 + b^2)(c^2 + d^2)} = \sqrt{a^2 + b^2} \cdot \sqrt{c^2 + d^2}$$
$$= \theta(a + bi)\theta(c + di)$$

15 $\cos 5x + i \sin 5x = \cos^5 x + 5i \cos^4 x \sin x - 10 \cos^3 x \sin x - 10i \cos^2 x \sin^3 x$
$+ 5 \cos x \sin^4 x + i \sin^5 \lvert x$

17 If we can write $1 + i$ in polar form, we shall be able to use DeMoivre's theorem.

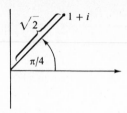

Thus $1 + i = \sqrt{2}e^{(\pi/4)i}$. We want to solve

$$z^3 = \sqrt{2}e^{(\pi/4)i} = \sqrt{2}e^{(\pi/4 + 2\pi)i} = \sqrt{2}e^{(\pi/4 + 4\pi)i}$$

Thus $z = 2^{1/6}e^{(\pi/12)i}$, $2^{1/6}e^{(9/12)\pi i}$, $2^{1/6}e^{(17/12)\pi i}$. These answers are correct though they can be put back in the other form. For instance,

$$2^{1/6}e^{(\pi/12)i} = 2^{1/6}\left(\cos\frac{\pi}{12} + i \sin\frac{\pi}{12}\right)$$

19 If u_1 and u_2 are the nth roots of unity, then $u_1{}^n = u_2{}^n = 1$, and thus $(u_1u_2)^n = u_1{}^nu_2{}^n = 1 \cdot 1 = 1$.

21

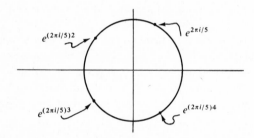

We must have

$$z^4 + z^3 + z^2 + z + 1 = (z - e^{2\pi i/5})(z - e^{4\pi i/5})(z - e^{6\pi i/5})(z - e^{8\pi i/5})$$

$$= (z - e^{2\pi i/5})(z - \overline{e^{2\pi i/5}})(z - e^{4\pi i/5})(z - \overline{e^{4\pi i/5}})$$

$$= \left(z - \cos\frac{2\pi}{5} - i \sin\frac{2\pi}{5}\right)\left(z - \cos\frac{2\pi}{5} + i \sin\frac{2\pi}{5}\right)$$

$$\left(z - \cos\frac{4\pi}{5} - i \sin\frac{4\pi}{5}\right)\left(z - \cos\frac{4\pi}{5} + i \sin\frac{4\pi}{5}\right)$$

$$= \left[\left(z - \cos\frac{2\pi}{5}\right)^2 + \sin^2\frac{2\pi}{5}\right]\left[\left(z - \cos\frac{4\pi}{5}\right)^2 + \sin^2\frac{4\pi}{5}\right]$$

23 What does raising something to the ith power or $\sqrt{-1}$th power mean? Alternatively, if $f(z) = z^i$, is f a function? If a proper definition of z^i is given, then multiple values are allowed.

Section 2.3

1 Both H and K are subrings.

3 $M_2(S)$ is the 2×2 matrices over a ring S. We see then that (for example)

$$
\left(\begin{pmatrix} a & b \\ c & d \end{pmatrix} + \begin{pmatrix} e & f \\ g & h \end{pmatrix} \right) \begin{pmatrix} l & m \\ p & q \end{pmatrix} = \begin{pmatrix} a+e & b+f \\ c+g & d+h \end{pmatrix} \begin{pmatrix} l & m \\ p & q \end{pmatrix}
$$

$$
= \begin{pmatrix} (a+e)l + (b+f)p & (a+e)m + (b+f)q \\ (c+g)l + (d+h)p & (c+g)m + (d+h)q \end{pmatrix}
$$

$$
= \begin{pmatrix} al+bp & am+bq \\ cl+dp & cm+dq \end{pmatrix} + \begin{pmatrix} el+fp & em+fq \\ gl+hp & gm+hq \end{pmatrix}
$$

$$
= \begin{pmatrix} a & b \\ c & d \end{pmatrix} \begin{pmatrix} l & m \\ p & q \end{pmatrix} + \begin{pmatrix} e & f \\ g & h \end{pmatrix} \begin{pmatrix} l & m \\ p & q \end{pmatrix}
$$

5 $R_1 = M_2(\mathbb{Z}_2)$ has 16 elements, since we have two choices at each of four locations. $|R_n| = 2^{4^n}$. (Why?)

7 $\mathcal{U}(M_2(\mathbb{Z}_2))$ is the set of nonsingular elements of $M_2(\mathbb{Z}_2)$. Thus,

$$
\mathcal{U}(M_2(\mathbb{Z})) = \left\{ \begin{pmatrix} 1 & 0 \\ 0 & 1 \end{pmatrix}, \begin{pmatrix} 1 & 1 \\ 0 & 1 \end{pmatrix}, \begin{pmatrix} 1 & 0 \\ 1 & 1 \end{pmatrix}, \begin{pmatrix} 0 & 1 \\ 1 & 0 \end{pmatrix}, \begin{pmatrix} 0 & 1 \\ 1 & 1 \end{pmatrix}, \begin{pmatrix} 1 & 1 \\ 1 & 0 \end{pmatrix} \right\}
$$

This is a noncommutative group (why?) of order 6, and thus isomorphic to Sym (3).

8 $|\mathrm{GL}\,(2,\mathbb{Z}_p)|$ can be determined by noting that $ad - bc = 0$ means that $(ka,kb) = (c,d)$ for some k in $\mathbb{Z}_p$ if $(a,b) \neq (0,0)$. Thus we first choose (a,b) to be not equal to $(0,0)$, and there are $p^2 - 1$ such choices. Then, to avoid $ad - bc = 0$, we avoid the p multiples (ka,kb). This leaves $p^2 - p$ choices, and $|\mathrm{GL}\,(2,\mathbb{Z}_p)| = (p^2 - 1)(p^2 - p)$.

11 (a) Let f_r and f_s be elements of $\bar{R}$. Then, for $f_r + f_s$ we have $(f_r + f_s)(x) = f_r(x) + f_s(x) = rx + sx = (r + s)x = f_{r+s}(x)$. Therefore, $f_r + f_s = f_{r+s}$. The additive identity is f_0, and f_{-r} is the additive inverse for f_r. Similarly, $f_r \circ f_s = f_{rs}$, so by Proposition 1 we have a ring.

13 Again Proposition 1 is the key. We earlier showed that matrix multiplication corresponds to the composition of functions.

15 (b) $f_{(x_1, y_1)}(X, Y) = (X_1 X, \ Y_1 Y) = (X, Y) \begin{pmatrix} X_1 & 0 \\ 0 & Y_1 \end{pmatrix}$, so $f_{(x_1, y_1)}$ corresponds to

$$
\begin{pmatrix} X_1 & 0 \\ 0 & Y_1 \end{pmatrix}.
$$

21 $-i = (ij)j = rj + sij + tj^2 = rj + s(r + si + tj) + tj^2 = (sr - t) + s^2 i + (st + r)j$. Thus $-1 = s^2$ by equating the coefficients of i.

Section 2.4

2 $0 \times \mathbb{Q} = \{(0,x) \,|\, x \in \mathbb{Q}\}$ is an ideal of $\mathbb{Q} \times \mathbb{Q}$. $S_1 = \{(x,x) \,|\, x \in \mathbb{Q}\}$ and $S_2 = \{(0,2n) \,|\, n \in \mathbb{Z}\}$ are subrings but not ideals. Why?

5 We proceed by induction on n. First $(a+b)^1 = \binom{1}{0}a^1b^0 + \binom{1}{1}a^0b^1 = a+b$. We assume $(a+b)^k = \sum_{i=0}^{k} \binom{k}{i}a^ib^{k-i}$ and examine $(a+b)^{k+1}$.

$$(a+b)^{k+1} = (a+b)\sum_{i=0}^{k} \binom{k}{i}a^ib^{k-i}$$

$$= \sum_{i=0}^{k} \binom{k}{i}a^{i+1}b^{k-i} + \sum_{i=0}^{k} \binom{k}{i}a^ib^{k+1-i} \quad \text{since the ring is commutative}$$

$$= \sum_{j=1}^{k+1} \binom{k}{j-1}a^jb^{k+1-j} + \sum_{i=0}^{k} \binom{k}{i}a^ib^{k+1-i}$$

$$= b^{k+1} + \sum_{l=1}^{k} \left[\binom{k}{l-1} + \binom{k}{l}\right]a^lb^{k+1-l}$$

$$= b^{k+1} + \sum_{l=1}^{k} \binom{k+1}{l}a^lb^{k+1-l} \quad \text{since } \binom{k}{l-1} + \binom{k}{l} = \binom{k+1}{l}$$

$$= \sum_{l=0}^{k+1} \binom{k+1}{l}a^lb^{k+1-l}$$

9 $-6 = 1 - 7 = (1 + \sqrt{7})(1 - \sqrt{7}) = \phi(1 + \sqrt{3})\phi(1 - \sqrt{3})\phi(1-3) = \phi(-2) = -2$. Thus ϕ is not a ring homomorphism.

10 $(a,b)\begin{pmatrix} -b & -b \\ a & a \end{pmatrix} = (0,0) = (ra,rb)\begin{pmatrix} -b & -b \\ a & a \end{pmatrix}$. If $(a,b) = (0,0)$, what is a suitable M?

14 $\mathbb{Z}_p$ is a commutative ring with 1, so all we need to establish is that every nonzero element is a unit. We shall show that $\mathbb{Z}_p$ has no zero divisors, and by Exercise 13 this will suffice. Assume $\bar{a}\bar{b} = \bar{0}$ in $\mathbb{Z}_p$ with $\bar{a} \neq \bar{0} \neq \bar{b}$; then $ab \equiv 0 \pmod{p}$ or $p \mid ab$. Since p is a prime, this means $p \mid a$ or $p \mid b$. Alternatively, $\bar{a} = \bar{0}$ or $b = \bar{0}$, and we have a contradiction.

Alternative proof: $\bar{a} \neq \bar{0} \Rightarrow \gcd(a,p) = 1 \Rightarrow ma + np = 1 \Rightarrow \bar{a}^{-1} = \bar{m}$.

16 The kernel of ϕ is $\mathbb{Z}$. {Why? What is the graph of $f(x) = [x]$?} However, the only ideals in $\mathbb{R}$ are (0) and $\mathbb{R}$. Thus $\mathbb{Z}$ is not an ideal, and $f(x) = [x]$ is not a ring homomorphism. The reader should also supply a more direct proof.

Section 2.5

1 The maximal ideals in $2\mathbb{Z}$ are the $2p\mathbb{Z}$, where p is prime. $\mathbb{Z}_{30}$ contains three maximal ideals, while $\mathbb{Q}_p$ has but one.

2 (a) Let $xr + i$ be an element of $\langle x,I \rangle$, and let r' be any element in R. Then $(xr + i)r' = xrr' + ir'$, which has the correct form since ir' is in I. The rest is even easier.

(b) We could now define $\langle x,I \rangle = \{r_1xr_2 + i \mid r_1,r_2 \in R, i \in I\}$.

3 (a) We wish to show that $(r_1 + I)(r_2 + I) = (r_2 + I)(r_1 + I)$ for all r_1 and r_2 in R. We know $r_1r_2 - r_2r_1 \in I$, so $r_1r_2 - r_2r_1 + I = I$ or $r_1r_2 + I = r_2r_1 + I$. Thus $(r_1 + I)(r_2 + I) = r_1r_2 + I = r_2r_1 + I = (r_2 + I)(r_1 + I)$.

(c)   We can strengthen the theorem by specifying only that R be a ring with 1. Then, to make the quotient ring commutative, we specify that $M \supseteq I = \langle r_1 r_2 - r_2 r_1 | r_1 r_2 \in \mathbb{R} \rangle$. For an example, we could have

$$R = \left\{ \begin{pmatrix} a & b \\ 0 & c \end{pmatrix} \middle| a, b, c \in \mathbb{R} \right\}$$

and $M = \left\{ \begin{pmatrix} a & b \\ 0 & 0 \end{pmatrix} \middle| a, b \in \mathbb{R} \right\}$. What is I?

6   Let w be any complex number. Then $M_w = \{f(z) | f(w) = 0\}$ is a maximal ideal of $\mathbb{C}(z)$. Since over $\mathbb{C}$ we have $f(z) = a_0(z - w_1)(z - w_2) \cdots (z - w_n)$, we have $M_w = \{f(z) | (z - w) \cdot g(z) = f(z)\} = \langle z - w \rangle$.

8   (a) I_1 is prime; I_2 isn't.
    (c) I_1 is prime. Since $(0,1)(3,0) = (0,0)$ without $(0,1) = (0,0)$ or $(3,0) = (0,0)$, we see that $I_2 = \{(0,0)\}$ is not prime.

10   If P is a prime ideal in R (a commutative ring with 1), then we want to show R/P has no zero divisors. Assume $(a + P)(b + P) = 0 + P$, where $a + P \neq 0 + P$, and $b + P \neq 0 + P$. Thus $ab + P = P$, so $ab \in P$, which implies $a \in P$ or $b \in P$ and, in turn, that $a + P = P$ or $b + P = P$, which is the desired contradiction.

13   Let D be a finite integral domain. Let x be any nonzero element of D. The function $f_x : D \to D$ given by $f_x(a) = ax$ is 1-1, since the cancellation laws hold. Since D is finite, f_x is also onto, and $f_x(y) = 1$ has a solution in D. This means that y exists, so that $xy = 1$, and thus $y = x^{-1}$ exists. (This yields another proof that $\mathbb{Z}_n$ is a field when n is a prime.)

CHAPTER 3

Section 3.1

1   We verify one of the distributive laws. In this context, $f(v_1 + v_2)$ becomes

$$f((a_1 x^2 + b_1 x + c_1) + (a_2 x^2 + b_2 x + c_2))$$
$$= f((a_1 + a_2)x^2 + (b_1 + b_2)x + (c_1 + c_2))$$
$$= f(a_1 + a_2)x^2 + f(b_1 + b_2)x + f(c_1 + c_2)$$
$$= fa_1 x^2 + fb_1 x + fc_1 + fa_2 x^2 + fb_2 x + fc_2$$
$$= f(a_1 x^2 + b_1 x + c_1) + f(a_2 x^2 + b_2 x + c_2)$$

which is $fv_1 + fv_2$ for $P_2(F)$.

2   We define $g + h$ by $(g + h)(x) = g(x) + h(x)$ for all $x \in [0,1]$. Additive closure is given by $(g + h)' = g' + h'$ or " the derivative of the sum equals the sum of the derivatives." The rest of the axioms are similar.

4   (a) and (c) are subspaces. For instance, for (a) we can take $v_1 = (a_1, b_1, c_1)$ and $v_2 = (a_2, b_2, c_2)$, where $a_1 + b_1 + c_1 = a_2 + b_2 + c_2 = 0$. Then

$$v_1 + v_2 = (a_1 + a_2, b_1 + b_2, c_1 + c_2)$$

is also in S, since $a_1 + a_2 + b_1 + b_2 + c_1 + c_2 = 0$. Also, $\alpha v_1 = (\alpha a_1, \alpha b_1, \alpha c_1)$ is in S, since $\alpha a_1 + \alpha b_2 + \alpha c_2 = \alpha(a_1 + b_1 + c_1) = 0$.

6 (b) is not a subspace. For an example, we note that $(\frac{1}{2}, 0, 0, 0, \ldots)$ is in S while $3(\frac{1}{2},0,0,0,\ldots) = (\frac{3}{2},0,0,0,\ldots)$ is not in S.

8 In the next section we shall show that $-a = -1 \cdot a$, where -1 is from F, and $-a$ is the additive inverse for a. Since $(ab)^{-1} = b^{-1}a^{-1}$ in an arbitrary group, we have $-(a+b) = -b+(-a)$ for $(V,+)$. Then $-b+(-a) = -(a+b) = -1(a+b) = -1 \cdot a + (-1 \cdot b)$. If we first add b to the right and then add a to the right, we have $-b+(-a)+b+a = 0$. Then we add b and a successively on the left to obtain $b + a = a + b$.

9 The key is that for a positive integer n we define $n \cdot a = a + a + \cdots + a$ (n copies). If m is negative, we say $m \cdot a = -(-m \cdot a)$, where $-m \cdot a$ is defined above. We also define $0 \cdot a = 0_A$.

11 Define $g + h$ by $(g + h)(s) = g(s) + h(s)$, and $\alpha \cdot g$ by $(\alpha \cdot g)s = \alpha g(s)$. We want to verify the distributive law $\alpha \cdot (v_1 + v_2) = \alpha \cdot v_1 + \alpha \cdot v_2$. For any s in S, we have

$$(\alpha \cdot (g + h))(s) = \alpha((g + h)(s)) = \alpha(g(s) + h(s))$$

$$= \alpha g(s) + \alpha h(s) = (\alpha \cdot g)(s) + (\alpha \cdot h)(s)$$

Thus $\alpha(g + h) = \alpha \cdot g + \alpha \cdot h$.

Section 3.2

1 $-\alpha \cdot v + \alpha \cdot v = (-\alpha + \alpha) \cdot v = 0_F \cdot v = 0_V$ by part a. Thus, in the group $(V,+)$, we have that $-\alpha \cdot v$ is an inverse of $\alpha \cdot v$. Therefore, $-\alpha \cdot v = -(\alpha \cdot v)$.

3 (a) $v + w = v = v + 0_V$, and then the left cancellation law yields $w = 0_V$.
(d) Use Exercise 1, Sec. 1.3.

5 The span of S is $\{a_0 \cdot 1 + a_1 \cdot x^2 + a_2 \cdot x^4 + \cdots + a_n \cdot x^{2n} | a_i \in F\}$, which is the set of all polynomials in x^2.

8 $0 = a_0 \cdot 1 + a_1 \cdot x^2 + a_2 \cdot x^4 + \cdots + a_n \cdot x^{2n}$. The only way this can happen (using our definition of polynomials in Sec. 2.6) is for $a_0 = a_1 = a_2 = \cdots = a_n = 0$, which shows S to be linearly independent. (We could also use Exercise 9 for a slightly different proof). If we take polynomials over $\mathbb{R}$ and consider polynomials as functions, then we can proceed as follows: First take $x = 0$ so that $a_0 = 0$. Then divide out an x^2 so that we obtain $a_1 + a_2 \cdot x^2 + \cdots + a_n \cdot x^{2n-2} = 0$. Again set $x = 0$ so that $a_1 = 0$. Keep repeating until $0 = a_0 = a_1 = a_2 = \cdots = a_n$.

10 If $\{1, \sqrt{2}\}$ is a basis, then the dimension of F over $\mathbb{Q}$ is 2. But $\{1, \sqrt{2}\}$ is a spanning set, since $a + b\sqrt{2} = a \cdot 1 + b \cdot \sqrt{2}$ for all $a + b\sqrt{2}$ in F. If $a + b\sqrt{2} = 0$ with a and b in $\mathbb{Q}$, then $\sqrt{2} = -a/b$ in $\mathbb{Q}$, which is a contradiction of the irrationality of $\sqrt{2}$ unless $b = 0$. If $b = 0$, then $a = 0$, and $\{1, \sqrt{2}\}$ is linearly independent.

12 First we show that $S = \{(1,0,0), (0,1,0), (0,0,1)\}$ is a spanning set. A typical element of $\mathbb{R}^3$ is (a,b,c). But $(a,b,c) = a(1,0,0) + b(0,1,0) + c(0,0,1)$, and S is a spanning set. If $a(1,0,0) + b(0,1,0) + c(0,0,1) = (0,0,0) = 0_{\mathbb{R}^3}$, then $(a,b,c) = (0,0,0)$, so $a = b = c = 0$. Thus S is linearly independent as well as a spanning set. Hence S is a basis having three elements, and, since $|S| = 3$, $\mathbb{R}^3$ has dimension 3 over $\mathbb{R}$.

Section 3.3

1 Let $v_1 = (a_1,b_1)$ and $v_2 = (a_2,b_2)$, and check that $T(v_1 + v_2) = T(v_1) + T(v_2)$ while $T(cv_1) = cT(v_1)$.

3 $\mathbb{R}$ is a vector space over any subfield of $\mathbb{R}$, and by the way we defined T it looks like $\mathbb{Q}$ is the subfield to choose.

4 In part *a*, T is a linear transformation. For part *c* we let $v_1 = (2,0,0)$ and $v_2 = (-3,0,0)$. Then $T(v_1) = (2,0,0)$, $T(v_2) = (|-3|,0,0) = (3,0,0)$, but $T(v_1 + v_2) = T(-1,0,0) = (1,0,0)$. Thus T is not a linear transformation.

5 The function in (*a*) is a linear transformation, unlike the function in (*c*).

8 (*a*) A typical element of W is $\sum \alpha_i c_i$, so that $T(\sum \alpha_i b_i) = \sum \alpha_i c_i$, and thus T maps V onto W.

11 Assume T is a ring homomorphism from $\mathbb{Q}(\sqrt{2})$ to $\mathbb{Q}(\sqrt{3})$. Then $T(1) = 1$, since a ring homomorphism preserves multiplicative identities (correspondence theorem, part *g*). We then have that $T(2) = T(1+1) = T(1) + T(1) = 1 + 1 = 2$. Then $T(\sqrt{2}) = q_1 + q_2\sqrt{3}$, so

$$2 = T(2) = T(\sqrt{2})T(\sqrt{2}) = (q_1 + q_2\sqrt{3})^2 = q_1{}^2 + 2q_1q_2\sqrt{3} + 3q_2{}^2$$

which contradicts the fact that $\sqrt{3}$ is irrational.

13 T is 1-1 if and only if $\ker(T) = \{0\}$. But $\dim[\ker(T)] = 0$ if and only if $\ker(T) = \{0\}$.

17 Here V is the vector space of all polynomials of degree 4 or less, so that $\dim(V) = 5$ since $\{1,x,x^2,x^3,x^4\}$ is a basis of V. T is differentiation. The only polynomials whose derivative is 0 are the constants, so $\ker(T) = \{c \mid c \in \mathbb{R}\}$, and the nullity of T is 1. However, T takes $P_4(\mathbb{R})$ onto $P_3(\mathbb{R})$, so the rank of T is 4.

Section 3.4

1 The B''-representation for $(1,1,1)$ is $(\tfrac{1}{2}, \tfrac{1}{2}, \tfrac{1}{2})$, whereas the B'-representation for $(1,1,1)$ is $(0,0,1)$.

2 *Definition*: Let B be a basis, and let B be ordered so that $B = (b_1,b_2,b_3,...)$. Then we can write any v in V uniquely as $v = \alpha_1 b_1 + \alpha_2 b_2 + \alpha_3 b_3 + \cdots$. If this is done, the *B-representation* of v is $(\alpha_1,\alpha_2,\alpha_3,...)$.

6
$$T(b_1) = T(1,0,0) = (0,1,0) = 0(1,1,1) + 1(1,1,0) - 1(1,0,0)$$
$$T(b_2) = T(0,1,0) = (1,0,0) = 0(1,1,1) + 0(1,1,0) + 1(1,0,0)$$
$$T(b_3) = T(0,0,1) = (0,0,-1) = -1(1,1,1) + 1(1,1,0) + 0(1,0,0)$$

Thus,
$$M_T = \begin{pmatrix} 0 & 1 & -1 \\ 0 & 0 & 1 \\ -1 & 1 & 0 \end{pmatrix}$$

8 (*a*) $T_a(k_1 + k_2) = (k_1 + k_2)a = k_1 a + k_2 a = T_a(k_1) + T_a(k_2)$, since k_1, k_2, and a are all in the field K. Similarly, $T_a(fk) = (fk)a = f(ka) = fT_a(k)$ $\forall f$ in F, since f, k, and a are all in the field K.

(*b*) Let $\{1,i\}$ be a basis for $\mathbb{C}$ over $\mathbb{R}$. Then $T_{2i+3} = \begin{pmatrix} 3 & 2 \\ -2 & 3 \end{pmatrix}$.

10 (a) $M_I = \begin{pmatrix} 1 & 2 & 3 \\ 0 & 1 & 0 \\ 0 & 0 & 3 \end{pmatrix}$.

11 If we rotate $(1,0)$ $90°$ counterclockwise, we obtain $(0,1)$, so $T_a(1,0) = (0,1) = 0(1,0) + 1(0,1)$. Similarly, $(0,1)$ goes to $(-1,0)$, and $T_a(0,1) = (-1,0) = -1(1,0) + 0(0,1)$. Thus $M_a = \begin{pmatrix} 0 & 1 \\ -1 & 0 \end{pmatrix}$.

14 T_c takes $(1,0)$ to itself, and $(0,1)$ to $(0,0)$. Where does T_c take (a,b)? Thus, $T_c(1,0) = (1,0) = 1(1,0) + 0(0,1)$, and $T_c(0,1) = (0,0) = 0(1,0) + 0(0,1)$.

17 AB is undefined, as is CD. $AC = \begin{pmatrix} 10 & 13 \\ 12 & 15 \end{pmatrix}$; $DC = (45, 48)$.

19 Let b_i be any basis vector in B_V. Then $T(b_i) = \sum_{k=1}^{m} \alpha_{ik} c_k$, where $B_W = (c_1, c_2, \ldots, c_n)$, and $S(T(b_i)) = S(\sum_{k=1}^{m} \alpha_{ik} c_k) = \sum_{k=1}^{m} \alpha_{ik} S(c_k)$. However, $S(c_k) = \sum_{j=1}^{r} \beta_{kj} d_j$. Therefore, we have

$$ST(b_i) = \sum_{k=1}^{m} \alpha_{ik} \sum_{j=1}^{r} \beta_{kj} d_j = \sum_{k=1}^{m} \sum_{j=1}^{r} \alpha_{ik} \beta_{kj} d_j = \sum_{j=1}^{r} \sum_{k=1}^{m} \alpha_{ik} \beta_{kj} d_j$$

However, $M_T = (\alpha_{ij})$ and $M_S = (\beta_{ij})$, and the (i,j)th entry of $M_T M_S$ is $\sum_{k=1}^{m} \alpha_{ik} \beta_{kj}$. Thus the ith row of the matrix representing ST is the same as the ith row of $M_T M_S$, and we now just let $i = 1, 2, 3, \ldots$.

CHAPTER 4

Section 4.1

1 All lines through the origin except the vertical one can be written as $\{(x,kx)\,|\,x \in \mathbb{R}\}$, where k is the slope of the line. The vertical line is $\{(0,x)\,|\,x \in \mathbb{R}\}$. Since $(x,kx)\begin{pmatrix} a & b \\ c & d \end{pmatrix} = (ax + ckx, \; bx + dkx)$, we see that $\begin{pmatrix} a & b \\ c & d \end{pmatrix}$ takes the line $\{(x,kx)\,|\,x \in \mathbb{R}\}$ to $\{((a + ck)x, \; (b + dk)x)\,|\,x \in \mathbb{R}\}$, which is the line

$$\{(x, \, [(b + dk)/(a + ck)]x)\,|\,x \in \mathbb{R}\}$$

unless $a + ck = 0$. We thus need only show

$$\left((x, kx)\begin{pmatrix} a & b \\ c & d \end{pmatrix}\right)\begin{pmatrix} e & f \\ g & h \end{pmatrix} = (x, kx)\left[\begin{pmatrix} a & b \\ c & d \end{pmatrix}\begin{pmatrix} e & f \\ g & h \end{pmatrix}\right]$$

and

$$(x, kx)I = (x, kx)$$

and make the appropriate conclusions about $\{(x,kx)\,|\,x \in \mathbb{R}\}$. Then, to conclude, check out $\{(0,x)\,|\,x \in \mathbb{R}\}$.

3 (b) $G_l = \left\{ \begin{pmatrix} 1 & 0 \\ 0 & 1 \end{pmatrix}, \begin{pmatrix} -1 & 0 \\ 0 & -1 \end{pmatrix} \right\}$.

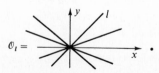

$\mathcal{O}_l =$

4   $G_5 = \left\{ I, \begin{pmatrix} 1 & 2 & 3 & 4 & 5 & 6 & 7 \\ 3 & 4 & 1 & 2 & 5 & 6 & 7 \end{pmatrix} \right\}$, $\mathcal{O}_5 = \{5,6\}$.

6   θ is a function from matrices of the form $\begin{pmatrix} a & b \\ 0 & c \end{pmatrix}$ to a group of functions on $\mathbb{R}$.
The operation for this group of functions is composition, and the identity I is the function such that $I(x) = x$ for all x in $\mathbb{R}$. Since

$$\begin{pmatrix} a & b \\ 0 & c \end{pmatrix} \begin{pmatrix} e & f \\ 0 & g \end{pmatrix} = \begin{pmatrix} ae & af + bg \\ 0 & ef \end{pmatrix}$$

and the composition of $(ax + b)/c$ and $(ex + f)/g$ is

$$\frac{a((ex + f)/g) + b}{c} = \frac{(ae)x + (af + cg)}{cg}$$

we see that $\theta(f)\theta(g) = \theta(f \circ g)$. Also, $\theta\begin{pmatrix} 1 & 0 \\ 0 & 1 \end{pmatrix} = $ the function such that x goes to

$(1 \cdot x + 0)/1 = x$. Thus, $\theta\begin{pmatrix} 1 & 0 \\ 0 & 1 \end{pmatrix} = I$. And ker (θ) turns out to be

$$\left\{ \begin{pmatrix} a & 0 \\ 0 & a \end{pmatrix} \middle| a \neq 0 \right\}$$

(Why?) $G_0 = \left\{ \begin{pmatrix} a & b \\ 0 & c \end{pmatrix} \right\}$ such that $(a \cdot 0 + b)/c = 0$, which is equivalent to $b = 0$,

so $G_0 = \left\{ \begin{pmatrix} a & 0 \\ 0 & c \end{pmatrix} \right\}$. In addition, $\mathcal{O}_0 = \mathbb{R}$. (Why?)

8   First note that $g^{-1}(t) = s$. Thus,

$$f \in G_s \Leftrightarrow f(s) = s \Leftrightarrow gf(s) = g(s) = t$$
$$\Leftrightarrow gf(g^{-1}(t)) = t = gfg^{-1}(t)$$
$$\Leftrightarrow gfg^{-1} \in G_t$$
$$\Leftrightarrow f \in g^{-1}G_t g$$

9   The correct answer is 12. The main cause for error in this problem is the substitution of a pyramid

for a regular tetrahedron. What would be the correct answer for the pyramid?

Section 4.2

2   Here we have $\theta(g) = f_g$.

$$\ker(\theta) = \{g \mid f_g = I\}$$
$$= \{g \mid f_g(x) = I(x) = x \qquad \forall x \text{ in } G\}$$
$$= \{g \mid gxg^{-1} = x \; \forall x \text{ in } G\} = Z(G)$$

4 (*b*) $Z(P)$ must have order 1, p, p^2, or p^3. Since P is nonabelian, p^3 is not a possibility. Proposition 1 eliminates 1 from the list. Were $|Z(P)| = p^2$, then $P/Z(P)$ would have order p and thus would be cyclic. Proposition 3 excludes this possibility, and we conclude that $|Z(P)| = p$.

6 For any group, $\{e\}$ is a conjugate class. Thus, if G has one conjugate class, then $G = \{e\}$. We start with the case that G has three conjugate classes. By the class equation, we have $|G| = \sum_{i=1}^{3} |\text{Cl}(a_i)| = |\{e\}| + |\text{Cl}(a_2)| + |\text{Cl}(a_3)|$. Thus $|G| = 1 + k + m$, where k and m divide $|G|$. The end result is that $|G| = 3$ or 6, with $G \simeq (\mathbb{Z}_3, \oplus)$ or Sym (3).

9 We can show that the interchange of any two rows or columns in the informal diagram does yield a collineation. Thus Coll (P,L) takes a to all the other eight points, and $|\text{Coll}(P,L)| = 9|G_a|$. We then look at b under G_a. Eventually we find that Coll (P,L) has order $9 \cdot 8 \cdot 6$.

10 $|G| = 168$.

Section 4.3

2 If we call a 60° rotation ρ, then the rotations are I, ρ, ρ^2, ρ^3, ρ^4, and ρ^5. Let a reflection through an opposite pair of vertices

be called v. Then v, $v\rho^2$, and $v\rho^4$ are the three reflections through opposite pairs of vertices. The other three reflections, $v\rho$, $v\rho^3$, and $v\rho^5$, are reflections about a line bisecting two opposite sides, such as

Thus we have the following:

Elements	Order	Description
I	1	0° rotation
ρ, ρ^5	6	60° rotation
ρ^2, ρ^4	3	120° rotation
ρ^3	2	180° rotation
$v, v\rho^2, v\rho^4$	2	vertex-vertex reflection
$v\rho, v\rho^3, v\rho^5$	2	reflection about an edge-edge bisector

3 $\chi(I) = 3^6$; $\chi(\rho) = \chi(\rho^5) = 3$; $\chi(\rho^2) = \chi(\rho^4) = 3^2$; $\chi(\rho^3) = 3^3$; $\chi(v) = \chi(v\rho^2) = \chi(v\rho^4) = 3^4$; $\chi(v\rho) = \chi(v\rho^3) = \chi(v\rho^5) = 3^3$. Thus the number of orbits is $\frac{1}{12}(3^6 + 2 \cdot 3 + 2 \cdot 3^2 + 3^3 + 3 \cdot 3^4 + 3 \cdot 3^3) = 92$. For instance, $\chi(\rho^2) =$ number of molecules left fixed by a 120° rotation. This yields all molecules of the form

where a and b can each be any of the three choices.

5   Label the triangular prism as follows:

Then the elements of G are I, $(123)(abc)$, $(132)(acb)$, $(1b)(2a)(3c)$, $(1c)(3a)(2b)$, and $(1a)(2c)(3b)$. $\chi(I) = 6^5$. Why? $\chi((123)(abc)) = 6^3$. Why? Also, $\chi((1b)(2a)(3c)) = 6^3$. The answer is 1,476.

6   Here the group is the cyclic group of order 2, and the answer is 528.

10 Assume G is doubly transitive on S, and we wish to show that G_t is transitive on $S - \{t\}$. Pick s_1 and s_2 in $S - \{t\}$. Then there is some element g in G taking (t,s_1) to (t,s_2). Rephasing, we have $g(t) = t$ and $g(s_1) = s_2$. Thus g is in G_t, and G_t is transitive on $S - \{t\}$, since s_1 and s_2 are arbitrary elements of $S - \{t\}$. To prove the converse, let (x_1,x_2) and (y_1,y_2) be ordered pairs of distinct elements in G. Pick g in G so that $g(x_1) = y_1$. Then $(g(x_1), g(x_2)) = (y_1, g(x_2))$. Since G_{y_1} is transitive on $S - \{y_1\}$, we have an h such that $h(y_1) = y_1$ and $h(g(x_2)) = y_2$. Thus $hg = h \circ g$ takes x_1 to y_1 and x_2 to y_2, and the result is that G is doubly transitive. This fact is useful for group characters (see Chap. 9), which is why we have given such a complete proof. Using this, you can show that GL $(2,\mathbb{R})$ is doubly transitive on the set of all lines in $\mathbb{R}^2$ through the origin.

11 The trick here is to define a new action of G on $S \times S$. We let $g(s_1,s_2) = (g(s_1), g(s_2))$, and this does define an action. Then, by Burnside's theorem, we have: number of orbits of this new action $= (1/|G|) \sum_{g \in G} \chi^*(g)$, $\chi^*(g) =$ number of (s_1,s_2) such that $g(s_1) = s_1$, $g(s_2) = s_2$, and thus $\chi^*(g) = (\chi(g))^2$. Also, an orbit of this new action corresponds to an orbit of G_t, where G_t acts on S. Why?

Section 4.4

1   Since $\bar{H} = \ker (\theta)$, $\bar{H}$ is normal G. If we had $\bar{H} \subseteq K \subseteq H$ with $K \lhd G$, then we would have $x^{-1}Kx = K \subseteq x^{-1}Hx$ for all x in H. Thus, $\bar{H} = \bigcap_x x^{-1}Hx \supseteq K \supseteq \bar{H}$, and $K = \bar{H}$.

2   Cayley's theorem says that G is isomorphic to a subgroup of Sym $(2m)$. If a is an element of G, then $\theta(a)$ is the permutation $(e,a,a^2,...)(g,ag,a^2g,...)(h,ah,a^2h,...) \cdots$. G has an element of order 2, by Exercise 4 of Sec. 1.3. Show that this element yields an odd permutation.

4   (a)   Let H have order 15, 20, or 30; since there are no normal subgroups, $\bar{H} = \ker (\theta)$ has order 1. Thus θ is a monomorphism. $S = \{$the cosets of $H\}$ has order 4, 3, or 2, so θ takes Alt (5) to Sym (4), Sym (3), or Sym (2), and there is no way to do this with a monomorphism. The case $|H| = 30$ is even easier, since, if H existed, it would have to be normal, being of index 2.

5   The first Sylow theorem gives us a subgroup H of order 5^4. If $|\bar{H}| = 1 = |\ker (\theta)|$, then we would have a monomorphism from G into Sym (2^4), but $|G| \nmid |$Sym $(2^4)|$.

8   We want to show $(a+b)^n = \sum_{k=0}^{n} \binom{n}{k} a^{n-k} b^k$. You may use

$$\binom{n}{k} + \binom{n}{k+1} = \binom{n+1}{k+1},$$

where $\binom{n}{k} = \dfrac{n!}{k!(n-k)!}$, and proceed by induction on n.

(b) $(a+b)^p = \sum_{k=0}^{p} \binom{p}{k} a^{p-k} b^k = a^p + 0 + 0 + \cdots + 0 + b^p$, since

$$\binom{p}{k} a^{p-k} b^k = 0$$

in a ring of characteristic p unless $k = 0$ or p.

10   Certainly $\bar{1} = \bar{1}^p = \theta(\bar{1})$ in $\mathbb{Z}_p$. Since $\mathbb{Z}_p$ is a commutative ring of characteristic p, we have $\theta(\bar{2}) = \theta(\bar{1} + \bar{1}) = \theta(\bar{1}) + \theta(\bar{1}) = \bar{1} + \bar{1} = \bar{2}$. Similarly, $\theta(\bar{3}) = \bar{3}$, and so on, until we have $\theta(a) = a$ for all a in $\mathbb{Z}_p$. But $a = \theta(a) = a^p$.

12   Let S be the set of all p-Sylow subgroups of G, and let Q act on S by conjugation. Since Q is a p-group, and $p \nmid |S| = 1 + kp$, we must have some Sylow subgroup fixed by Q. That is, $q^{-1}Pq = P \ \forall q \in Q$. If $x \in Q - P$, then the subgroup generated by x and P is a p-subgroup containing P properly, which contradicts the maximality of P as a Sylow subgroup. Thus $Q \subseteq P$.

13   A 2-Sylow subgroup of Sym (7) is generated by the elements (1234), (24), and (56) and is isomorphic to $D_8 \times C_2$.

15   If $p = q$, then the group is abelian. If $p < q$, then look at the subgroup of order q.

16   Let S_5 be a subgroup of order 5^2, and S_7 one of order 7^2. (Why do these exist?) Both S_5 and S_7 are normal subgroups. (Why?) $S_5 \cap S_7 = \{e\}$. Also, there are at least $5^2 \cdot 7^2$ elements of the form xy, where $x \in S_5$ and $y \in S_7$. Thus, $G \simeq S_5 \times S_7$.

CHAPTER 5

Section 5.1

1   If G_1 is a subgroup of G, and H_1 is a subgroup of H, then the subgroup $G_1 \times H_1 = \{(g_1, h_1) \mid g_1 \in G_1, h_1 \in H_1\}$. Show that if $\{G_i\}_{i=1}^{n}$ and $\{H_i\}_{i=1}^{m}$ are solvable series for G and H, then

$$G \times H = G_1 \times H_1 \supseteq G_2 \times H_1 \supseteq G_3 \times H_1 \supseteq \cdots$$

$$\supseteq \{e_G\} \times H_1 \supseteq \{e_G\} \times H_2 \supseteq \cdots \supseteq \{e_G\} \times \{e_H\}$$

is a solvable series for $G \times H$.

2   Show that G has order 27. Let

$$G_2 = \left\{ \begin{pmatrix} 1 & a & b \\ 0 & 1 & 0 \\ 0 & 0 & 1 \end{pmatrix} \middle| a, b \in \mathbb{Z}_3 \right\}$$

and show that $G_2 \triangleleft G$ and that G_2 is abelian.

3 Let θ be the canonical homomorphism from M_i' to M_i'/N. Since $M_{i+1} \lhd M_i = M_i'/N$, we have, by the correspondence theorem, that $M_{i+1}' = \theta^{-1}(M_{i+1}) \lhd \theta^{-1}(M_i) = M_i'$.

4 Were Sym (5) solvable, then Alt (5) as a subgroup of Sym (5) would be solvable also.

5 Since Alt $(4) \lhd$ Sym (4), and Sym (4)/Alt (4) is abelian, we see that $G' \subseteq$ Alt (4). Computing $x^{-1}y^{-1}xy$ for various x and y in G will show that Alt $(4) \subseteq G'$. The determination of $G'' =$ Alt $(4)'$ follows from the discussion of Alt (4) early in the section.

6 If $n = 2k$, then $D_{2n} = \{a^i b^j \mid i = 0, \ldots, 2k-1, \ j = 0,1\}$. Show $D_{2n}' = \langle a^2 \rangle$ and $D_{2n}'' = \{e\}$.

7 Assume $q \geq p$. Then G has a normal subgroup of order q, by Exercise 15 of Sec. 4.4. If you didn't read this section, just assume this result.

9 Let α be an automorphism of G. Then $\alpha(H) = H$, since H char G. Then $\alpha_{|H}$, the restriction of α to H, is an automorphism of H. (Note that $\alpha_{|H}$ need not be inner even if α is inner.) Thus $\alpha(K) = \alpha_{|H}(K) = K$, since K char H.

12 Let $f_x(g) = x^{-1}gx$, so f_x is an automorphism of G. Since $H \lhd G$ we have $f_x(H) = H$, and $f_{x|H}$ is an automorphism of H. Thus, $f_x(K) = f_{x|H}(K) = K$ for any $x \in G$. That is, $x^{-1}Kx = K$ for all x in G.

15. The key to this exercise is Exercise 10. This shows another way in which solvable is a generalization of abelian.

19 If P has order p^n, we are guaranteed a subgroup H of order p^{n-1} by the last exercise. $H \lhd P$ (why?). Thus, $P' \subseteq H$ (why?). (In fact, if $n \geq 2$, then we can guarantee that $|P'| \leq p^{n-2}$. See Chap. 9.)

Section 5.2

2 We use the division algorithm to find that $p_1 = cm_1 + r$ with $0 \leq r < m_1$.

$$\sum_{i=1}^{k} p_i g_i' - c \sum_{i=1}^{k} m_i g_i' = rg_1' + \sum_{i=2}^{k} (p_i - cm_1)g_i'$$

and the minimality of m_1 forces r to be 0.

3 Assume m_i is not divisible by m_1 so that $m_i = cm_1 + r$ with $0 < r < m_1$. Show that $\{g_1' + cg_i', g_2', g_3', \ldots, g_k'\}$ is a generating set and it represents 0 in such a way as to force a contradiction of the minimality of m_1.

6 If a has order m, and b has order n, then $mn(a + b) = 0$, so the order of $a + b$ divides mn and is finite. This shows that $T(A)$ is closed under addition.

11 Factoring gives $90,000 = 2^4 \cdot 3^2 \cdot 5^4$. There are five abelian groups of order 2^4, two of order 3^2, and five of order 5^4. This yields fifty possibilities for the group, since the group must be isomorphic to $A_2 \times A_3 \times A_5$.

12 $(\Phi(16), \cdot)$ has one element of order 1, three elements of order 2 (which are $\{7,9,15\}$), and four elements of order 4. Since there are no elements of order 8, $\Phi(16) \not\cong \mathbb{Z}_8$. Since there are elements of order 4 in $\Phi(16)$, we have $\Phi(16) \not\cong \mathbb{Z}_2 \oplus \mathbb{Z}_2 \oplus \mathbb{Z}_2$. This leaves $\Phi(16) \simeq \mathbb{Z}_4 \oplus \mathbb{Z}_2$.

16   (a)   A typical element of E_{p^n} is $(a_1,a_2,\ldots,a_n)$, and $a(a_1,a_2,\ldots,a_n)$ is defined as $(aa_1,aa_2,\ldots,aa_n)$. Since $a = 0, 1, 2, \ldots$, or $p - 1$, we see that

$$a(a_1,a_2,\ldots,a_n) = \underbrace{(a_1,a_2,\ldots,a_n) + \cdots + (a_1,a_2,\ldots,a_n)}_{a \text{ copies}}$$

Thus

$$\phi(a(a_1,a_2,\ldots,a_n))$$
$$= \phi((a_1,a_2,\ldots,a_n) + (a_1,a_2,\ldots,a_n) + \cdots + (a_1,a_2,\ldots,a_n))$$
$$= \phi(a_1,a_2,\ldots,a_n) + \phi(a_1,a_2,\ldots,a_n) + \cdots + \phi(a_1,a_2,\ldots,a_n)$$
$$= a\phi(a_1,a_2,\ldots,a_n)$$

and $\phi a = a\phi$.

    (b)   Since every homomorphism corresponds to a linear transformation, every automorphism corresponds to a nonsingular linear transformation. Fixing a basis will give us an isomorphism from the group of nonsingular linear transformations and GL $(n, \mathbb{Z}_p)$.

23   (b)   We want to show that $(1,1)$ is a generator of $\mathbb{Z}_{m_1} \oplus \mathbb{Z}_{m_2}$. If $k(1,1) = (k,k) = (0,0)$, then m_1 and m_2 both divide k. Since gcd $(m_1,m_2) = 1$, this implies that lcm $(m_1,m_2) = m_1 m_2$ divides k, and thus the smallest positive k such that $k(1,1) = (0,0)$ is $k = m_1 m_2$. Thus the order of $(1,1)$ is $m_1 m_2$, which shows that $\mathbb{Z}_{m_1} \oplus \mathbb{Z}_{m_2}$ is cyclic and thus isomorphic to $\mathbb{Z}_{m_1 m_2}$.

    (d)   Let $m_1 = p_1^{a_1} p_2^{a_2} \cdots p_u^{a_n}$, where the p_i's are distinct primes, and $\mathbb{Z}_{p_i}^{a_i}$ is the highest power involving p_i that is available. Then $m_2 = p_1^{b_1} p_2^{b_2} \cdots p_u^{b_n}$, where $\mathbb{Z}_{p_i}^{b_i}$ is the highest power involving p_i other than the factor $\mathbb{Z}_{p_i}^{a_i}$.

CHAPTER 6

Section 6.1

1   We verify a few of the steps—hopefully enough to persuade the reader to do the rest.

Step 2   We show $\sim$ is symmetric. Assume $(a,b) \sim (c,d)$ so that $ad = bc$. This implies $cb = da$ so that $(c,d) \sim (a,b)$.

Step 3   If $[a,b] = [a',b']$, then $ab' = a'b$. Also, $[a,b] + [c,d] = [ad + bc, bd]$, and $[a',b'] + [c,d] = [a'd + b'c, b'd]$. All will be well if

$$(ad + bc, bd) \sim (a'd + b'c, b'd),$$

which is true since $(ad + bc)b'd = bd(a'd + b'c)$.

Step 4   The distributive law is hardest, and it goes as follows: $[a,b]([c,d] + [e,f]) = [a,b][cf + de, df] = [a(cf + de), bdf]$. On the other hand,

$$[a,b][c,d] + [a,b][e,f] = [ac,bd] + [ae,bf] = [acbf + bdae, b^2 df]$$
$$= [ba(cf + de), bbdf]$$

However, $[x,y] = [bx,by]$, so the distributive law holds.

Step 5   The mapping $d \to [d,1]$ is 1-1 unless $[d',1] = [d,1]$, but this implies $d'1 = d1 = d' = d$. Also, $d + d' \to [d + d',1] = [1d + d'1, \ 1 \cdot 1] = [d,1] + [d',1]$, and $dd' \to [dd',1] = [dd',11] = [d,1][d',1]$. Thus the mapping is a ring monomorphism.

Step 6   If $\theta: D \to G$ is a monomorphism (inclusion being the most important case), then $\theta(d)^{-1}$ exists in G for all $d \in D^*$. Thus, let $\theta[a,b] = \theta(a) \cdot \theta(b)^{-1}$ for all $[a,b] \in F$.

2   The quotient field of $Z[i]$ is isomorphic to $\mathbb{Q}[i] = \{a + bi \,|\, a,b \in \mathbb{Q}\}$. The quotient field of $F[x]$ is the field of rational functions, that is, all $p(x)/q(x)$, where $q(x)$ is not the constant zero. Here we operate on the rational functions with the usual operations of addition and multiplication. We aren't concerned with $q(x)$ having roots, since we did not define $q(x)$ as a function. (See Sec. 2.6.)

4   If gcd $(m,n) = 1$, then there exist integers a and b such that $am + bn = 1$. Thus,

$$d = d^1 = d^{am+bn} = (d^m)^a(d^n)^b = (d'^m)^a(d'^n)^b$$
$$= (d')^{am+nb} = (d')^1 = d'$$

This goes fine, and would seem not to use the properties of an integral domain at all, until we remember that usually a or b is negative. Thus some sense must be made of d^{-k}, where k is a positive integer. However, D can be considered as a subring of its quotient field where this does make sense. (If $d = 0$, we see that $0 = d = d^m = d'^m = d'$.)

8   Let u_1 and u_2 be units in R. Then $u_1(u_2 r) = (u_1 u_2)r$ for all u_1, u_2 in $\mathscr{U}(R)$ and all r in R. Also, $1 \cdot r = r$ for all r in R. Thus $\mathscr{U}(R)$ acts on R. The equivalence relation induced by $\mathscr{U}(R)$ is $\sim$, so the orbits are the classes of associated elements.

9   The inverse of $a + bi$ in $\mathbb{C}$ is $a/(a^2 + b^2) - [b/(a^2 + b^2)]i$. If a and b are in $\mathbb{Z}$, and $a/(a^2 + b^2)$ and $b/(a^2 + b^2)$ are to be in $\mathbb{Z}$, then either $a = 0$ and $b = \pm 1$, or $b = 0$ and $a = \pm 1$. Thus $\mathscr{U}(\mathbb{Z}[i]) = \{1,-1,i,-i\}$, and the associates of $z = c + d$ are $\{z, -z = -c - di, zi = -d + ci,$ and $-zi = d - ci\}$.

10   Since we usually only look at irreducibles in domains, we might expect some strange results in $\mathbb{Z}_{12}$. The units are $\{1,5,7,11\}$. Thus $2 \sim 10 = 2 \cdot 5$, $3 \sim 9 = 7 \cdot 3$, and $4 \sim 8 = 5 \cdot 4$. The elements 4, 6, and 8 are not irreducibles, since, for instance, $6 = 2 \cdot 3$ and neither 2 nor 3 is a unit. However, 2, 3, 9, and 10 are irreducible.

12   $\mathscr{U}(F) = F^* = F - (0)$.

Section 6.2

1   The key to this exercise is the plenitude of units in a field. If $\delta(1) = n$, then, by Proposition 1, $\delta(x) = n$ for all $x \in F^*$, where F is our field. Since $n \in \mathbb{Z}^+ \cup (0)$, we have $\delta(x) = n$, and $n \le \delta(xy) = n$. Also, if $b \in F^*$, then for any $a \in F$ we can solve $a = bc$ and thus can let $r = 0$. So it makes no difference which positive integer n we choose.

3   (b) $\mathscr{U}(E_p) = \{a/b \,|\, p \nmid a,\ p \nmid b\}$.

   (c)   If $a/b \in E_p$, we can write $a/b = (a'/b)p^k$, where $k = 0,1,2,\ldots$, and $p \nmid a'$. Let $\delta(a/b) = \delta((a'/b)p^k) = k$.

7 If $u \in I$, then $uu^{-1} = 1 \in I$. (Note that this proposition could be extended to cover right or left ideals.)

8 We need only to produce an ideal that is not principal. Let $I = \langle x,y \rangle$ be the smallest ideal containing both x and y. Since each $x^i y^j$, for $i + j \geq 1$, must be in I, so must all polynomials with no constant term, which is another description of I. It is straightforward to check that I is an ideal. Assume some polynomial $f(x,y)$ generated I. Then we would have $f(x,y) | x$, so that $f(x,y)$ is either c_1 or $c_2 x$. Similarly, $f(x,y) | y$, so that $f(x,y)$ is either c_3 or $c_4 y$. Thus $f(x,y)$ is a constant, which is our desired contradiction.

9 (a) If $f(x)$ is in $I - \{0\}$, then the corresponding variety consists of a subset of the roots of $f(x)$.

   (b) If $I \subsetneqq \mathbb{C}[x]$, then $I = \langle g(x) \rangle$, since $\mathbb{C}[x]$ is a PID. Thus the variety consists precisely of the zeros of $g(x)$.

Section 6.3

2 (c) $\langle 60 \rangle \subseteq \langle 30 \rangle \subseteq \langle 6 \rangle \subseteq \langle 3 \rangle \subseteq \langle 1 \rangle = \mathbb{Z}$ is one chain. There are 12 possible chains of maximal length.

3 $\langle 2 \rangle \supsetneq \langle 4 \rangle \supsetneq \langle 8 \rangle \supsetneq \cdots \supsetneq 2^n \supsetneq \cdots$ again provides an infinite strictly descending chain of ideals. The only change is that $\langle 2^n \rangle = \{2^n a + 2^n b \sqrt{-3} \,|\, a,b \in \mathbb{Z}\}$. The hint shows that, if $I \neq (0)$, then the ideal I contains a positive integer n. But $a^2 + 3b^2 = n$ has, for a fixed n, only a finite number of solutions. Thus, n can be an element of only a finite number of ideals. Thus, if $n \in I$, then only a finite number of ideals contain I.

6 Since $a | ab/(a,b) = a\{b/(a,b)\}$, and $b | ab/(a,b)$, we see that $ab/(a,b)$ is a multiple of both a and b. Thus, $[a,b]k = ab/(a,b)$ for some integer k. Rewriting, we see that $([a,b]/a)k(a,b) = b$, so $k(a,b) | b$. Similarly, $k(a,b) | a$, and this says $k(a,b) | (a,b)$. Thus, $k | 1$ and k is a unit. Therefore, $ab \sim [a,b](a,b)$.

7 Let $k[a,b] = ab$ [k will turn out to be (a,b)]. Show that $k | a$ and $k | b$. If l also divides a and b, then show that $[k,l] | a$ and $[k,l] | b$. Thus $[k,l] = kg$. But $(a/[k,l])ga = [a,b]$, so $ga | [a,b]$. Similarly, $gb | [a,b]$. Thus, $g[a,b] | [a,b]$ and $g | 1$. Thus, $[k,l] \sim k$ and $l | k$.

11 Observe that, if k is any integer, then k^2 has the form $4n$ or $4n + 1$, depending on whether k is even or odd. Thus z^2 is odd, since if it were even, x^2 and y^2 would both have the form $4n + 1$, and z^2 can't have the form $4n + 2$. Thus, either x^2 or y^2 is odd, the other is even, and z^2 is odd. In particular, $1 + i \nmid z^2$, since $(1 + i)(1 - i) = 2 \nmid z^2$.

   If $a + bi | x + iy$ and $x - iy$, then $a + bi | (x + iy) \pm (x - iy)$, and $a + bi | 2x$ and $2y$. Thus $a + bi | x$ and $a + bi | y$ (eliminating $a + bi = 2$, $2i$, or $1 + i$, all of which would make z^2 even). Taking complex conjugates would show $a - bi | x$ and $a - bi | y$, so that $a^2 + b^2$ divides x^2, y^2, and, finally, z^2. If $p = a + bi$ divides z, then $p^2 | z^2$, and thus, by the above, $p^2 | x + iy$ or $p^2 | x - iy$. Thus,

$$z^2 = \underbrace{(x_1 + iy_1)^2 (x_2 + iy_2)^2 \cdots (x_n + iy_n)^2}_{x + iy} \underbrace{(x_1 - iy_1)^2 \cdots (x_n - iy_n)^2}_{x - iy}$$

12 We can assume x, y, and z are coprime and that x and z are odd. Factoring in $\mathbb{Z}(\sqrt{-2})$, we have $z^2 = (x + \sqrt{-2}y)(x - \sqrt{-2}y)$. We again arrive at $z^2 = \Pi_{i=1}^n p_i^2 \cdot \Pi_{i=1}^n \bar{p}_i^2$, and $(x + \sqrt{-2}y) = (u + \sqrt{-2}v)^2$. Finally, $x = u^2 - 2v^2$, $y = 2uv$, and $z = u^2 + 2v^2$.

15 Let (r_1, r_2) and (r_3, r_4) be two pairs of rationals on the unit circle. Define $(r_1, r_2)(r_3, r_4) = (r_1 r_3 - r_2 r_4, \ r_2 r_3 + r_1 r_4)$, which is the same result we would obtain by letting $(r_1, r_2) = r_1 + i r_2$. Show that this defines a group.

Section 6.4

1 $\mathbb{C}$ is a UFD.

4 If $u = a + b\sqrt{2}$ is a unit then $\theta(u) = \theta(a + b\sqrt{2}) = a^2 - 2b^2 = \pm 1$. This is a special case of *Pell's equation*. By trial and error $a = 1$ and $b = 1$ yields the smallest nontrivial solution. Show that $\{(1 + \sqrt{2})^n \mid n \in \mathbb{Z}\}$ yields all the units.

CHAPTER 7

Section 7.1

2 $\frac{1}{7} = 0.142857$ with remainder 1; thus $10^6/7 = 142,857 + \frac{1}{7}$. This yields $10^6 - 1 = 7(142,857)$ or $7 \mid 10^6 - 1$. Similarly, $13 \mid 10^{12} - 1$ and $17 \mid 10^{16} - 1$. You are welcome to try some similar computations in base b.

6 For $p = 11$, the primitive roots are 2, 6, 7, and 8. For $p = 5$, the primitive roots are 2 and 3.

7 The cyclic group $\Phi(p)$ has order $p - 1$. Thus $\Phi(p) \simeq (\mathbb{Z}_{p-1}, \oplus)$ since all cyclic groups of the same order are isomorphic. In $(\mathbb{Z}_{p-1}, \oplus)$, $\bar{k}$ is a generator if and only if $\gcd(k, p - 1) = 1$. This yields $\phi(p - 1)$ generators.

8 $\Phi(12)$ has order 4, and the elements $\{5, 7, 11\}$ each have order 2. The abelian groups of order 4 are $\mathbb{Z}_4$ and $\mathbb{Z}_2 \oplus \mathbb{Z}_2$, so $\Phi(12)$ must be isomorphic to one of these. Since $\mathbb{Z}_4$ has elements of order 4, we must have $\Phi(12) \simeq \mathbb{Z}_2 \oplus \mathbb{Z}_2$.

9

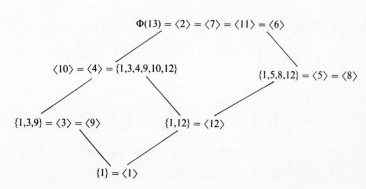

Section 7.2

1 In $\Phi(17)$ we have $R = \{1,2,4,8,9,13,15,16\}$. In $\Phi(8) = \{1,3,5,7\}$, the only residue is $1 = 1^2 = 3^2 = 5^2 = 7^2$. However, 8 is not a prime, so $\Phi(8)$ need not be cyclic.

2 If $p = 5$, 13, 17, 29, or 41, then $-\bar{1}$ is a residue modulo p. What pattern is present here? If $p = 7$, 17, 23, 31, or 41, then 2 is a residue modulo p. Here 2 is a residue for those p of the form $8k \pm 1$, but not for those of the form $8k \pm 3$.

4 $$\left(\frac{2,005}{103}\right) = -1; \left(\frac{2,020}{401}\right) = -1.$$

6 $$1 = \left(\frac{7}{p}\right) \Rightarrow \left(\frac{p}{7}\right) = \begin{cases} +1 & \text{if } p = 4k + 1 \\ -1 & \text{if } p = 4k + 1 \end{cases}$$

The residues of 7 are 1, 2, and 4. Thus, if p has the form $7k + 1$, 2, or 4, and simultaneously $4k + 1$, then 7 is a residue modulo p. If we put these together, p has the form $28k + 1$, 9, or 25. Alternatively, $p = 7k + 3$, 5, or 6 while having the form $4k + 3$ will also give 7 as a residue modulo p. These yield $p = 28k + 3$, 19, or 27. Collectively we have $p = 28k + 1$, 3, 9, 19, 25, or 27.

7 Since $9,993 = (100)^2 - 7$, we know that $p \mid 9,993$ implies 7 is a residue modulo p. Thus, by Exercise 6, we see that $p = 28k + 1$, 3, 9, 19, 25, or 27. Also, $p < \sqrt{9,993} < 100$ if p is the smallest prime dividing 9,993, assuming p factors Thus $p = 3$, 19, 29, 31, 37, 47, 53, 59, or 83. This is a shorter list of primes, but now these should be checked individually.

9 $\Phi(p)$ has order $p - 1$, which is even. If g is a primitive root modulo p, then $\Phi(p) = \{g, g^2, g^3, \ldots, g^{p-1} = 1\}$. The residues are $R = \{g^2, g^4, \ldots, g^{p-1}\}$. If g were a residue, we would have $R = \Phi(p)$, which is impossible.

10 In these last exercises we showed that every primitive root is a nonresidue. $\Phi(f)$ has $\phi(f - 1) = \phi(2^m + 1 - 1) = \phi(2^m) = 2^{m-1}$ primitive roots (cf. Exercise 4 of Sec. 7.1). We also have $(f - 1)/2 = (2^m + 1 - 1)/2 = 2^{m-1}$ nonresidues. Thus, here, every nonresidue is a primitive root. Finally,

$$\left(\frac{3}{f}\right) = \left(\frac{f}{3}\right) = \left(\frac{2^m + 1}{3}\right).$$

If $2^m + 1$ is prime, $3 \nmid 2^m + 1$, so m is even and $m = 2k$. So

$$\left(\frac{3}{f}\right) = \left(\frac{2^{2k} + 1}{3}\right) = \left(\frac{4^k + 1}{3}\right) = \left(\frac{2}{3}\right) = -1.$$

Section 7.3

1 If $1 = \left(\frac{-2}{p}\right) = \left(\frac{-1}{p}\right)\left(\frac{2}{p}\right)$ then, if $1 = \left(\frac{-1}{p}\right) = \left(\frac{2}{p}\right)$, we have $p = 4k + 1$ and $p = 8k + 1$ or 7. For $-1 = \left(\frac{-1}{p}\right) = \left(\frac{2}{p}\right)$, we have $p = 4k + 3$ as well as $p = 8k + 3$ or 5. Thus we must have either $p = 8k + 1$ or $8k + 3$.

2 Imitate closely the proof of Fermat's two-square theorem, using Exercise 1.

3 If $x = z - 2w$ and $y = w$, then $x^2 + 2y^2 = z^2 + 6w^2 - 4zw$.

4   If integers x and y exist such that $7y^2 - 1 = x^2$, then what happens in $\mathbb{Z}_7$?

5   Let $m = a^2 + b^2 = (a + bi)(a - bi)$ and $n = (c + di)(c - di)$, so that $mn = (a + bi)(c + di)(a - bi)(c - di) = (ac - bd)^2 + (ad + bc)^2$.

6   The term " opposite parity " means that if x is odd then y is even, while if x is even then y is odd.

7   Look at $ea_2 a_3 \cdots a_n$, where $A = \{e, a_2, a_3, \dots, a_n\}$, and pair each element with its inverse when possible. Multiplying will leave $eb_1 b_2 \cdots b_k$, where each b_i is its own inverse. Since $b_i = b_i^{-1} \Leftrightarrow b_i^2 = e$, we shall obtain e if there are no elements of order 2, and t if t is the unique element of order 2. Assume $b_1, b_2, \dots, b_k$ are the elements of order 2, and that $k \geq 2$. Look at $\langle b_1, b_2 \rangle = \{e, b_1, b_2, b_1 b_2\}$, which is isomorphic to Klein's four group. The product $e \cdot b_1 \cdot b_2 \cdot b_1 b_2 = e$. If $\{he, hb_1, hb_2, hb_1 b_2\}$ is a coset of $\langle b_1, b_2 \rangle$ in $\langle b_1, b_2, \dots, b_k \rangle$, then the product of the elements in this coset is likewise e. Continuing through all these cosets yields $eb_1 b_2 \cdots b_k = e = \Pi_{a \in A} a$.

8   -1 is the only element of order 2 in $\mathbb{Z}_p$, since $x^2 = 1$ can have at most two solutions over a field.

11   (a) $\Phi(29) = \{2, 2^2, 2^3, \dots, 2^{27}, 2^{28} = 1\}$; thus, the four solutions of $x^4 = 1$ are 2^7, 2^{14}, 2^{21}, and 2^{28}. Simplifying, we have $2^7 = 12$, $2^{14} = 28 = -1$, $2^{21} = -12 = 17$, and $2^{28} = 1$.

   (b) $1 + x + x^2 + \cdots + x^6 = (x^7 - 1)/(x - 1)$, so all the solutions of $x^7 - 1$ will work except $x = 1$.

CHAPTER 8

1   Division rings have no proper left ideals, and finite rings can only have a finite number of left ideals. $M_2(\mathbb{R})$ will be shown to be artinian in Exercise 6.

2   The ring $\mathbb{Z}$ is not semisimple since it does not satisfy the dcc. However, it has no nonzero nilpotent elements and, therefore, no nonzero nilpotent left ideals.

3   Assume $d^2 \mid m$ for some positive integer $d > 1$. Then $m = d^2 k$, and the elements $\overline{dk}$ are nilpotent since $m \mid (dk)^2$, and $\overline{(dk)}^2 = \overline{0}$ in $\mathbb{Z}_m$. Conversely, if $\bar{r}^k = \overline{0}$ in $\mathbb{Z}_m$, then $(\bar{r}^{k-1})^2 = 0$; so if we can show $\mathbb{Z}_m$ has no nonzero nilpotent elements with $\bar{x}^2 = \overline{0}$, then $\mathbb{Z}_m$ has no nonzero nilpotent elements. If $m \mid x^2$ in $\mathbb{Z}$ without $m \mid x$, then for some prime p we have $p^l \mid x^2$ without $p^l \mid x$; this implies $l \geq 2$. Thus $p^2 \mid m$, and this is a necessary and sufficient condition for Z_m to have nilpotent elements.

4   $M_2(\mathbb{R})$ is semisimple since it has no nilpotent left ideals. For instance, if $\begin{pmatrix} a & b \\ c & d \end{pmatrix} \in L$, where L is a left ideal, then $\begin{pmatrix} 1 & 1 \\ 0 & 0 \end{pmatrix} \begin{pmatrix} a & b \\ c & d \end{pmatrix} = \begin{pmatrix} a & b \\ 0 & 0 \end{pmatrix} \in L$, and $\begin{pmatrix} a & b \\ 0 & 0 \end{pmatrix}^n = \begin{pmatrix} a^n & * \\ 0 & 0 \end{pmatrix} \neq \begin{pmatrix} 0 & 0 \\ 0 & 0 \end{pmatrix}$ if $a \neq 0$. Similar computations hold for $b \neq 0$, $c \neq 0$, or $d \neq 0$, and we end up with no nonzero nilpotent left ideals. Exercise 12, along with Proposition 1, will complete the proof of semisimplicity.

8   We need only produce inverses for each nonzero element. If $x \neq 0$ and $x \in R$, then $Rx = \{rx \mid r \in R\}$ is a left ideal. The presence of x in Rx gives us that $Rx = R$,

and thus $rx = 1$ for some r in R. A similar procedure yields y such that $yr = 1$. But $y = yrx = x$, and we have $rx = xr = 1$.

12 For (d) and (f), note that we define

$$f \begin{pmatrix} a_{11} & a_{12} & a_{13} & \cdots & a_{1n} \\ a_{21} & a_{22} & a_{23} & \cdots & a_{2n} \\ \cdots\cdots\cdots\cdots\cdots\cdots \\ a_{n1} & a_{n2} & a_{n3} & & a_{nn} \end{pmatrix} = \begin{pmatrix} fa_{11} & fa_{12} & \cdots & fa_{1n} \\ fa_{21} & fa_{22} & \cdots & fa_{2n} \\ \cdots\cdots\cdots\cdots\cdots \\ fa_{n1} & fa_{n2} & & fa_{nn} \end{pmatrix}$$

With this definition we have $f(AB) = (fA)B = A(fB)$ for any A and B in $M_n(F)$. We could also identify f with the matrix $\begin{pmatrix} f & & \bigcirc \\ & f & \\ & & \ddots & \\ \bigcirc & & & f \end{pmatrix}$. Similarly, we can take f in $P(F)$ to be the constant polynomial f.

13 We already have additive closure for L, so we need only check that if $i \in L$, then $\alpha i \in L \ \forall \alpha \in F$. However, $\alpha i = \alpha(1_A i) = (\alpha 1_A)i$, and since $\alpha 1_A$ is in A, we have $\alpha i = (\alpha 1_A)i$ is in L.

14 $L = \left\{ \begin{pmatrix} 0 & a \\ 0 & 0 \end{pmatrix} \middle| a \in \mathbb{R} \right\}$ is the largest nilpotent left ideal in R. L is two-sided.

17 If $a^n = e$ and $b^m = e$, then $(ab)^{mn} = e$, since G is abelian. Now if $H = \langle x_1, x_2, \ldots, x_n \rangle$, and the order of x_i is m_i, then the order of H is at most $m_1 m_2 \cdots m_n$. Thus G is locally finite.

18 If $x^2 = e$ for all x in G, we can return to Exercise 2 of Sec. 1.3 to see that G is abelian.

20 Since $x^3 = x$ for all x, we can have no nonzero nilpotent ideals. Thus we have that R is semisimple and, as such, $R \simeq \oplus \sum_{i=1}^{k} M_{n_i}(\Delta_i)$. Were some n_i larger than 1, we would have

$$\begin{pmatrix} 0 & 1 & 0 & \cdots & 0 \\ 0 & 0 & 0 & \cdots & 0 \\ \cdots\cdots\cdots\cdots\cdots \\ 0 & 0 & 0 & \cdots & 0 \end{pmatrix}^3 = \begin{pmatrix} 0 & 1 & 0 & \cdots & 0 \\ 0 & 0 & 0 & \cdots & 0 \\ \cdots\cdots\cdots\cdots\cdots \\ 0 & 0 & 0 & \cdots & 0 \end{pmatrix}$$

which is false. Thus we have $R \simeq \Delta_1 \oplus \Delta_2 \oplus \cdots \oplus \Delta_k$, where each Δ_i is a division ring. If $1 + 1 = 2 \neq 0$ in Δ_i, then $2^3 = 2$, which tells us that $8 = 2$ or $6 = 0$. Thus the characteristic of Δ_i is 2 or 3. Let $Y = \{0,1\}$ or $\{0,1,2\}$, and assume $x \in \Delta_i - Y$. Then $\langle x, Y \rangle$ is a field, and $x^3 - x$ has at most three roots, among which are 0, 1, and -1. Thus we must have $\Delta_i = \mathbb{Z}_2$ or $\mathbb{Z}_3$, as all other fields (and division rings) have at least four elements. Thus $R \simeq \Delta_1 \oplus \Delta_2 \oplus \cdots \oplus \Delta_k$, where Δ_i is isomorphic to $\mathbb{Z}_2$ or $\mathbb{Z}_3$.

CHAPTER 9

Section 9.1

1 $a = \begin{pmatrix} 1 & 2 & 3 & 4 \\ 2 & 3 & 4 & 1 \end{pmatrix}$; then $\theta(a) = \begin{pmatrix} 0 & 1 & 0 & 0 \\ 0 & 0 & 1 & 0 \\ 0 & 0 & 0 & 1 \\ 1 & 0 & 0 & 0 \end{pmatrix}$. For $b = \begin{pmatrix} 1 & 2 & 3 & 4 \\ 1 & 4 & 3 & 2 \end{pmatrix}$, we have

$$\theta(b) = \begin{pmatrix} 1 & 0 & 0 & 0 \\ 0 & 0 & 0 & 1 \\ 0 & 0 & 1 & 0 \\ 0 & 1 & 0 & 0 \end{pmatrix}.$$

2   If $g = \begin{pmatrix} 1 & 2 & \cdots & n \\ g(1) & g(2) & \cdots & g(n) \end{pmatrix}$, then b_i goes to $b_{g(i)} = 0 \cdot b_1 + 0 \cdot b_2 + \cdots + 1 \cdot b_{g(i)}$
$+ \cdots + 0 \cdot b_n$. Thus the ith row of the corresponding matrix is all zeros except
for a 1 in the $[i, g(i)]$th position.

3   For $\mathbb{Z}_4$,

$$\bar{0} \leftrightarrow \begin{pmatrix} 1 & 0 & 0 & 0 \\ 0 & 1 & 0 & 0 \\ 0 & 0 & 1 & 0 \\ 0 & 0 & 0 & 1 \end{pmatrix} \qquad \bar{1} = \begin{pmatrix} 0 & 1 & 0 & 0 \\ 0 & 0 & 1 & 0 \\ 0 & 0 & 0 & 1 \\ 1 & 0 & 0 & 0 \end{pmatrix}$$

$$\bar{2} \leftrightarrow \begin{pmatrix} 0 & 0 & 1 & 0 \\ 0 & 0 & 0 & 1 \\ 1 & 0 & 0 & 0 \\ 0 & 1 & 0 & 0 \end{pmatrix} \qquad \bar{3} = \begin{pmatrix} 0 & 0 & 0 & 1 \\ 1 & 0 & 0 & 0 \\ 0 & 1 & 0 & 0 \\ 0 & 0 & 1 & 0 \end{pmatrix}$$

4   Let x be a nonidentity element in G, and assume that the (i,i)th position in the
matrix representing x is 1. If g is the ith element in the list of elements of g, then
$g = 0 + 0 + \cdots + 0 + xg + 0 + \cdots + 0$, which contradicts that $x \neq e$.

5   Since α is a homomorphism, we have

$$\begin{pmatrix} A(gh) & 0 \\ B(gh) & C(gh) \end{pmatrix} = \alpha(gh) = \alpha(g)\alpha(h) = \begin{pmatrix} A(g) & 0 \\ B(g) & C(g) \end{pmatrix} \begin{pmatrix} A(h) & 0 \\ B(h) & C(h) \end{pmatrix}$$

$$= \begin{pmatrix} A(g)A(h) & 0 \\ B(g)A(h) + C(g)B(h) & C(g)C(h) \end{pmatrix}$$

Thus $\alpha'(gh) = A(gh) = A(g)A(h) = \alpha'(g)\alpha'(h)$, and similarly for α''.

6   If char $(F) = p$, then if $p \mid |G|$ we would have $|\bar{G}| = \bar{0}$ in $\mathbb{Z}_p \subseteq F$, and thus
$1/|G|$ wouldn't be defined. If G is infinite, $\sum_{g \in G} B(g)A(g^{-1})$ won't be defined.

8   The homomorphism $\theta: \mathbb{Z} \to GL\,(2,G)$ given by $\theta(n) = \begin{pmatrix} 1 & n \\ 0 & 1 \end{pmatrix}$ is reducible, since
it is already reduced. If θ is decomposable, we would have

$$\begin{pmatrix} a & b \\ c & d \end{pmatrix}^{-1} \begin{pmatrix} 1 & n \\ 0 & 1 \end{pmatrix} \begin{pmatrix} a & b \\ c & d \end{pmatrix} = \begin{pmatrix} x & 0 \\ 0 & y \end{pmatrix}$$

so that $\begin{pmatrix} a + cn & b + dn \\ c & d \end{pmatrix} = \begin{pmatrix} ax & by \\ cx & dy \end{pmatrix}$, implying $d = c = 0$, which is impossible.

15   Suppose $G \simeq C_p \times C_p \times C_p = \{a^i b^j c^k\}$. Define $\alpha_{r,s,t}$ by $\alpha_{r,s,t}(a^i b^j c^k) = \varepsilon^{ri + sj + tk}$,
where $\varepsilon = e^{2\pi i/p}$.

19   Let P be a nonabelian p-group of order p^3. Since $p^3 = N_1 + N_2 p^2 + N_3 p^4 + \cdots$,

we must have $N_i = 0$, for $i = 3, 4, \ldots$. We must also have $N_2 \geq 1$, since P is non-abelian. $p^3 = N_1 + N_2 p^2$ implies $p^2 | N_1$, which, together with $N_2 \geq 1$ and $N_1 | p^3$, forces $N_1 = p^2$. This leads to $N_2 = p - 1$ and $N_1 + N_2 = p^2 + p - 1$.

22 The conjugate classes for D_{2n}, where $n = 2k$, are $\{e\}$, $\{a, a^{-1}\}$, $\{a^2, a^{-2}\}$, ..., $\{a^{k-1}, a^{k+1}\}$, $\{a^k\}$, $\{b, ba^2, ba^4, \ldots, ba^{2n-2}\}$, and $\{ba, ba^3, ba^5, \ldots, ba^{2n-1}\}$. Thus there are $k + 3 = n/2 + 3$ conjugate classes, and the same number of inequivalent irreducible representations. Also, $D'_{2n} = \langle a^2 \rangle$ (see Sec. 1.14), and $|D_{2n}/D'_{2n}| = 4 = N_1$. Thus,

$$2n = 4 \cdot 1^2 + \sum_{i=1}^{n/2-1} n_i^2 \geq 4 \cdot 1^2 + \sum_{i=1}^{n/2-1} 2^2$$

$$= 4 \cdot 1 + \left(\frac{n}{2} - 1\right) 4 = 2n$$

and each $n_i = 2$, where n_i is the degree of a nonlinear representation.

Section 9.2

1 If $A = (a_{ij})$ and $B = (b_{ij})$, then $AB = (c_{ij})$, where $c_{ij} = \sum_{k=1}^{n} a_{ik} b_{kj}$. $\mathrm{Tr}\,(AB) = \sum_{i=1}^{n} c_{ii} = \sum_{i=1}^{n} \sum_{k=1}^{n} a_{ik} b_{ki} = \mathrm{Tr}\,(BA) = \sum_{i=1}^{n} \sum_{k=1}^{n} b_{ik} a_{ki}$.

2 If G acts on S, $\theta(g) = \begin{pmatrix} s_1 & s_2 & \cdots & s_k \\ g(s_1) & g(s_2) & \cdots & g(s_k) \end{pmatrix}$. Then if π is the permutation representation of G, we have $\pi(g) = $ matrix with 0s except for 1s in the $[s_i, g(s_i)]$th positions. $\mathrm{Tr}\,(\pi(g))$ is the number of 1s in the $[s(i), s(i)]$th position, and, for each of these, $g(s_i) = s_i$, so that s_i is a fixed point of g. Thus, $\mathrm{Tr}\,(\pi(g)) = \chi(g)$.

5

C_2	e	a
χ_1	1	1
χ_2	1	-1

Sym (3)	I	(ab)	(abc)
χ_1	1	1	1
χ_2	1	-1	1
χ_3	2	0	-1

6 The character table for Sym (4) is

	I	(ab)	(abc)	$(abcd)$	$(ab)(cd)$
χ_1	1	1	1	1	1
χ_2	1	-1	1	-1	1
χ_3	2	0	-1	0	2
χ_4	3	1	0	-1	-1
χ_5	3	-1	0	1	-1

7 The columns representing elements of order 2 should have integral entries, and these entries should differ from those in the first column by even numbers.

CHAPTER 10

Section 10.1

1   $[\mathbb{C}, \mathbb{R}] = 2$, since $\{1, i\}$ is a basis for $\mathbb{C}$ over $\mathbb{R}$. Also, $[\mathbb{R}: \mathbb{Q}] = \infty$.

2   Assume

$$\sum_{\substack{1 \leq j \leq n \\ 1 \leq j \leq m}} f_{ij} l_i k_j = 0 \qquad \text{with} \qquad f_{ij} \in F$$

Then, since each $f_{ij} \in L$, we have $\sum_{j=1}^{m} \left(\sum_{i=1}^{n} f_{ij} l_i \right) k_j = 0$, which implies $\sum_{i=1}^{n} f_{ij} l_i = 0$ for $j = 1, 2, \ldots, m$, since the $\{k_j\}$ are linearly independent. But the $\{l_i\}$ are linearly independent also, so $f_{ij} = 0$ for all i and j.

3   If l is an arbitrary element of L, then since the $\{l_i\}$ span L over K, we have $l = \sum_{i=1}^{n} \alpha_i l_i$ with $\alpha_i \in K$. However, each $\alpha_i = \sum_{j=1}^{m} \beta_{ij} k_j$ with $\beta_j \in F$, and this yields $l = \sum_{j=1}^{m} \left(\sum_{i=1}^{n} \beta_{ij} k_j \right) l_i = \sum_{i, j} \beta_{ij} l_i k_j$; thus, the $\{l_i k_j\}$ span L over F.

6   Multiplicative closure follows from $[p_1(a)/q_1(a)][p_2(a)/q_2(a)] = p_1(a)p_2(a)/q_1(a)q_2(a)$, and the inverse of $p(a)/q(a)$ is $q(a)/p(a)$ if $p(a) \neq 0$.

8   If $a^3 + 2b^3 + 4c^3 - 6abc = 0$ for relatively prime integers a, b, and c, then a^3 must be even. Thus a is even, so $a = 2a'$, which yields $8a'^3 + 2b^3 + 4c^3 - 12a'bc = 0$ and $b^3 + 2c^3 + 4a'^3 - 16a'bc = 0$. This yields in turn that b and then c are even, so that a, b, and c are not relatively prime.

11   A minimal polynomial for $\sqrt[5]{7}$ over $\mathbb{Q}$ is $x^5 - 7$. Similarly, a minimal polynomial for $\sqrt{2} + \sqrt{3}$ over $\mathbb{Q}$ is $x^4 - 10x^2 + 1$. Thus, $[\mathbb{Q}(\sqrt[5]{7}): \mathbb{Q}] = 5$.

12   If $\alpha = e^{2\pi i/5}$, then $\alpha^5 = 1$, so that α satisfies $x^5 - 1 = (x - 1)(x^4 + x^3 + x^2 + x + 1)$. Obviously α is not a root of $x - 1$, so α must be a root of $x^4 + x^3 + x^2 + x + 1 = p(x)$. We have seen that $p(x)$ is irreducible over $\mathbb{Q}$ (Exercise 17 of Sec. 2.6), so that $[\mathbb{Q}(e^{2\pi i/5}): \mathbb{Q}] = 4$. Every polynomial in $\mathbb{R}[x]$ factors into a product of quadratic and linear polynomials (Exercise 27 of Sec. 2.2), and a quick look at $p(x)$ says it has no linear factor in $\mathbb{R}$. In fact,

$$x^4 + x^3 + x^2 + x + 1 = \left(x^2 + \frac{1 + \sqrt{5}}{2} x + 1 \right)\left(x^2 + \frac{1 - \sqrt{5}}{2} x + 1 \right)$$

α is a root of $x^2 + [(1 + \sqrt{5})/2]x + 1$, and $[\mathbb{R}(e^{2\pi i/5}): \mathbb{R}] = 2$.

14   Assume $[\mathbb{A}: \mathbb{Q}] = m$, and look at $2^{1/(m+1)} = \alpha$, so that α is a root of $x^{m+1} - 2$. By Eisenstein's criterion, $x^{m+1} - 2$ is irreducible over $\mathbb{Q}$. Thus $[\mathbb{Q}(\alpha): \mathbb{Q}] = m + 1 > [\mathbb{A}: \mathbb{Q}]$, contradicting the fact that $\alpha \in \mathbb{A}$.

17   Since $e^{2\pi i/360}$ satisfies $x^{360} - 1$, we have $(\cos 1° + i \sin 1°)^{360} - 1 = 0 = (\cos 1° + i\sqrt{1 - \cos^2 1°})^{360} - 1$. Expanding and taking the real part will give a polynomial of degree 360 in $\mathbb{Q}[x]$ satisfied by $\cos 1°$. This shows that $\cos 1°$ is in $\mathbb{A}$ and, similarly, $\sin 1° \in \mathbb{A}$.

20   $\sqrt{1 + \sqrt{2}}$ is algebraic over $\mathbb{Q}$ but π^3 isn't, since, if $\pi^3 \in \mathbb{A}$, then $x^3 - \pi^3$ would have its solutions in $\mathbb{A}$. But π is transcendental and thus not in $\mathbb{A}$.

24   If $\alpha \in T$, then α is algebraic over F. Thus α is a root of a minimal polynomial

$p(x) = a_0 + a_1 x + a_2 x^2 + \cdots + a_n x^n$ with $a_i \in F$. Since $a_0 \neq 0$ (else we would have a polynomial of smaller degree satisfied by α), we have

$$0 = a_0 + a_1 \alpha + a_2 \alpha^2 + \cdots + a_n \alpha^n$$

Thus

$$-a_0 = \alpha(a_1 + a_2 \alpha + \cdots + a_n \alpha^{n-1})$$

and

$$\alpha^{-1} = \frac{-1}{a_0}(a_1 + a_2\alpha + \cdots + a_n \alpha^{n-1})$$

Thus every nonzero α in T has an inverse, and T is a field.

25 (d) $\begin{vmatrix} 2-x & 3 \\ -3 & 2-x \end{vmatrix} = x^2 - 4x + 13$ is irreducible over $\mathbb{R}$ and thus is both the minimal and characteristic polynomial for $2 + 3i$.

(e) Let $(1, \sqrt[3]{2}, \sqrt[3]{4})$ be an ordered basis for $\mathbb{Q}(\sqrt[3]{2})$ over $\mathbb{Q}$. Then the characteristic polynomial for $a + b\sqrt[3]{2} + c\sqrt[3]{4}$ is

$$\begin{vmatrix} a-x & b & c \\ 2c & a-x & b \\ 2b & 2c & a-x \end{vmatrix} = -x^3 + 3ax^2 + 3(2bc - a^2)x + a^3 - 6abc + 2b^3 + 4c^2$$

This can be used to compute $(a + b\sqrt[3]{2} + c\sqrt[3]{4})^{-1}$ by the procedure in Exercise 24.

Section 10.2

1 If g is constructible, then so is $|g|/2$, and we can set up

3 If a regular 9-gon (nonagon) were constructible, then $\cos (360/9)° = \cos 40°$ would be constructible. Bisecting $40°$ would imply $20°$ is constructible, which we proved impossible in proving the impossibility of trisection of general angles.

5 The angles $15°$ and $21°$ are constructible, unlike $20°$.

7 (a) We identify the plane with the complex plane. Then the problem is constructing $\alpha = e^{2\pi i/p}$. We see that $\alpha^p = 1$, so that α is a root of $x^p - 1$. However, $x^p - 1 = (x-1)(x^{p-1} + x^{p-2} + \cdots + x + 1)$, and α is a root of $x^{p-1} + x^{p-2} + \cdots + x + 1$, which is irreducible over $\mathbb{Q}$ (see Exercise 17 of Sec. 2.6). Thus $[\mathbb{Q}(a): \mathbb{Q}] = p - 1$. But $p - 1 = 2^n$ only if $p = 2^n + 1$, so that p is a Fermat prime. (If p is to be prime, n must itself be a power of 2. Why?)

Section 10.3

1 We have $y^2 + y - 1 = x^2 + x + 1 + 1/x + 1/x^2$, so that $y = (-1 \pm \sqrt{5})/2$ is a root. But $x^2 - yx + 1 = 0$ yields

$$x = \frac{1 - \sqrt{5}}{4} \pm \frac{\sqrt{\sqrt{5} - 2}}{2}$$

3 Let D be an integral domain, and let $p(x) \in D[x]$. If F is the quotient field of D, then $D[x]$ can be considered as a subring of $F[x]$, and we can view $p(x)$ as an element in $F[x]$. In F, $p(x)$ will have at most n roots, and restricting the possibilities to D cannot increase the number of roots.

5 What are possible roots of $x^2 + 1$ in $\mathcal{2}$?

7 We see that $x^3 - 8 = (x - 2)(x^2 + 2x + 4)$ is a complete factorization of $x^3 - 8$ into irreducible elements in $\mathbb{Q}[x]$. Since $2 \in \mathbb{Q}$, we have $\mathbb{Q}((-2 \pm \sqrt{4 - 16})/2) = \mathbb{Q}(-1 \pm \sqrt{3}) = \mathbb{Q}(\sqrt{3})$ is the splitting field for $x^3 - 8$ over $\mathbb{Q}$.

9 $[S: \mathbb{Q}] = 8$, where S is the splitting field of $x^4 - 2$ over $\mathbb{Q}$.

10 If $p(x) = ax^3 + bx^2 + cx + d$ with $a > 0$, then a rough graph of $p(x)$ is as follows:

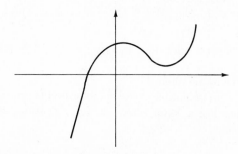

The graph eventually enters the first and third quadrants and thus must cross the x axis at least once. This yields a real root.

11 If x_1 and x_2 are in E_S, then $\alpha(x_1 \pm x_2) = \alpha(x_1) \pm \alpha(x_2) = x_1 \pm x_2 \ \forall \alpha \in S$. Also,

$$\alpha(x_1 x_2) = \alpha(x_1)\alpha(x_2) = x_1 x_2 \quad \text{and} \quad \alpha(x^{-1}) = (\alpha(x))^{-1} = (x)^{-1} = x^{-1}$$

$\forall x_1, x_2, x \in E_S$ and $\alpha \in S$.

13 (b) $E_{(I, \alpha)} = \{q_0 + q_3\sqrt{3} \mid q_i \in \mathbb{Q}\} = \mathbb{Q}(\sqrt{3})$. Define β in an analogous manner, so that $E_{(I, \beta)} = \mathbb{Q}(\sqrt{2})$. Check that $\alpha^2 = \beta^2 = I$ and that $\alpha\beta = \beta\alpha$, so G is isomorphic to Klein's four group.

Section 10.4

1 $G_{(E_H)}$ is the subgroup of automorphisms of E fixing the subfield E_H. However E_H is defined as the subfield of elements fixed by H, so certainly $H \subseteq G_{(E_H)}$. But $[H: G_{(E_H)}] = [E_H: E_H] = 1$, so that $H = G_{(E_H)}$.

3 $G = G(\mathbb{Q}(\alpha, i)/\mathbb{Q})$ has five elements of order 2, and two of order 4. Thus $G \not\simeq C_8$ or $C_2 \times C_2 \times C_2$, where C_n denotes a cyclic group of order 4. $G \not\simeq C_4 \times C_2$, since $C_4 \times C_2$ has four elements of order 4. This disposes of the abelian groups of order 8, and the only nonabelian groups of this order are D_8 and $\mathcal{2}$ (the quaternions). In $\mathcal{2}$ there is but a single element of order 2, and thus $G \simeq D_8$.

4 We first factor

$$x^5 - 7 = (x - 7^{1/5})(x^5 - \varepsilon 7^{1/5})(x - \varepsilon^2 7^{1/5})(x - \varepsilon^3 7^{1/5})(x - \varepsilon^4 7^{1/5})$$

so $x^5 - 7$ splits in $\mathbb{Q}(\varepsilon, 7^{1/5})$, where $7^{1/5}$ denotes the real fifth root of 7. Since $[\mathbb{Q}(\varepsilon): \mathbb{Q}] = 4$ (Exercise 12 of Sec. 10.1) and $[\mathbb{Q}(7^{1/5}): \mathbb{Q}] = 5$, we see that $4 \cdot 5 \mid [\mathbb{Q}(\varepsilon, 7^{1/5}): \mathbb{Q}] = [\mathbb{Q}(\varepsilon, 7^{1/5}): \mathbb{Q}(7^{1/5})][\mathbb{Q}(7^{1/5}): \mathbb{Q}] \leq 4 \cdot 5$, so that

$$[\mathbb{Q}(\varepsilon, 7^{1/5}): \mathbb{Q}] = 20.$$

Since the splitting field for $x^5 - 7$ must contain both $7^{1/5}$ and $\varepsilon 7^{1/5}$, and thus ε, we see that $K = \mathbb{Q}(\varepsilon, 7^{1/5})$ is the splitting field.

7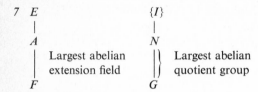

The largest abelian extension field corresponds to the largest abelian quotient group G/N, and this is given by $N = G'$, the commutator subgroup of G. See Sec. 1.14 for more details about G'.

9   $x^{p^m - 1} - 1$ is far from irreducible, so we do not expect $[\mathbb{Z}_p(\alpha): \mathbb{Z}_p] = p^m - 1$, where α is a root. Since E is a vector space over $\mathbb{Z}_p$, we see that E has dimension m, since $|E| = p^m$ and

$$E = \{a_1 b_1 + z_2 b_2 + \cdots + z_m b_k \mid z_i \in \mathbb{Z}_p, (b_1, \ldots, b_k) \text{ a basis for } E \text{ over } \mathbb{Z}_p\}$$

11 (a) $[\mathbb{Q}(\varepsilon): \mathbb{Q}] = 2$, since ε is a root of the polynomial $x^2 + x + 1$, which is irreducible over $\mathbb{Q}$. But $\mathbb{Q}(\varepsilon)$ is a splitting field for $x^2 + x + 1$, and thus $G(\mathbb{Q}(\varepsilon)/\mathbb{Q})$ is a group of order 2 and therefore cyclic.

(c)   See Exercise 13b in Sec. 10.3.

13   Since F has characteristic p, it is true that $x^p + x^p = (x + y)^p$ for all x and y in F. Thus $0 = x^p - 1 = x^p - 1^p = (x - 1)^p$, and $x = 1$ is the only root of $x^p - 1$.

15   Let $p(x) \in \mathbb{R}[x]$ so that $p(x) = l_1(x) l_2(x) \cdots l_k(x) q_1(x) q_2(x) \cdots q_m(x)$, where the l_i are linear polynomials, and the q_i are irreducible quadratic polynomials (Exercise 27 of Sec. 2.2). $\mathbb{R}$ is a splitting field for the l_i, and $\mathbb{C}$ is a splitting field for the q_i. Thus the splitting field for $p(x)$ is $\mathbb{C}$ or $\mathbb{R}$. The Galois group has order 2 or 1, thus is abelian, and thus is solvable.

18   The Galois group of $f(x)$ is a permutation group on the roots $r_1, r_2, \ldots, r_p$ and is transitive (corollary to Theorem 2 of Sec. 10.3). The following diagram holds:

$$
\begin{array}{ll}
\mathbb{Q}(r_i, r_j) & \text{Subgroup fixing } r_i \text{ and } r_j \\
\mid & \mid \\
\mathbb{Q}(r_i) & \text{Subgroup fixing } r_i \\
\mid & \mid \\
\mathbb{Q} & G
\end{array}
$$

Thus, $\mathbb{Q}(r_i, r_j) = $ the splitting field of $f(x) \Leftrightarrow I = $ subgroup fixing r_i and r_j, which is the condition for solvability.

19 If $p(x)$ is a cubic or quartic over F, then the splitting field S over F of $p(x)$ satisfies $[S:F] \leq 3! = 6$, or $[S:F] \leq 4! = 24$, respectively. The Galois group permutes the (at most) four roots, and thus $G(S/F) \subseteq \text{Sym}\,(4)$. However, Sym (4) is solvable, so $G(S/F)$ is solvable.

20 Let $y = x^2$, and observe that $ay^4 + by^3 + cy^2 + dy + f$ is solvable over F by the last exercise. If S is the splitting field for this quartic, then let S' be the splitting field for $x^2 - \alpha$, for α a root of $ay^4 + 6y^3 + cy^2 + dy + f$. Then $[S':S] \leq 2$, so S' is solvable over S, S is solvable over F, and thus S' is a solvable extension of F.

CROSS-REFERENCES TO ELEMENTARY MATHEMATICS

are related to homomorphisms; trigonometry, logarithms, absolute value, casting out nines.

NOTATION

page	symbol	definition
5	a^{-1} (or $-a$)	inverse element for a
8	$\lvert S \rvert$	the number of elements in S
14	$\mathbb{Z}_n$	$\{\bar{0}, \bar{1}, ..., \overline{n-1}\}$ with addition modulo n
18	V	Klein's Four Group
18–19	$\langle g \rangle$	the cyclic subgroup generated by g
23	$f: A \to B$	f is a function from A to B
23	$f \circ g$	composition of the function g and f
25	I, I_A	the identity function on the set A
26	$\begin{pmatrix} a & b \\ c & d \end{pmatrix}$	2×2 matrix
28	$GL(2,\mathbb{R})$	the 2×2 general linear group
30	$SL(2,\mathbb{R})$	the 2×2 special linear group
31	Sym (A)	group of permutations of A
32	Sym (n)	group of permutations of $\{1, 2, ..., n\}$
32	$\begin{pmatrix} 1 & 2 & ... & n \\ a & b & ... & k \end{pmatrix}$	a permutation of $\{1, 2, ..., n\}$
36, (37)	$D_8, (D_{2n})$	dihedral group of order 8, $(2n)$
38	$\sim$	equivalence relation
39	$\bar{t}$	equivalence class containing t
41	Hg or $(H + g)$	right coset of H
43	$[G:H]$	index of H in G
44	$Z(G)$	the center of G
51	gcd	greatest common divisor
55	$n\mathbb{Z}$	$\{0, n, -n, 2n, -2n, 3n, -3n, ...\}$
57	$A \times B$ (or $A \oplus B$)	direct product (or sum) of A and B
58, (135)	$\langle S \rangle$	subgroup generated by S, (span of S)
60	aH (or $a + H$)	left coset of H
60	$\lhd, N \lhd G$	N is a normal subgroup of G
63	G/N	quotient group of G over N
64	$\mathcal{N}: G \to G/N$	canonical homomorphism of G onto G/N
65	ker (ϕ)	the kernel of ϕ
71	G'	the commutator subgroup of G
73	G_{ab}	G abelianized, $G \mid G'$
79	Alt (n)	the alternating group on $\{1, 2, ..., n\}$
85–86	$(R, +, \cdot)$	ring
89	$\mathscr{B}$	Boolean ring
91	$\begin{pmatrix} n \\ k \end{pmatrix}$	$n!/k!\,(n-k)!$
93	$C[0,1]$	the ring of continuous functions on $[0,1]$
95	$\mathbb{C}$	the complex numbers
96	$\lvert z \rvert = \sqrt{z\bar{z}}$	norm of z
99, 195	$G = \mathbb{Z}[i]$	gaussian integers
102–103	$M_2(S)$	2×2 matrices over S
105	$\mathscr{Q}$	the quaternions
106	H	the group of quaternions
120	$R[x]$	the ring of polynomials in x over R

page	symbol	definition
125	$C(p(x))$	content of the polynomial $p(x)$
128	$(V, +, \cdot\, F)$	the vector space of n-tuples over F
129	F^n	vector space, V, over F
130	$P_n(x)$	all polynomials over F of degree less than $n + 1$
147	(a_{ij})	matrix
155	G_t	subgroup stabilizing t
156	$\mathcal{O}_t$	orbit containing t
159	f_g	conjugation by g
160	Inn (G)	the group of inner automorphisms of G
160	Cl (a)	the conjugate class containing a
163	$X(g)$	the number of elements fixed by g
183	$G^{(k)}$	kth commutator subgroup of G
184, 186	Aut (G)	the group of automorphisms of G
195	ED	euclidean domain
198	PID	principal ideal domain
199	UFD	unique factorization domain
200	acc	ascending chain condition
200	dcc	descending chain condition
211	$\Phi(n)$	$\{\bar{1}, ..., \bar{k}, ..., n - 1 \mid \gcd(k,n) = 1\}$
211	$\phi(n)$	Euler's phi function
216	$\left(\dfrac{a}{p}\right)$	Legendre symbol
232	$GL(V)$	group of all nonsingular linear transformations on V
245	N_i	number of inequivalent, irreducible representations of degree i
247	X_j	jth irreducible character of G
254	$F(a)$	the field F with a adjoined
254	$[K : F]$	the degree of K over F
259	$\mathbb{A}$	the field of algebraic numbers
261	$\mathbb{G}$	the field of constructible numbers
270	F_H	subfield of elements fixed by H
270	$G(K/F)$	group of K over F

SUBJECT INDEX

vector space (= linear transformation), 139

Ideal, 108
  maximal, 116, 207
  prime, 119, 207
Idempotent, 91
Identity element, e, 5
Image, 22
Impossibility theorems:
  constructing the square root of two, 263
  squaring the circle, 264
  trisecting an angle, 263
Improper subgroup, 16
Improper subset, 2
Independence, linear, 136
Index of a subgroup, 43
Inner automorphism, 160
Integers, $\mathbb{Z}$, 2
  algebraic, 250
  Gaussian, 99, 195, 197–198
  (*See also* Integral domain)
Integral domain, 22, 192–210, 228
Intersection, 1
Inverse element in a group, 5, 8, 10
Inverse function, 24
Inverse image, $(\phi^{-1}(X) = \{g \mid \phi(g) \in X\},$ 64
Inverse matrix, 27
Involution, 14
Irreducible, 195
Irreducible representation, 236
Isomorphic groups, 46–51
Isomorphic rings, 109–110
Isomorphic vector spaces, 139
Isomorphism:
  first isomorphism theorem, 70, 182
  of groups, 46–51
  of rings, 109–110
  second isomorphism theorem, 69, 181
  of vector spaces, 139

Jordan, C., 84, 85

k-cycle, 79
kth commutator subgroup, 183–184
Kernel, 65, 70, 114, 216
Klein's four group, V, 18, 19, 58, 76
Kronecker, L., 267, 268
Kummer, E., 85, 115

Lagrange, J. L., 44, 81–82
Lagrange's theorem, 42–44, 158, 161
Lattice of subgroups, 17
Least common multiple (lcm), 75, 202–204
Left artinian ring, 224
Left coset, 41, 42
Left idea, 228, 229
Left R-module, 134, 191
Legendre, A. M., 216
Legendre, symbol, 216
Lie, S., 7
Lindemann, F., 259
Linear property, 51
  in an integral domain, 202–204
Linear transformation, 139
Linearly dependent set, 136
Linearly independent set, 136
Liouville, J., 259, 269
Locally finite, 230

McKay, J., 162
Maschke's theorem, 238
Map (= function) (*see* Functions)
Matrix, 26
  inverse, 27
  multiplication, 26, 148
  nonsingular, 26
  ring, 101–106
Maximal ideal, 116, 207
Minimum polynomial for $a = p(x)$ of smallest degree ≥ 1 so that $p(a) = 0$, $p(x) \in F[x]$, 256–258, 260
Module, 134, 191
Monomorphism:
  of a group, 46, 240
  of a ring, 109
  of a vector space, 139